MATERIALS SCIENCE AND TECHNOLOGIES

ANALYTICAL MODELS OF THERMAL STRESSES IN COMPOSITE MATERIALS III

MATERIALS SCIENCE AND TECHNOLOGIES

Additional books in this series can be found on Nova's website
under the Series tab.

Additional E-books in this series can be found on Nova's website
under the E-book tab.

MATERIALS SCIENCE AND TECHNOLOGIES

ANALYTICAL MODELS OF THERMAL STRESSES IN COMPOSITE MATERIALS III

LADISLAV CENIGA

Nova Science Publishers, Inc.
New York

Copyright © 2012 by Nova Science Publishers, Inc.

All rights reserved. No part of this book may be reproduced, stored in a retrieval system or transmitted in any form or by any means: electronic, electrostatic, magnetic, tape, mechanical photocopying, recording or otherwise without the written permission of the Publisher.

For permission to use material from this book please contact us:
Telephone 631-231-7269; Fax 631-231-8175
Web Site: http://www.novapublishers.com

NOTICE TO THE READER

The Publisher has taken reasonable care in the preparation of this book, but makes no expressed or implied warranty of any kind and assumes no responsibility for any errors or omissions. No liability is assumed for incidental or consequential damages in connection with or arising out of information contained in this book. The Publisher shall not be liable for any special, consequential, or exemplary damages resulting, in whole or in part, from the readers' use of, or reliance upon, this material. Any parts of this book based on government reports are so indicated and copyright is claimed for those parts to the extent applicable to compilations of such works.

Independent verification should be sought for any data, advice or recommendations contained in this book. In addition, no responsibility is assumed by the publisher for any injury and/or damage to persons or property arising from any methods, products, instructions, ideas or otherwise contained in this publication.

This publication is designed to provide accurate and authoritative information with regard to the subject matter covered herein. It is sold with the clear understanding that the Publisher is not engaged in rendering legal or any other professional services. If legal or any other expert assistance is required, the services of a competent person should be sought. FROM A DECLARATION OF PARTICIPANTS JOINTLY ADOPTED BY A COMMITTEE OF THE AMERICAN BAR ASSOCIATION AND A COMMITTEE OF PUBLISHERS.

Additional color graphics may be available in the e-book version of this book.

LIBRARY OF CONGRESS CATALOGING-IN-PUBLICATION DATA

ISBN: 978-1-61324-710-5

Published by Nova Science Publishers, Inc. † New York

Contents

Preface	**ix**
About the Author	xi

1 Outline of principles   **1**
- 1.1 Cell model 1
- 1.2 Mathematical techniques 2
- 1.3 Reason of thermal stresses 3
- 1.4 Temperature range 3
- 1.5 Finite matrix 4
- 1.6 Subscripts and notation 4
- 1.7 Recommendations of author 4
  - 1.7.1 Categories of readers 4
  - 1.7.2 Programming techniques 6

2 Cell model   **9**
- 2.1 Geometric boundary condition for cell matrix 9
  - 2.1.1 Summary 12
- 2.2 Particle volume fraction 12
  - 2.2.1 General formulae 12
  - 2.2.2 Analysis of interval boundaries $v \to 0$, $v \to v_{max}$ 13
  - 2.2.3 Rectangular-based prismatic and cubic cells 14
  - 2.2.4 Hexagonal-based prismatic cell 15
- 2.3 Determination of distance $r_c = r_c(\varphi, \nu)$ 16
  - 2.3.1 Rectangular-based prismatic and cubic cells 16
    - 2.3.1.1 Interval $\nu \in \langle 0, \nu^* \rangle$ 16
    - 2.3.1.2 Interval $\nu \in \langle \nu^*, \pi/2 \rangle$ 18
  - 2.3.2 Hexagonal-based prismatic cell 19
- 2.4 Real composite material 20

3 Thermal stresses in elastic solid continuum   **21**
- 3.1 Selected topics of Mechanics of Solid Continuum 21
  - 3.1.1 Coordinate system 21

3.1.2	Displacements of the infinitesima spherical cap. Thermal-stress induced radial displacement	22
3.1.3	Cauchy's equations	24
3.1.4	Saint-Venant's equations	26
3.1.5	Equilibrium equations	26
3.1.6	Hooke's law	28
3.2	Mathematical techniques	29
3.2.1	Determination of tangential and shear stresses	29
3.2.2	Determination of radial stress	33
3.2.3	Different solutions regarding exponent λ	36
3.2.4	Elastic energy	37
3.2.5	Castigliano's theorem	38
3.3	Mathematical solutions	38
3.4	Reason of thermal stresses	41
3.4.1	Coefficient $\beta_p, \beta_e, \beta_m$	41
3.4.2	Phase-transformation induced radial strain	43
3.4.3	Analysis of the conditions $\beta_{q_1} \neq \beta_{q_2}, \beta_{q_1} = \beta_{q_2}$	44
3.4.4	Formulae for determination of the radial stresses p_1, p_2	46
3.4.5	Temperature dependence of thermal stresses	47
3.4.6	Temperature range	48

4 Boundary conditions — **49**

4.1	Spherical particle	49
4.2	Spherical envelope	50
4.3	Cell matrix	51
4.3.1	Mandatory boundary conditions	52
4.3.2	Additional boundary conditions	53
4.4	One-particle-(envelope)-matrix system	53
4.5	Supplement	54
4.5.1	Condition $\beta_p = \beta_e \neq \beta_m$	54
4.5.2	Condition $\beta_p \neq \beta_e = \beta_m$	55

5 Thermal stresses in multi-particle-(envelope)-matrix system I — **57**

5.1	Multi-particle-matrix system	57
5.1.1	Thermal stresses in spherical particle	57
5.1.2	Thermal stresses in cell matrix	58
5.1.3	Radial stress p_1	64
5.2	Multi-particle-envelope-matrix system	65
5.2.1	Condition $\beta_p \neq \beta_e = \beta_m$	65
5.2.1.1	Thermal stresses in spherical particle	65
5.2.1.2	Thermal stresses in spherical envelope	65
5.2.1.3	Thermal stresses in cell matrix	66
5.2.1.4	Radial stress p_1	69

	5.2.2	Condition $\beta_p \neq \beta_e \neq \beta_m$	69
		5.2.2.1 Thermal stresses in spherical particle	69
		5.2.2.2 Thermal stresses in spherical envelope	69
		5.2.2.3 Thermal stresses in cell matrix	71
		5.2.2.4 Radial stresses p_1, p_2	73
	5.2.3	Condition $\beta_p = \beta_e \neq \beta_m$	73
		5.2.3.1 Thermal stresses in spherical particle	73
		5.2.3.2 Thermal stresses in spherical particle	74
		5.2.3.3 Thermal stresses in cell matrix	75
		5.2.3.4 Radial stress p_2	76

6 Thermal stresses in one-particle-(envelope)-matrix system — 77

6.1 Thermal stresses in spherical particle and envelope — 77

6.2 Thermal stresses in infinit matrix — 77

    6.2.1 Conditions $\beta_m \neq \beta_p$, $\beta_p \neq \beta_e \neq \beta_p$, $\beta_p = \beta_e \neq \beta_p$ — 77

    6.2.2 Condition $\beta_p \neq \beta_e = \beta_p$ — 79

7 Thermal stresses in multi-particle-(envelope)-matrix system II — 81

8 Related phenomena — 83

8.1 Introduction — 83

    8.1.1 Analytical fracture mechanics — 83

    8.1.2 Analytical model of energy barrier — 85

    8.1.3 Analytical model of strengthening — 86

    8.1.4 Analytical-computational and analytical-experimental-computational methods of lifetime prediction — 86

        8.1.4.1 Resistive and contributory effects of thermal stresses — 87

        8.1.4.2 Analytical-computational method — 87

        8.1.4.3 Analytical-computational-experimental method — 88

8.2 Analytical fracture mechanics — 88

    8.2.1 General analysis — 88

    8.2.2 Cell model — 92

        8.2.2.1 Conditions of crack formation — 94

            A. Multi-particle-matrix system — 94

            B. Multi-particle-envelope-matrix system — 94

            C. Conclusions — 95

        8.2.2.2 Crack parameters — 96

            A. Multi-particle-matrix system — 96

            B. Multi-particle-envelope-matrix system — 103

        8.2.2.3 Modificatio of stress-deformation field — 106

    8.2.3 Crack formation I — 107

        8.2.3.1 Determination of $W_{cp}^{(ij)}$, $W_{cm}^{(ij)}$ in multi-particle-matrix system — 107

viii   Ladislav Ceniga

A. Spherical particle 107
B. Cell matrix 108
8.2.3.2   Multi-particle-envelope-matrix system 109
A. Spherical particle 109
B. Spherical envelope 110
C. Cell matrix 112
8.2.4   Crack formation II 114
8.2.4.1   Determination of $W_{cp}^{(ij)}$, $W_{cm}^{(ij)}$ in multi-particle-matrix
system 114
A. Method 1 117
B. Method 2 121
8.2.4.2   Determination of $W_{cp}^{(ij)}$, $W_{ce}^{(ij)}$, $W_{cm}^{(ij)}$ in multi-particle-
envelope-matrix system 125
A. Method 1 129
B. Method 2 136
8.2.5   Transformations concerning cracking in planes $x_i x_j$, $x_j x_k$, $x_k x_i$ . . 142
8.3   Analytical model of energy barrier 144
8.3.1   General analysis 144
8.3.2   Multi-particle-matrix system 147
8.3.3   Multi-particle-envelope-matrix system 148
8.4   Analytical model of strengthening 150
8.4.1   General analysis 150
8.4.2   Strengthening in multi-particle-(envelope)-matrix system 151
8.4.2.1   Multi-particle-matrix system 152
8.4.2.2   Multi-particle-envelope-matrix system 154
8.5   Analytical-computational and analytical-experimental-computational
methods of lifetime prediction 157
8.5.1   Resistive and contributory effects of thermal stresses 158
8.5.2   Analytical-computational method 161
8.5.3   Analytical-computational-experimental method 164

9   Appendix   167
9.1   Elastic modulus and thermal expansion coefficien 167
9.1.1   Transformations concerning subscripts ij and q 168
9.2   Coefficien c 178

Bibliography   195

Index   197

Preface

This book is dedicated with love to my dearest parents.

This book is the third volume of the trilogy *Analytical models of thermal stresses in composite materials I, II, III*, presenting, in each of the volumes, only original results created by the author. The fact that the author proceeds from fundamental equations of solid continuum mechanics which are presented in Sections 3.1.3–3.1.6, 3.2.4 confirm originality of the results and accordingly the establishment of a new scientifi school with an interdisciplinary character. As an imagination considered for the analytical models, an **elastic** solid continuum is represented by multi-particle-matrix and multi-particle-envelope-matrix systems which consist of components represented by **spherical** particles periodically distributed in an **infinit** matrix, without and with a **spherical** envelope on a surface of each of the spherical particles, respectively. The multi-particle-matrix and multi-particle-envelope-matrix systems with different distributions of the spherical particles are considered to represent as model systems for the determination of the thermal stresses in real composite materials with finit dimensions, i.e. two- and three-component materials, respectively, which are define in Section 2.1 (see Items 1–4). The particle and envelope radii, R_1 and $R_1 < R_2$, respectively, the envelope thickness $t = R_2 - R_1 > 0$, the interparticle distance d and the particle volume fraction v represent parameters of the multi-particle-(envelope)-matrix system as well as microstructural parameters of the two- and three-component materials. The thermal stresses are thus functions of the microstructural parameters. Additionally, the thermal stresses which originate during a cooling process are a consequence of different thermal expansion coefficient of material components as well as a consequence of different dimensions of crystal lattices which are mutually transformed due to a phase transformation originating at least in one of the material components.

Volumes I, II present analytical models of the thermal stresses which are considered for the composite materials with **isotropic** components with **isotropic** crystal lattices. In contrast to Volumes I, II, this book presents analytical models of the thermal stresses which are considered for the composite materials with **anisotropic** components with **anisotropic** crystal lattices. The anisotropy is considered to be uniaxial or triaxial.

As usual in solid continuum mechanics, a stress-strain state of a system can be determined using different mathematical techniques resulting in different solutions for the thermal stresses in the components of the multi-particle-(envelope)-matrix system. The different solutions for at least one of the components result in different energy of the system. Due to the different solutions, such combination of different solutions for the component of the system is considered to exhibit minimal energy of the system as expressed by the Castigliano's theorem (see Section 3.2.5). These different solutions for the anisotropic component of the multi-particle-(envelope)-matrix system which are obtained by mathematical techniques analysed in Section 3.2.2 are analysed in Section 3.2.3. On the one hand, a number of these different solutions is more than sufficient and then this book exhibits permanent validity without a need of the updating in future. On the other hand, in case of interest, a reader can increase a number of these different solutions by such mathematical techniques which are based on those in Section 3.2.2.

With regard to completeness of this topic, the analytical model of the thermal stresses in one-particle-matrix and one-particle-envelope-matrix systems, both with anisotropic components, is determined (see Chapter 6).

Additionally, the mathematical technique for the determination of the thermal stresses in the multi-particle-(envelope)-matrix system with isotropic and anisotropic components is also presented (see Chapter 7).

In addition to the analytical models of the thermal stresses in the composite (two- and three-component) materials, this book also includes

1. an analytical model of crack formation (see Section 8.2),

2. an analytical model of an energy barrier (see Section 8.3),

3. an analytical model of micro- and macro-strengthening (see Section 8.4),

4. analytical-computational and analytical-experimental-computational methods of lifetime prediction (see Section 8.5).

These analytical models which are based on curve and surface integrals of energy density exhibit a general character regarding the stress-strain state which induces the energy density. Strictly speaking, the energy density is induced by the thermal stresses or is induced by stresses of a different reason than a difference in thermal expansion coefficient of material components. In case of the thermal stresses, the energy density is considered to represent elastic energy density. The same is also valid for the lifetime prediction methods which are, along with these analytical models, applicable to the two- and three-component materials with isotropic, uniaxial and triaxial anisotropic components.

The trilogy *Analytical models of thermal stresses in composite materials I, II, III* thus represents **an integrated scientifi work** with an interdisciplinary character. As presented

Preface

in Section 1.7.1, the trilogy is suitable for several categories of readers i.e. senior undergraduates and PhD. students in mechanical engineering, solid continuum mechanics, applied physics, materials engineering, as well as for researchers and practicing engineers working at universities, scientifi institutes and in industry. [1].

About the Author

Born in 1965 in Košice, Slovak Republic, graduated from the Mechanical Engineering Faculty of the Technical University in Košice (1988) (Department of Mechanical Engineering Technology), and from the Faculty of Sciences of the P. J. Šafárik University in Košice (1993) (Department of Physics of Solids), both with distinction, awarded the prize of the Chancellor of the Technical University in Košice for excellent study results, aimed at heat and chemical treatment and magnetic properties of amorphous alloys (1994–2000), defending a PhD. thesis in Physics of Condensed Matters and Acoustics at the Institute of Experimental Physics of the Slovak Academy of Sciences in Košice (1999), since 2000 employed at the Institute of Materials Research of the Slovak Academy of Sciences in Košice, Dr. Ladislav Ceniga currently works on analytical models of thermal stresses and related thermal-stress induced phenomena in composite materials [2].

Dr. Ladislav Ceniga

Košice, Slovak Republic
February 2011

---

[1] This work was supported by the Slovak Research and Development Agency under the contract No. COST-0022-06.

[2] E-mail addresses: lceniga@yahoo.com, lceniga@hotmail.com, ladislav_ceniga@yahoo.com, ladislav_ceniga@hotmail.com, lceniga@imr.saske.sk

Chapter 1

Outline of principles

1.1 Cell model

The analytical model of the thermal stresses presented in this book is applicable to the types of real composite materials which are define in Items 1-4, Section 2.1. With regard to analytical modelling, the real two- and three-component composite materials with finit dimensions are replaced by infinit multi-particle-matrix and multi-particle-envelope-matrix systems, respectively. The multi-particle-matrix and multi-particle-envelope-matrix systems are **imaginarily** divided into identical cells (see Fig. 2.1a,b), where each cell contains a central spherical particle (with the radius R_1) without and with a spherical envelope (with the radii $R_1 < R_2$ and with the thickness $t = R_2 - R_1 > 0$) on the particle surface, respectively.

The thermal stresses are thus investigated within the cell. Due to the matrix infinit , the analytical model of the thermal stresses which is related to a certain cell is also valid for any cell of the same shape. Additionally, such imaginary dividing of the infinit matrix is required regarding the particles distribution that the cells could fulfi the infinit matrix perfectly. As shown in Fig. 2.1a,b, rectangular-based and hexagonal-based prismatic cells are considered in this book. Finally, corresponding mathematical techniques similar to those presented in Sections 2.2, 2.3 can be applied to such cells to fulfi the infinit matrix perfectly, exhibiting a different shape than the rectangular-based and hexagonal-based prismatic cells.

The inter-particle distances d_1, d_2, d_3 along the axes x_1, x_2, x_3 of the Cartesian system $(Ox_1x_2x_3)$ (see Fig. 3.1) respectively, the radii R_1, R_2 and the particle volume fraction $v \in (0, v_{max})$ are mutually connected within relationships define in Section 2.2, where v_{max} depends on the particle distribution. These parameters of the cell are also microstructural parameters of the types of real composite materials which are define in Items 1-4, Section 2.1.

1.2 Mathematical techniques

In general, the stress-deformation state of a solid continuum is determined at an arbitrary point of the solid continuum, where a position of the arbitrary point is define by suitable coordinates. The arbitrary point is consequently replaced by an infinitesima part of the solid continuum, where a shape of the infinitesima part corresponds to a shape of the solid continuum. Finally, an effect of the rest of the solid continuum on the infinitesima part is replaced by forces/stresses which act on sides of the infinitesima part.

With regard to the spherical particle and the spherical envelope, the infinitesima spherical cap at the point P with a position determined by the spherical coordinates (r, φ, ν) (see Fig. 3.1) regarding the Cartesian system $(Ox_1x_2x_3)$ represents the infinitesima part of the solid continuum, where $r = |OP|$ (see Section 3.1.1). The stress-deformation state which is characterized by the radial stress (strain) $\sigma'_{11} = \sigma_r$ ($\varepsilon'_{11} = \varepsilon_r$), the tangential stresses (strains) $\sigma'_{22} = \sigma_\varphi$, $\sigma'_{33} = \sigma_\nu$ ($\varepsilon'_{22} = \varepsilon_\varphi$, $\varepsilon'_{33} = \varepsilon_\nu$), and the shear stresses (strains) $\sigma'_{12} = \sigma_{r\varphi}$, $\sigma'_{13} = \sigma_{r\nu}$ ($\varepsilon'_{12} = \varepsilon_{r\varphi}$, $\varepsilon'_{13} = \varepsilon_{r\nu}$) is thus determined within the Cartesian system $(Px'_1x'_2x'_3)$ (see Fig. 3.1), where the axes $x'_1 = x_r$ and $x'_2 = x_\varphi$, $x'_3 = x_\nu$ defin radial and tangential directions, respectively. The stresses σ'_{11}, σ'_{12}, σ'_{13} and σ'_{22}, σ'_{21} and σ'_{33}, σ'_{31} act along the axes x'_1 and x'_2 and x'_3, respectively, where $\sigma'_{12} = \sigma'_{21}$, $\sigma'_{13} = \sigma'_{31}$.

The conventional subscripts r, φ, ν, $r\varphi$, $r\nu$, $\varphi\nu$ in a connection with the spherical coordinates (r, φ, ν) (see Fig. 3.1) are replaced by the subscripts 11, 22, 33, 12, 13, 23, respectively, due to the mathematical techniques in Section 3.2 (see e.g. Eqs. (3.23)–(3.34))

The determination of the stress-deformation state results from the Cauchy's, Saint-Venant's and equilibrium equations (see Sections 3.1.3–3.1.5), and from the Hooke's law for an anisotropic solid continuum (see Section 3.1.6) which represent fundamental equations of solid continuum mechanics. The Cauchy's, Saint-Venant's and equilibrium equations are determined for the infinitesima spherical cap which exhibits a radial displacement only in the Cartesian system $(Px'_1x'_2x'_3)$ as analysed in Sections 3.1.2.

As presented in Section 3.2.1, suitable mathematical techniques are used for the determination of the tangential and shear stresses as functions of the radial stress. Consequently, as presented in Section 3.2.2, suitable mathematical techniques are used for the determination of different solutions for the radial stress (see Section 3.2.3) which results in different solutions for the tangential and shear stresses. These different solutions for each of components of the multi-particle-matrix and multi-particle-envelope-matrix systems (see Section 3.3) result in different elastic energy of these systems represented by elastic energy accumulated in the cell (see Eqs. (3.74), (3.75)). Accordingly, the Castigliano's theorem is required to be considered due to different values of the cell elastic energy (see Section 3.2.5).

Integration constants included in formulae for the thermal stresses, strains, thermal-stress induced elastic energy density and thermal-stress induced energy in components of the multi-particle-matrix and multi-particle-envelope-matrix systems are determined by boundary conditions analysed in Chapter 4.

1.3 Reason of thermal stresses

The thermal stresses originate at the temperature $T < T_r$, where T_r is relaxation temperature below that the stress relaxation as a consequence of thermal-activated processes does not occur, where the analysis concerning the relaxation temperature of the multi-particle-matrix and multi-particle-envelope-matrix systems is presented in Section 3.4.1 (see Items 4, 5).

With regard to the multi-particle-matrix system, the thermal stresses originate as a consequence of the difference $\beta_m - \beta_p \neq 0$. With regard to the multi-particle-envelope-matrix system, the thermal stresses originate as a consequence of the differences $\beta_e - \beta_p \neq 0$ and/or $\beta_m - \beta_e \neq 0$. The coefficien β_q (see Eqs. (3.100)–(3.102)) $(q = p,e,m)$ includes the thermal expansion coefficien $\alpha'_{1q} = \alpha'_{1q}(\varphi, \nu)$ (see Eq. (9.11)) along the radial direction as well as the radial strain $\varepsilon_{11tq} = \varepsilon_{11tq}(\varphi, \nu)$ (see Eq. (3.104)) which is induced by a phase transformation at the temperature $T_{tq} \leq T_r$.

If $\beta_m = \beta_p$, the thermal stresses are a consequence of the phase transformation which originates at least in one of components of the multi-particle-matrix system. In this case, the thermal stresses originate at the temperature T_{tq}. The same is also valid for the multi-particle-envelope-matrix system provided that $\beta_p = \beta_e = \beta_m$. Finally, the phase-transformation induced radial strain ε_{11tq} results from a difference in dimensions of mutually transforming crystal lattices of a component of the multi-particle-matrix and system multi-particle-envelope-matrix systems.

The same analysis as presented in Section 1.3 is also valid for the one-particle-(envelope)-matrix system (see Chapter 6) which is required to be determined due to the analysis in Section 2.2.2.

1.4 Temperature range

The analytical model of the thermal stresses in the multi-particle-matrix and multi-particle-envelope-matrix systems presented in this book is related to thermal-stress induced strains which are elastic, i.e. the stress-strain relationships are define by the Hooke's law. Accordingly, the validity of the analytical model is required to consider yield stresses of components of the multi-particle-matrix and multi-particle-envelope-matrix systems as well as a maximal stress of the thermal-stress fiel which is induced in the cell due to $\beta_m \neq \beta_p$, and $\beta_p \neq \beta_e = \beta_m$ (see Sections 5.2.1) or $\beta_p \neq \beta_e \neq \beta_m$ (see Sections 5.2.2) or $\beta_p = \beta_e \neq \beta_m$ (see Sections 5.2.3), respectively.

The analytical model of the thermal stresses presented in this book is valid for the temperature $T \in \langle T_{ys}, T_r \rangle$, where the determination of the critical fina temperature of a cooling process, T_{ys}, is presented in Section 3.4.6.

1.5 Finite matrix

With regard to a finit matrix, the analytical model is required to consider a shape and dimensions of the finit matrix, and a position of each cell in the finit matrix. Additionally, boundary conditions for the radial displacement u_{1m} or for the radial stress σ_{11m}, both related to a surface of each cell, are required to be determined for each cell separately.

On the one hand, let the boundary conditions are defined An application of the analytical model to real composite materials results in numerical dependencies of the thermal stresses and corresponding quantities of the related phenomena analysed in Chapter 8. On the other hand, such application which considers the microstructural parameters v, R_1 and v, R_1, R_2 of real two- and three-component (composite) materials would be probably time-consuming, respectively.

1.6 Subscripts and notation

As an example for the understanding of a calculation of subscripts presented in this book, the subscript $1 + i$ $1 + j$ $1 + k\varphi m$, related to the angle φ and to material parameters of the cell matrix (m), is transformed to the subscripts $112\varphi m$ and $212\varphi m$ for $i, j = 0$, $k = 1$ and $i, k = 1$, $j = 0$, respectively. Analogically, the subscript $1 + i$ $1 + jm$ is transformed to the subscripts $11m$ and $21m$ for $i, j = 0$ and $i = 1$, $j = 0$, respectively.

Similarly, the subscript $i1 + j$ $1 + ke$ related to material parameters of the spherical envelope (e) is transformed to the subscripts $312e$ and $121e$ for $i = 3$, $j = 0$, $k = 1$ and $i, j = 1$, $k = 0$, respectively. Analogically, the subscript $1 + i + 2je$ is transformed to the subscripts $1e$ and $3e$ and $4e$ for $i, j = 0$ and $i = 0$, $j = 1$ and $i, j = 1$, respectively.

Finally, the subscript $1 + ijq$ related to the material parameters of the spherical particle $(q = p)$, the spherical envelope $(q = e)$ and the cell matrix $(q = m)$ is transformed to the subscript $12q$ and $23q$ for $i = 0$, $j = 2$, and $i = 1$, $j = 3$, respectively. Additionally, the subscript $11 + iq$ is transformed to the subscripts $11q$ and $12q$ for $i = 0$ and $i = 1$, respectively.

1.7 Recommendations of author

1.7.1 Categories of readers

With regard to practicability, this book as well as the trilogy is devoted to the following categories of readers:

1. senior undergraduates, PhD students and senior researchers in

    (a) mechanical engineering and solid continuum mechanics,

    (b) applied physics,

    (c) Materials Engineering,

Outline of principles

2. practicing engineers in materials engineering, working at universities, scientifi institutes and in industry.

This book as well as the trilogy exhibit an interdisciplinary character which resulting
from an application of principles of solid continuum mechanics to composite materials with
anisotropic (Volume I) and isotropic (Volumes I, II).

This book is thus intended to be used as a textbook on applied mechanics regarding
readers of **Category 1a**. The readers are able

- to determine parameters of any cell which fulfi the infinit matrix perfectly (see
  Item 2, Section 2.1.1),

- to determine analytical models of thermal stresses in the multi-particle-**multi-
  envelope**-matrix and one-particle-**multi-envelope**-matrix systems with anisotropic
  components on the conditions $\beta_p \neq \beta_e \neq \beta_m$, $\beta_p \neq \beta_e = \beta_m$, $\beta_p = \beta_e \neq \beta_m$
  (see Eqs. (3.100)–(3.102)),

- consequently to defin corresponding boundary conditions for the determination of
  integration constants included in solutions for the thermal stresses in the spherical
  particle, the spherical envelopes and the cell matrix,

- and to modify the results concerning the analytical fracture mechanics (see Section 8.2), and the analytical models of the energy barrier (see Section 8.3) and the
  analytical model of the micro-/macro-strengthening (see Section 8.4) for the multi-
  particle-**multi-envelope**-matrix system.

The readers of **Categories 1a, 1b** are able to incorporate the thermal-stress induced
field in the multi-particle-matrix, multi-particle-envelope-matrix and multi-particle-**multi-
envelope**-matrix systems with anisotropic components into the Eshelby's model. The Eshelby's model and its development [1]–[3] defin the disturbance of an applied stress-fiel
in a solid continuum, where the applied stress-fiel is disturbed due to the presence of
inclusions in a solid continuum. This incorporation thus define a stress-strain state in a
composite material loaded by thermal and mechanical stresses.

With regard to **Category 1b** formulae for thermal-stress induced elastic energy density in components of the multi-particle-matrix, multi-particle-envelope-matrix and multi-
particle-**multi-envelope**-matrix systems with anisotropic components can be substituted to
formulae for the energy barrier (see Section 8.3). Consequently, formulae for this thermal-
stress induced energy barrier can be incorporated into analytical models which are created
by physicists and which describe an interaction of energy barriers with dislocations or magnetic domain walls.

Readers of **Categories 1c, 2** are able to determine parameters of the crack formation
analysed in Section 8.2 (critical particle radii, functions describing crack shapes), micro-
and macro-strengthening analysed in Section 8.4 and lifetime analysed in Section 8.5.
Strictly speaking, readers of **Categories 1c, 2** are able to design and develop microstructure
of a real composite material with such parameters

- which do not correspond to a limit state characterized by the critical particle radius (see Section 8.2),

- which result in such micro- and macro-strengthening to be required against mechanical loading (see Section 8.4),

- which result in the resistive effect of the thermal stresses against mechanical loading (see Section 8.5).

Additionally, with regard to completeness of the topic of the thermal stresses, analytical models of the thermal stresses in one-particle-matrix and one-particle-envelope-matrix systems are presented in Chapter 6.

The monographs, presented in Section *Bibliography* and written in Slovak and Czech, represent common textbooks on material engineering [4], mechanics of solid continuum [5]–[7] and applied mathematics [8], and accordingly equivalent monographs are offered by any public library as well as libraries at universities and scientifi institutes.

Finally, the cracking parameters, the energy barrier along with the micro- and macro-strengthening are determined by integration of energy density. This integration is determined numerically by the programming techniques analysed in Section 1.7.2.

1.7.2 Programming techniques

Numerical determination of the thermal stresses, and the related phenomena in Chapter 8 includes determination of derivatives and integrals.

With regard to software creation for the determination of the thermal stresses, a derivative of the parameter Q regarding the variable η can be calculated numerically by the formula

$$\frac{\partial Q}{\partial \eta} \approx \frac{Q(\eta + \Delta \eta) - Q(\eta)}{\Delta \eta}, \tag{1.1}$$

where $Q = \kappa, \psi$, and $\eta = \varphi, \nu$. The angle steps $\Delta\varphi$, $\Delta\nu$ can be taken as small as possible, e.g. $\Delta\varphi = \Delta\nu = 10^{-6}$ [deg].

Similarly, with regard to the elastic energy W representing a definit integral of the coefficien Ω as a function of the angles $\varphi, \nu \in \langle 0, \pi/2 \rangle$, the definit integral can be calculated numerically by the formula

$$W = \int_{0}^{\pi/2} \int_{0}^{\pi/2} \Omega \, d\varphi \, d\nu \approx \sum_{j=0}^{m} \left(\sum_{i=0}^{n} W(i \times \Delta\varphi; j \times \Delta\nu) \, \Delta\varphi \right) \Delta\nu, \tag{1.2}$$

where n, m are integral parts of the real numbers $\pi/(2\,\Delta\varphi)$, $\pi/(2\,\Delta\nu)$, respectively. With regard to an experience of the author, the steps $\Delta\varphi = \Delta\nu = 0.01 - 0.1$ [deg], dependent on a computer hardware, is sufficient

Outline of principles

With regard to the determination of the particle radius R_1, in case of the R_1-independent and φ, ν-dependent function $(\varepsilon_{22q})_{r=R_1}$ $(q = p,m)$, the definit integral in Eq. (3.65) can be calculated numerically by the formula

$$\int_0^{\pi/2} \int_0^{\pi/2} (\varepsilon_{22q})_{r=R_1} \, d\varphi \, d\nu \approx \sum_{j=0}^{m} \left(\sum_{i=0}^{n} [\varepsilon_{22q} \, (i \times \Delta\varphi; j \times \Delta\nu)]_{r=R_1} \, \Delta\varphi \right) \Delta\nu, \quad (1.3)$$

considering the same conditions concerning the steps $\Delta\varphi$, $\Delta\nu$ as related to Eq. (1.2).

With regard to quantitative determination of the critical parameters R_{1c}, t_c (see Section 8.5), the average values $\overline{p_1}$ and $\overline{p_2}$ of the radial stresses p_1, p_2 given by Eqs. (8.125) are required to be derived. Considering the same conditions concerning the steps $\Delta\varphi$, $\Delta\nu$ as related to Eq. (1.2), the definit integral given by Eq. (8.125) can be calculated numerically by the formula

$$\overline{p_{1+i}} = \left(\frac{2}{\pi} \right)^2 \int_0^{\pi/2} \int_0^{\pi/2} p_{1+i} \, d\varphi \, d\nu \approx \left(\frac{2}{\pi} \right)^2 \sum_{k=0}^{m} \left(\sum_{j=0}^{n} p_{1+i} \, (j \times \Delta\varphi; k \times \Delta\nu) \, \Delta\varphi \right) \Delta\nu,$$

$$i = 0, 1. \tag{1.4}$$

Finally, the integrals in Sections 8.3, 8.4 can be numerically determined by the procedure described in Eqs. (1.2)–(1.4), where the step $\Delta\eta$ $(\eta = \varphi, \nu)$ is replaced by Δx. The step Δx

With regard to an experience of the author, the step $\Delta x = (x_2 - x_1)/1000$ which depends on a computer hardware is sufficient where x_1, x_2 are integration boundaries, and $x_2 > x_1$.

Chapter 2

Cell model

2.1 Geometric boundary condition for cell matrix

The analytical models of the thermal stresses and corresponding quantities presented of the related phenomena (see Chapter 8) in this book are applicable to real composite materials represented by

1. grains and precipitates, the latter aperiodically distributed in the grains or at the grain boundaries without or with a continuous component on a surface of each of the precipitates, where the grains and the precipitates along with the continuous component exhibit mutually different thermal expansion coefficients

2. two types of aperiodically distributed grains with different thermal expansion coefficients and without the continuous component on a surface of the grains of both types,

3. two types of aperiodically distributed grains with identical thermal expansion coefficients, where one type of the grains is covered by the continuous component with a different thermal expansion coefficien than those of the grains of both types,

4. one type of grains covered by a continuous component with a different thermal expansion coefficien than those of the grains.

With regard to the analytical models of the thermal stresses, the real composite materials with finit dimensions are replaced by a multi-particle-(envelope)-matrix system with infinit dimensions. The multi-particle-matrix and multi-particle-envelope-matrix systems consist of by periodically distributed spherical particles without and with a spherical envelope on a surface of each of the spherical particle surfaces which are embedded in an infinit matrix, respectively. To derive the thermal stresses acting in the multi-particle-(envelope)-matrix system, the infinit matrix is **imaginarily** divided into identical cells with such a shape which corresponds to a distribution of the particles. Consequently, the thermal stresses are investigated within the cell representing a part of the solid continuum related to one spherical particle. Considering the matrix infinit , formulae for thermal stresses related to a certain cell are consequently valid for any cell of the infinit matrix (see Fig. 2.1a,b).

As presented in Items 1, 2 and 3, the precipitates, one of the two types of the grains, and the grains covered by the continuous component are considered to represent the periodically distributed spherical particles. Consequently, with regard to Items 1, 2 and 3, the grains embedding the precipitates, the other type of the grains, and the grains without the continuous component are considered to represent the infinit matrix resulting in the cell matrix.

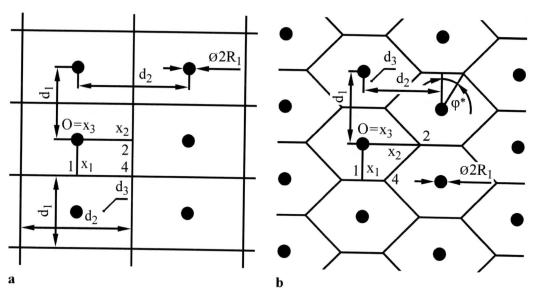

Figure 2.1: The infinit matrix divided into the imaginary (a) rectangular-based and (b) hexagonal-based prismatic cells with the central spherical particles of the radius R_1 at the point O of the Cartesian system $(Ox_1x_2x_3)$, and the inter-particle distance d_i along the axis x_i ($i = 1, 2, 3$). Along with the inter-particle distance d_i, the hexagonal-based prismatic cell is additionally define by the angle $\varphi^* \in (0, \pi/2)$. Considering the multi-particle-envelope-matrix system, the spherical envelope on the particle surface with the radii $R_1 < R_2$ is not presented regarding simplificatio of the figure

With regard to Item 4, the grain covered by the continuous component represents the spherical particle covered by the spherical envelope with such envelope radius so that the spherical envelope fill the cell maximally. Consequently, the rest of the cell, i.e. the cell matrix, is represented by the grains, neglecting the presence of the continuous component on their surfaces (see Section 2.4).

With regard to Items 1–4, boundaries of the grains representing the infinit matrix, and consequently the cell matrix, are considered to represent a coherent solid continuum. Finally, the continuous component on the particle surfaces is considered to represent the spherical envelope.

With regard to the periodical distribution of the spherical particles with the radius R_1 shown in Fig. 2.1a,b, the infinit matrix is **imaginarily** divided into rectangular-based (*a*) and hexagonal-based (*b*) prismatic cells, both define by the inter-particle distance d_i along the axis x_i ($i = 1,2,3$) and the latter additionally define by the angle $\varphi^* \in (0, \pi/2)$.

Moreover, the thermal stresses are sufficien to be investigated in the interval $\varphi, \nu \in \langle 0, \pi/2 \rangle$ regarding the spherical coordinates (r, φ, ν) (see Figs. 2.1a,b, 3.1). This is a consequence of symmetry of the multi-particle-matrix and multi-particle-envelope-matrix systems due to the matrix infinit and the periodical distribution of the spherical particles and the spherical envelopes.

The dimensions R_1, R_2, d_i ($i = 1,2,3$) in Fig. 2.1a,b are related to the temperature $T \in \langle T_f, T_r \rangle$, where T_f is fina temperature of a cooling process, and the relaxation temperature T_r is analysed in Section 3.4.1 (see Items 4, 5, p. 41).

The spherical particles in the individual planes $x_1' x_2'$, mutually parallel with the inter-plane distance d_3 along the axis x_3', are mutually located along the axis x_3'.

Considering zero displacement of the points O, the centres *1, 2, 4* of the abscissae *O1O*, *O2O*, *O4O* in the rectangular-based prismatic cell, respectively, exhibit zero displacement along these abscissae, as also valid for the arbitrary point C on the cell surface along the axis x_1' representing a radial direction (see Figs. 2.2, 2.3, 3.1). If a condition a non-zero radial displacement on a surface of a certain cell is assumed, then this condition related to the surface of a certain cell is required to be valid - due to the matrix infinit - for a surface of any cell as well as for surfaces of neighbouring cells. Accordingly, if the condition of the non-zero radial displacement on the cell surface is assumed, then the space between neighbouring cells is replaced by the vacuum what is physically unacceptable.

The condition of zero displacement on the cell surface is also valid for the points *1, 2, 4* of the abscissae *O1O*, *O22O*, *O4*, respectively, and for the arbitrary point C on a surface of the hexagonal-based prismatic cell (see Figs. 2.1b, 2.3). The same physically unacceptable situation arises provided that cubic and regular-hexagonal-based prismatic cells, transformed from the rectangular-based and hexagonal-based prismatic cells on the conditions $d_i = d$ ($i = 1,2,3$) and $d_1 = d_3 = d$, $d_2 = 2d\sqrt{3}/3$, $\varphi^* = \pi/6$, respectively, are replaced by spherical cells connected at one point related to a pair of the spherical cells, where the matrix between three spherical cells is replaced by the vacuum.

The analytical model of the thermal stresses acting in the multi-particle-(envelope)-matrix system with a finit matrix is required to consider a position, a shape and dimensions of each cell in the finit matrix. Additionally, boundary conditions, related to the radial displacement u_{1m} or the radial stress σ_{11m} both on surfaces of each cell, are required to be define for each cell separately. Accordingly, even though the boundary conditions are defined the application of the analytical model to a real composite material, resulting in numerical dependencies of the thermal stresses and corresponding quantities (see Chapter 8) on the particle volume fraction v and on the particle and envelope radii, R_1 and $R_1 < R_2$, respectively, would be probably time-consuming.

On the one hand, as a hypothetic conclusion based on the tendency of a system to exhibit such a stress-deformation state which corresponds to minimal energy [7, p. 219], provided that the multi-particle-(envelope)-matrix system is loaded by the thermal stresses only, the system represented by the real composite materials define in Items 1–4 might be assumed to exhibit such particle distribution of those shown in Fig. 2.1a,b, which results in lower thermal-stress induced elastic energy of the cell, W_c (see Eqs. (3.74), (3.75)). On the other hand, with regard to the quantitative determination of the elastic thermal-stress loading in the real composite materials, such particle distribution of those shown in Fig. 2.1a,b is considered within the presented analytical models, which results in lower thermal-stress induced elastic energy of the cell.

2.1.1 Summary

Accordingly, the following conclusions, including a geometric boundary condition for the cell matrix define in Item 1, are considered:

1. zero radial displacement of an arbitrary point in the cell matrix on the cell surface, $(u_{1m})_{r=r_c} = 0$, accordingly represents a geometric boundary condition for the cell matrix, simultaneously valid for cells of such a shape which fil the infinit matrix perfectly, where the φ, ν-dependent distance r_c as a function of the experimentally determined parameters v, R_{1ex}, d_{1ex}, d_{2ex}, d_{3ex} (see Section 2.2.1) is derived in Section 2.3,

2. conversely, such an imaginary dividing of the infinit matrix is required regarding particles distribution that cells can fil the infinit matrix perfectly,

3. corresponding mathematical techniques similar to those presented in Sections 2.3.1, 2.3.2 can be applied to such cells which fil the infinit matrix perfectly, exhibiting a different shape than the rectangular-based and hexagonal-based prismatic cells.

2.2 Particle volume fraction

2.2.1 General formulae

With regard to a real composite material, provided that the particles and the envelopes, exhibiting general shapes and aperiodic distribution, are both considered to exhibit the spherical shape and periodic distribution, the multi-particle-(envelope)-matrix system is characterized at the fina temperature T_f of a cooling process by the average particle and envelope radii, R_{1ex} and $R_{1ex} < R_{2ex}$, respectively, by the average inter-particle distance d_{iex} $(i = 1,2,3)$ along the axis x_i of the Cartesian system $(Ox_1x_2x_3)$ (see Fig. 2.1a,b), and by the particle volume fraction v, where at least four of the parameters R_{1ex}, d_{1ex}, d_{2ex}, d_{3ex}, v are determined by experimental methods, e.g. using microscopy [1]. Provided that

---

[1] With regard to the microscopy, the envelope thickness t is usually determined, and thus the envelope radius has the form $R_{2ex} = R_{1ex} + t$ (see Eq. (2.5)).

the inter-particle distance d_{iexT} ($i=1,2,3$) is determined at the temperature $T \in \langle T_f, T_r \rangle$, the inter-particle distance d_{iex} at the temperature T_f has the form

$$d_{iex} = d_{iexT}\left(1 + \beta_{imT}\right), \quad i = 1, 2, 3, \tag{2.1}$$

where the coefficien β_{imT} along the axis x_i ($i=1,2,3$) (see Figs. 2.1, 3.1) is given by Eqs. (3.100)–(3.102), (9.2)–(9.4), (9.11), replacing the temperature T_r in the integrals of Eqs. (3.100)–(3.102) by the temperature T, where an analysis concerning the relaxation temperature T_r is presented in Section 3.4.1 (see Items 4, 5, p. 41). The subscripts $i=1$; $i=2$; $i=3$ in β_{imT}, x_i are related to the conditions $\varphi = 0$, $\nu = \pi/2$; $\varphi = \nu = \pi/2$; $\varphi = \nu = 0$ for the angles $\varphi, \nu \in \langle 0, \pi/2 \rangle$ (see Fig. 3.1) included in Eqs. (9.2)–(9.4), respectively.

With regard to four parameters of $R_{1ex}, d_{1ex}, d_{2ex}, d_{3ex}, v$ determined by experimental methods, the fift parameter can be derived from the formula

$$v = \frac{V_p}{V_c} = \frac{4\pi R_{1ex}^3}{3V_c} \in (0, v_{max}), \tag{2.2}$$

where $V_p = 4\pi R_{1ex}^3/3$ is volume of the spherical particle, and V_c is volume of the rectangular-based and hexagonal-based prismatic cells resulting from the particles distribution in Fig. 2.1a,b. The hexagonal-based prismatic cell is additionally define by the angle φ^*.

Consequently, the parameters $R_{1ex}, R_{2ex}, d_{1ex}, d_{2ex}, d_{3ex}, v$ can exhibit different numerical values related to cells in different areas of the multi-particle-(envelope)-matrix system, provided that the cells fil the infinit matrix perfectly, as expressed in Item 2 in Section 2.1.1.

With regard to a physical point of view, the multi-particle-(envelope)-matrix system 'is transformed' on the condition $v \to 0$ to a one-particle-(envelope)-matrix system containing one spherical particle which is embedded in the infinit matrix.

2.2.2 Analysis of interval boundaries $v \to 0, v \to v_{max}$

As presented in Section 3.2.4, elastic energy of the infinit multi-particle-(envelope)-matrix system is represented by the thermal-stress induced elastic energy of the cell, W_c (see Eqs. (3.74), (3.75)).

The dependence $W_c - v$, resulting from dependencies of the thermal stresses on v, is required to exhibit limits $W_{c0} = \lim_{v \to 0} W_c$, $W_{cmax} = \lim_{v \to v_{cmax}} W_c$, where v_{max} is given by Eqs. (2.4), (2.7), (2.10) considering $t = 0$ for the multi-particle-matrix system. Accordingly, the analytical model of the thermal stresses inducing the elastic energy $W_{c0} = \lim_{v \to 0} W_c$ related to the multi-particle-(envelope)-matrix system defines regarding $v \to 0$, a stress-deformation state in the one-particle-(envelope)-matrix system.

The limits $W_{c0} = \lim_{v \to 0} W_c$, $W_{cmax} = \lim_{v \to v_{max}} W_c$ are meant to be determined numerically, considering material parameters of a real composite material, and consequently the limit $W_{cmax} = \lim\limits_{v \to v_{max}} W_c$ is compared with a numerical value of $W^{(1)}$ related to the same real composite material, where $W^{(1)} = W_p^{(1)} + \left(W_e^{(1)}\right) + W_m^{(1)} {}^2$ is elastic energy accumulated in the one-particle-(envelope)-matrix system (see Section 6). Additionally, the limit $W_{c0} = \lim\limits_{v \to 0} W_c$ exists provided that the positive dependence $W_c - r_c$ is a decreasing function of the v-dependent term $r_c = r_c(v)$, or provided that the term $r_c \to \infty$ for $v \to 0$ (see Section 2.3) is reduced in fractions included in a formula for W_c, otherwise $\lim\limits_{v \to 0} W_c \to \infty$. With regard to a real composite material with $v \not\to 0$, the investigation concerning the condition $\lim\limits_{v \to 0} W_c \to \infty$ is irrelevant.

With regard to the Castigliano's theorem (see Section 3.2.5), such a solution or such a combination of solutions for components of the multi-particle-(envelope)-matrix system are considered to result in minimal energy W_c. When $W_{c0}^{(i)}$ ($i = 1, \dots, n$) is minimal elastic energy of the set $\left\{W_{c0}^{(1)}, \dots, W_{c0}^{(n)}\right\}$ related to n combinations of solutions, and provided that $W_{c0}^{(i)} \leq W^{(1)}$, consequently the combination of solutions resulting in the elastic energy $W_{c0}^{(i)} = \lim\limits_{v \to 0} W_c^{(i)}$ is simultaneously considered for the determination of the elastic thermal-stress loading in the one-particle-(envelope)-matrix system, regarding $v \to 0$. Accordingly, the analytical model for the one-particle-(envelope)-matrix system arises from that for the multi-particle-(envelope)-matrix system.

On the one hand, the condition $W_{c0}^{(i)} > W^{(1)}$ results in a discontinuity of solutions related to the multi-particle-(envelope)-matrix system for $v \to 0$ and to the one-particle-(envelope)-matrix system. On the other hand, with regard to a real composite material with $v \not\to 0$, the investigation of the discontinuity is irrelevant, such as in the case $\lim\limits_{v \to 0} W_c \to \infty$.

2.2.3 Rectangular-based prismatic and cubic cells

Considering the volume $V_c = d_{1ex}d_{2ex}d_{3ex}$ of the rectangular-based prismatic cell, the particle volume fraction v has the form

$$v = \frac{4\pi R_{1ex}^3}{3 d_{1ex}d_{2ex}d_{3ex}} \in (0, v_{max}), \tag{2.3}$$

where the parameter v_{max}, determined by the condition $R_{1ex} = (d_{ex}/2) - t$, is derived as

$$v_{max} = \frac{4\pi}{3 d_{1ex}d_{2ex}d_{3ex}} \left(\frac{d_{ex}}{2} - t\right)^3, \tag{2.4}$$

where d_{ex} is the minimum of the set $\{d_{1ex}, d_{2ex}, d_{3ex}\}$, considering the thickness t of the spherical envelope in the form

$$t = R_{2ex} - R_{1ex}, \tag{2.5}$$

---

[2]The superscript (1) is related to the *one*-particle-(envelope)-matrix system.

and accordingly $t = 0$ for the multi-particle-matrix system. Similarly, on the condition $d_{iex} = d_{ex}$ $(i = 1,2,3)$ corresponding to a cubic cell, we get

$$v = \frac{4\pi}{3} \left(\frac{R_{1ex}}{d_{ex}} \right)^3 \in (0, v_{max}), \tag{2.6}$$

$$v_{max} = \frac{\pi}{6} \left(1 - \frac{2t}{d_{ex}} \right)^3 = \frac{\pi}{6} \left(1 - \frac{t}{R_{1ex} + t} \right)^3, \tag{2.7}$$

transforming to $v_{max} = \pi/6$ for the multi-particle-matrix system.

2.2.4 Hexagonal-based prismatic cell

Considering the volume $V_c = d_{1ex}d_{2ex}d_{3ex}$ of the hexagonal-based prismatic cell [3], the particle volume fraction v and the parameter v_{max} are given by Eqs. (2.3) and (2.4), respectively, where d_{ex} is the minimum of the set $\{d_{1ex}, d_{24}, d_{3ex}\}$. The parameter d_{24} represents altitude of the triangle $O24$ related to the basis 24 derived in the form [4]

$$d_{24} = \frac{d_{1ex} \left(2d_{2ex} - d_{1ex} \tan \varphi^* \right)}{2 \left[d_{1ex}^2 + 4 \left(d_{2ex} - d_{1ex} \tan \varphi^* \right)^2 \right]^{1/2}}. \tag{2.8}$$

On the conditions $d_{1ex} = d_{3ex} = d_{ex}$, $d_{2ex} = 2d_{ex}\sqrt{3}/3$ corresponding to a regular hexagon, thus for $\varphi^* = \pi/6$, we get

$$v = \frac{2\pi\sqrt{3}}{3} \left(\frac{R_{1ex}}{d_{ex}} \right)^3 \in (0, v_{max}), \tag{2.9}$$

where the parameter v_{max}, determined by the condition $R_{1ex} = (d_{ex}/2) - t$, is derived as

$$v_{max} = \frac{\pi\sqrt{3}}{12} \left(1 - \frac{2t}{d_{ex}} \right)^3 = \frac{\pi\sqrt{3}}{12} \left(1 - \frac{t}{R_{1ex} + t} \right)^3, \tag{2.10}$$

transforming to $v_{max} = \pi\sqrt{3}/12$ for the multi-particle-matrix system, thus for $t = 0$, where the envelope thickness t is given by Eq. (2.5).

---

[3]Replacing d_i by d_{iex} $(i = 1,2,3)$ (see Fig. 2.1b), the volume V_c of the hexagonal-based prismatic cell is given by $V_c = 4d_{3ex} (S_{O14} + S_{O24})$, where $S_{O14} = |O1| \times |14| /2$ and $S_{O24} = |O1| \times |O2| /2$ [8, p. 117] represent surface area of the triangles $O14$ and $O24$, respectively, considering the lengths $|O1| = d_{1ex}/2$, $|O2| = d_{2ex} - |14|$, $|14| = (d_{1ex} \tan \varphi^*) /2$ of the abscissae $O1$, $O2$, 14, respectively. To avoid misunderstanding with regard to Fig. 2.1b, the abscissa $O4$ is not drawn.

[4]Considering the surface area $S_{O24} = d_{24} \times |24| /2 = |O1| \times |O2| /2$ of the triangle $O24$ with the basis $|24|$ and the altitude d_{24}, we get $d_{24} = |O1| \times |O2| / |24|$, where $|24|^2 = |O1|^2 + (d_{2ex} - 2 |14|)^2$.

2.3 Determination of distance $r_c = r_c(\varphi, \nu)$

With regard to the geometric boundary condition $(u_{1m})_{r=r_c} = 0$ define in Item 1 in Section 2.1.1, the parameter r_c, representing a φ, ν-dependent distance from the particle centre O to the points C_1 or C_3 on the cell surfaces *1456* or *3657* with the normals x_1 or x_3 (see Figs. 2.2, 2.3), respectively, is required to be determined.

With regard to Figs. 2.2, 2.3, C_1 and C_3 are points of intersection of the cell surfaces *1456* and *3657* with the axis x_1' for $\nu = \angle(\overline{OC_3}, x_3) \in \langle 0, \nu^* \rangle$ and $\nu = \angle(\overline{OC_1}, x_3) \in \langle \nu^*, \pi/2 \rangle$ (see Figs. 2.2, 2.3), respectively. The axis x_1' define a radial direction, regarding the Cartesian system $(Px_1'x_2'x_3')$ (see Fig. 3.1), and the angle $\nu^* = \angle(\overline{O9}, \overline{O3})$ is given by Eqs. (2.31)–(2.37).

Finally, the φ, ν-dependent distance $r_c = |\overline{OC_3}|$ or $r_c = |\overline{OC_1}|$ for $\nu \in \langle 0, \nu^* \rangle$ or $\nu \in \langle \nu^*, \pi/2 \rangle$, representing length of the abscissae $\overline{OC_3}$ or $\overline{OC_1}$ in the radial direction, respectively, is derived as a function of the experimentally determined parameters v, R_{1ex}, $d_{1ex}, d_{2ex}, d_{3ex}$ (see Section 2.2.1, 2.3.1, 2.3.2), for the rectangular-based and hexagonal-based prismatic cells, and resulting from the formulae for the particle volume fraction v given by Eqs. (2.3), (2.4), (2.8).

2.3.1 Rectangular-based prismatic and cubic cells

2.3.1.1 Interval $\nu \in \langle 0, \nu^* \rangle$

The distance $r_c = |\overline{OC_3}|$ between the particle centre O and the point C_3 on the surface *3657*, i.e. for $\nu = \angle(\overline{OC_3}, x_3) \in \langle 0, \nu^* \rangle$, is derived by the following mathematical procedure.

With regard to Eq. (2.3), the length $|\overline{O3}|$ and $|\overline{39}|$ of the abscissae $\overline{O3}$ and $\overline{39}$, respectively, and consequently the angle φ^* are derived as

$$|\overline{O3}| = \frac{d_{3ex}}{2} = \frac{4\pi R_{1ex}^3}{3vd_{1ex}d_{2ex}}, \tag{2.11}$$

$$|\overline{39}| = \frac{1}{2c_\varphi}, \tag{2.12}$$

$$\nu^* = \arctan\left(\frac{|\overline{39}|}{|\overline{O3}|}\right) = \arctan\left(\frac{1}{c_\varphi d_{3ex}}\right), \tag{2.13}$$

where the coefficien c_φ is given by Eqs. (2.26)–(2.28) and (2.31)–(2.33) for the rectangular-based prismatic and cubic cells, respectively.

Considering the sine rule (see Fig. 2.2) [8, p. 63]

$$\frac{|\overline{C_39}|}{\sin[\angle(\overline{OC_3}, \overline{O9})]} = \frac{|\overline{O9}|}{\sin[\angle(\overline{OC_3}, \overline{C_39})]} \tag{2.14}$$

and the condition

$$|\overline{O9}| = r_c \cos[\angle(\overline{OC_3}, \overline{O9})] + |\overline{C_39}| \cos[\angle(\overline{C_39}, \overline{O9})], \tag{2.15}$$

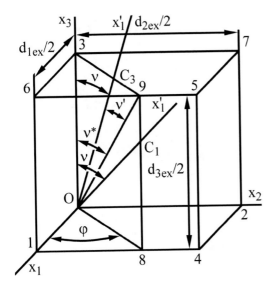

Figure 2.2: The points C_3 and C_1 as intersections of the axis x_1', representing a radial direction (see Fig. 3.1), with the surfaces *3657* and *1456* of one eighth of the rectangular-based prismatic cell (see Fig. 2.1a) for $\nu \in \langle 0, \nu^* \rangle$ and $\nu \in \langle \nu^*, \pi/2 \rangle$, respectively.

we get

$$r_c = \frac{|\overline{O9}|}{\cos\left[\angle\left(\overline{OC_3},\overline{O9}\right)\right]} \left\{ 1 - \frac{\sin\left[\angle\left(\overline{OC_3},\overline{O9}\right)\right] \cos\left[\angle\left(\overline{C_39},\overline{O9}\right)\right]}{\sin\left[\angle\left(\overline{OC_3},\overline{C_39}\right)\right]} \right\}. \quad (2.16)$$

The length $|\overline{O9}|$ of the abscissa $\overline{O9}$ and the angles $\angle\left(\overline{OC_3},\overline{O9}\right)$, $\angle\left(\overline{C_39},\overline{O9}\right)$, $\angle\left(\overline{OC_3},\overline{C_39}\right)$, regarding Eqs. (2.11), (2.12), are derived as

$$|\overline{O9}| = \sqrt{|\overline{O3}|^2 + |\overline{39}|^2} = \frac{\sqrt{1 + (c_\varphi d_{3ex})^2}}{2 c_\varphi}, \quad (2.17)$$

$$\angle\left(\overline{OC_3},\overline{O9}\right) = \nu' = \nu^* - \nu, \quad \nu' \in (0, \nu^*), \quad (2.18)$$

$$\angle\left(\overline{C_39},\overline{O9}\right) = \frac{\pi}{2} - \nu^*, \quad (2.19)$$

$$\angle\left(\overline{OC_3},\overline{C_39}\right) = \frac{\pi}{2} + \nu^* - \nu'. \quad (2.20)$$

Considering the following relationships [8, p. 72]

$$\sin\left(\frac{\pi}{2} \pm \alpha\right) = \cos\alpha, \quad (2.21)$$

$$\cos\left(\frac{\pi}{2} \pm \alpha\right) = \mp \sin\alpha, \quad (2.22)$$

$$\sin(\alpha \pm \beta) = \sin\alpha \cos\beta \pm \cos\alpha \sin\beta, \quad (2.23)$$

$$\cos(\alpha \pm \beta) = \cos\alpha \cos\beta \mp \sin\alpha \sin\beta, \quad (2.24)$$

and with regard to Eqs. (2.3), (2.17)–(2.20), the distance $r_c = \left|\overline{OC_3}\right|$ (see Fig. 2.3) has the form

$$r_c = \left|\overline{OC_3}\right| = \frac{d_{3ex}\sqrt{1 + (c_\varphi d_{3ex})^2}}{2\left(\sin \nu' + c_\varphi d_{3ex} \cos \nu'\right)} = \frac{2\pi R_{1ex}^3 \sqrt{1 + (c_\varphi d_{3ex})^2}}{3\upsilon d_{1ex} d_{2ex}\left(\sin \nu' + c_\varphi d_{3ex} \cos \nu'\right)},$$
$$\nu' \in (0, \nu^*\rangle, \tag{2.25}$$

where the coefficien c_φ and consequently the angle φ^* are derived as

$$c_\varphi = \frac{\cos \varphi}{d_{1ex}}, \quad \varphi \in \langle 0, \varphi^*\rangle, \tag{2.26}$$

$$c_\varphi = \frac{\sin \varphi}{d_{2ex}}, \quad \varphi \in \left(\varphi^*, \frac{\pi}{2}\right), \tag{2.27}$$

$$\varphi^* = \arctan\left(\frac{d_{2ex}}{d_{1ex}}\right). \tag{2.28}$$

On the condition $d_{iex} = d_{ex}$ $(i = 1,2,3)$ corresponding to a cubic cell, regarding Eq. (2.6), we get

$$r_c = \left|\overline{OC_3}\right| = R_{1ex} f_c, \tag{2.29}$$

where the function f_c, regarding Eq. (2.6), and consequently the angle ν^* along with the coefficien c_φ are derived as

$$f_c = \frac{\sqrt{1 + c_\varphi^2}}{2\left(c_\varphi \cos \nu' + \sin \nu'\right)} \left(\frac{4\pi}{3\upsilon}\right)^{1/3}, \quad \nu' \in (0, \nu^*\rangle, \tag{2.30}$$

$$\nu^* = \arctan\left(\frac{1}{c_\varphi}\right), \tag{2.31}$$

$$c_\varphi = \cos \varphi, \quad \varphi \in \left\langle 0, \frac{\pi}{4}\right\rangle, \tag{2.32}$$

$$c_\varphi = \sin \varphi, \quad \varphi \in \left(\frac{\pi}{4}, \frac{\pi}{2}\right). \tag{2.33}$$

2.3.1.2 Interval $\nu \in \langle \nu^*, \pi/2\rangle$

Similarly, the φ, ν-dependent distance $r_c = \left|\overline{OC_1}\right|$ between the particle centre O and the point C_1 on the surface *1456*, i.e. for $\nu = \angle\left(\overline{OC_1}, x_3\right) \in \langle \nu^*, \pi/4\rangle$, has the form

$$r_c = \frac{1}{2c_\varphi \sin \nu}, \quad \nu \in \left\langle \nu^*, \frac{\pi}{2}\right\rangle, \tag{2.34}$$

where the angle ν^* and the coefficient c_φ are given by Eqs. (2.13) and (2.26)–(2.28), respectively.

On the condition $d_{iex} = d_{ex}$ ($i = 1, 2, 3$) corresponding to a cubic cell, regarding Eq. (2.6), we get
$$r_c = |\overline{OC_1}| = R_{1ex} f_c, \qquad (2.35)$$
where the function f_c, regarding Eq. (2.6), is derived as
$$f_c = \frac{1}{2 c_\varphi \sin \nu} \left(\frac{4\pi}{3v} \right)^{1/3}, \quad \nu \in \langle \nu^*, \frac{\pi}{2} \rangle, \qquad (2.36)$$
and the angle ν^* along with the coefficien c_φ are given by Eqs. (2.31)–(2.33).

2.3.2 Hexagonal-based prismatic cell

The φ, ν-dependent distance r_c between the particle centre O and the points C on the surfaces *3657* and *1456* of the hexagonal-based prismatic cell with the parameters $d_{1ex}, d_{2ex}, d_{3ex}, \varphi^*$ (see Fig. 2.3) is derived by Eqs. (2.13), (2.25), (2.26), (2.34), considering the coefficien c_φ for $\varphi \in (\varphi^*, \pi/2\rangle$ in the form

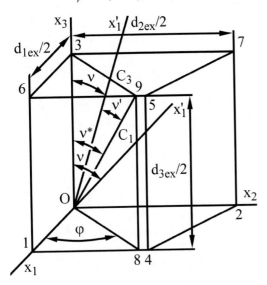

Figure 2.3: The points C_3 and C_1 as intersections of the axis x'_1, representing a radial direction (see Fig. 3.1), with the surfaces *3657* and *1456* of one eighth of the hexagonal-based prismatic cell (see Fig. 2.1b) for $\nu \in \langle 0, \nu^* \rangle$ and $\nu \in \langle \nu^*, \pi/2 \rangle$, respectively.

$$c_\varphi = \frac{d_{1ex} \sin \varphi + 2 (d_{2ex} - d_{1ex} \tan \varphi^*) \cos \varphi}{d_{1ex} (2 d_{2ex} - d_{1ex} \tan \varphi^*)}, \quad \varphi \in \left(\varphi^*, \frac{\pi}{2} \right). \qquad (2.37)$$

On the conditions $d_{1ex} = d_{3ex} = d_{ex}$, $d_{2ex} = 2 d_{ex} \sqrt{3}/3$ corresponding to a regular hexagon, thus for $\varphi^* = \pi/6$, regarding Eq. (2.9), we get
$$r_c = R_{1ex} f_c, \qquad (2.38)$$
where the function f_c, regarding Eq. (2.9), has the form
$$f_c = \frac{\sqrt{1 + c_\varphi^2}}{2 (c_\varphi \cos \nu' + \sin \nu')} \left(\frac{2\pi \sqrt{3}}{3v} \right)^{1/3}, \quad \nu' \in \langle 0, \nu^* \rangle, \qquad (2.39)$$

$$f_c = \frac{1}{2c_\varphi \sin \nu} \left(\frac{2\pi\sqrt{3}}{3v} \right)^{1/3} , \quad \nu \in \left\langle \nu^*, \frac{\pi}{2} \right\rangle , \tag{2.40}$$

and the angle ν^* is given by Eq. (2.31), considering the coefficien c_φ derived as

$$c_\varphi = \cos\left(\varphi - \frac{z\pi}{3} \right), \quad \varphi \in \left\langle \frac{\pi(2z-1)}{6}, \frac{\pi(2z+1)}{6} \right\rangle , \quad z = 0, 1. \tag{2.41}$$

2.4   Real composite material

Provided that particles in a real composite material are represented by precipitates without or with a continuous phase on a surface of each of the precipitates as define in Item 1 in Section 2.1, both considered to be spherical with the radii R_{1ex} and $R_{1ex} < R_{2ex}$, respectively, in contrast to the particles distribution characterized by $d_{1ex}, d_{2ex}, d_{3ex}$ (see Fig. 2.1a,b), the real composite material is characterized by one inter-particle distance $d_{ex} = d_{1ex} = d_{2ex} = d_{3ex}$, and accordingly the cubic cell and the φ, ν-dependent distance r_c for $v \in (0, v_{max})$ given by Eqs. (2.29)–(2.33), (2.35), (2.36) are recommended to be considered in formulae for the thermal stresses and corresponding quantities (see Chapter 8), where R_{1ex}, R_{2ex}, d_{ex} are determined by experimental methods. With regard to Items 2, 3 in Section 2.1, the same conditions are also valid provided that a real material is represented

1. by two types of grains, A and B, with the volume fractions v_A and $v_B = 1 - v_A$, respectively, characterized by different properties without continuous phases on a surface of the grains of both types,

2. by two types of grains with identical properties and one type of the grains is covered by a continuous phase with different properties than those of the grains of both types.

With regard to Item 1 and to the analytical models of the thermal stresses, provided that $v_A < v_{max}$ and simultaneously $v_B < v_{max}$, the grains A, B can represent the spherical particle, the infinit matrix, respectively, or vice versa, resulting in the elastic energy of the cell, $W_{cAB} = W_{pA} + W_{mB}$ or $W_{cBA} = W_{pB} + W_{mA}$ (see Eq. (3.74)), substituting $v = v_A$ or $v = v_B$ in Eqs. (2.30), (2.36), respectively. Considering the Castigliano's theorem (see Section 3.2.5) [7, p. 219], and provided that $W_{cAB} < W_{cBA}$ or $W_{cAB} > W_{cBA}$, the $A - B$ or $B - A$ multi-particle-matrix system is considered, respectively. Finally, provided that $v_Q > v_{max}$ ($Q = A, B$), the grains Q are considered to represent the infinit matrix.

Provided that a real composite material is represented by one type of grains covered by a continuous phase with different properties than those of the grains as define in Item 4 in Section 2.1, both considered to be spherical with the radii R_{1ex} and $R_{2ex} = R_{1ex} + t$, respectively, the prismatic cell with the regular-hexagonal basis and the φ, ν-dependent distance r_c for $v \rightarrow v_{max}$ given by Eqs. (2.10), (2.38)–(2.41) are recommended to be considered for the analytical models of the thermal stresses.

Chapter 3

Thermal stresses in elastic solid continuum

3.1 Selected topics of Mechanics of Solid Continuum

Along with the analysis concerning the radial displacement of an infinitesima spherical cap and the elastic energy (see Sections 3.1.2, 3.2.4), Section 3.1, presenting fundamental equations of a solid continuum including the Castigliano's theorem (see Sections 3.1.3–3.1.5, 3.2.5)), represents a short introduction to the scientifi branch Mechanics of Solid Continuum, assuming elasticity of the solid continuum resulting from the Hooke's law which considers a linear strain-stress dependence (see Section 3.1.6). Finally, the differential equations (3.1)–(3.10), (3.14)–(3.16) along with the Hooke's law (see Eqs. (3.17)–(3.22), (9.1)–(9.10)), represent initial equations to which the mathematical techniques presenting in Section 3.2 are applied.

3.1.1 Coordinate system

As usual in Mechanics of Solid Continuum representing a fundamental basis of Applied Mechanics, a stress-deformation state of a solid continuum is investigated within an infinitesima part of the solid continuum at an arbitrary point with a position determined by a suitable system of coordinates. Additionally, the infinitesima part has a suitable shape regarding the geometry of the solid continuum. An influenc of the solid continuum on the infinitesima part is represented by displacements of the infinitesima part along individual directions, and by forces acting along individual directions on surfaces of the infinitesima part [5, p. 12–14].

Thermal stresses are investigated at the arbitrary point P of the solid continuum along the axes x_1', x_2', x_3' of the Cartesian system $(Px_1'x_2'x_3')$ (see Fig. 3.1). With regard to the spherical coordinates (r, φ, ν), the axes $x_1' = x_r$ and $x_2' = x_\varphi \parallel x_1 x_2$, $x_3' = x_\nu$ represent radial and tangential directions related to the spherical surface in the point P, respectively, where $r = |OP|$ is radius of the spherical surface, and $P = (r, \varphi, \nu)$ for $r \in \langle 0, r_c \rangle$,

$\varphi, \nu \in \langle 0, \pi/2 \rangle$ (see Section 2.1). Finally, the infinitesima part of the solid continuum at the point P is represented by an infinitesima spherical cap described by the parameters dr, $d\varphi$, $d\nu$ (see Figs. 3.2, 3.3).

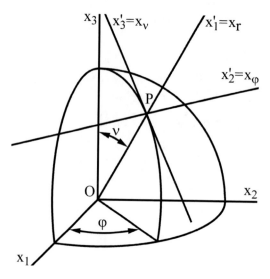

Figure 3.1: The radial and tangential axes $x_1' = x_r$ and $x_2' = x_\varphi \parallel x_1 x_2$, $x_3' = x_\nu$, respectively, and the point P with a position determined by the spherical coordinates (r, φ, ν) regarding the Cartesian system $(Ox_1 x_2 x_3)$, where O is the centre of the spherical particle, and x_2', x_3' are tangents to the spherical surface with the radius $r = |OP|$.

3.1.2 Displacements of the infinitesima spherical cap. Thermal-stress induced radial displacement

Let the multi-particle-matrix system is considered. The thermal stresses in the multi-particle-matrix system (see Fig. 2.1) originate as a consequence of the condition $\beta_p \neq \beta_m$. A detailed analysis of the coefficien β_q (see Eqs. (3.100)–(3.102)) is presented in Section 3.4.1, where the subscript $q = p$ and $q = m$ is related to the spherical particle and cell matrix, respectively. The thermal stresses originate at the temperature $T \in \langle T_f, T_r \rangle$, where T_f is fina temperature of a cooling process, and the relaxation temperature T_r is analysed in Section 3.4.1.

If $T \in \langle T_f, T_r \rangle$ and $\beta_m - \beta_p > 0$, then the cell matrix is pushed by the spherical particle, and the spherical particle is pushed by the cell matrix. If $T \in \langle T_f, T_r \rangle$ and $\beta_m - \beta_p < 0$, then the cell matrix is pulled by the spherical particle, and the spherical particle is pulled by the cell matrix.

An analysis of the pulling or pushing, i.e. an analysis of displacement of the infinites imal spherical cap, is as follows. As presented in Section 2.1, the multi-particle-matrix system is symmetric, where this symmetry results from the matrix infinit and from the periodical distribution of the spherical particles. Due to this symmetry, the pulling or pushing of an arbitrary point at the particle-matrix boundary is realized along a normal to this surface (i.e. to the particle-matrix boundary) at this arbitrary point.

Let P (see Fig. 3.1) is such arbitrary point at the particle-matrix boundary, thus for $r = R_1$. The point P as well as the spherical the infinitesima spherical cap at the point P thus exhibit a displacement along a normal to the surfaces S_r and S_{r+dr} of the spherical

Thermal stresses in elastic solid continuum

the infinitesima spherical cap at the radii r and $r + dr$, where $S_r = r\,d\varphi \times r\,d\nu$, $S_{r+dr} = (r + dr)\,d\varphi \times (r + dr)\,d\nu$ represent an area of S_r and S_{r+dr}, respectively. The normal to the surfaces S_r, S_{r+dr} is represented by the axis x'_1 which define the radial direction regarding the Cartesian system $(Ox_1x_2x_3)$ (see Fig. 3.1). The infinitesima spherical cap in the point P at the particle-matrix boundary exhibits a displacement along the axis x'_1, i.e. a radial displacement.

The condition $\beta_p \neq \beta_m$ is a reason of the radial stress p_1 acting at the particle-matrix boundary along the axis x'_1, i.e. along the radial direction. A condition for the determination of p_1 is derived in Section 3.4.4 (see Eqs. (3.82), (3.109)). The radial stress p_1 is a reason of the fact that this analysis concerning the radial displacement of the infinitesima spherical cap in the point P at the particle-matrix boundary (i.e. for $r = R_1$) is also valid for each point of the axis x'_1 (i.e. for $r \in \langle 0, r_c \rangle$), where r_c is determined in Section 2.3.

The same is also valid for the multi-particle-envelope-matrix system. In this case, p_1 and p_2 acting at the particle-envelope and matrix envelope boundaries are a consequence of the conditions $\beta_p \neq \beta_e$ and $\beta_e \neq \beta_m$. With regard to $\beta_p \neq \beta_e$ and $\beta_e \neq \beta_m$, the radial stresses p_1 and p_2 are determined by Eqs. (3.110) and (3.111), respectively, where β_e is a thermal expansion coefficien of the spherical envelope (see Eqs. (3.100)–(3.102)).

The infinitesima spherical cap in the arbitrary point P (see Fig. 3.1) with a position described by the spherical coordinates (r, φ, ν) exhibits the **radial displacement** $u_1 = u_r$ only, along the axis x'_1 (see Fig. 3.1).

Additionally, the following analysis concerning the radial displacement $u_1 = u_r$ is required to be considered. This analysis is based on a concept of imaginary separation.

Let the multi-particle-matrix system is considered. Let the spherical particles and the infinit matrix are imaginarily separated, and then spherical hollows are periodically distributed in the infinit matrix.

Let $T \in \langle T_f, T_r \rangle$ represent temperature of the separated spherical particles and of the infinit matrix with the spherical hollows. If the temperature T increases or decreases within the interval $\langle T_f, T_r \rangle$, then the components which are imaginarily separated expand or contract, respectively. The expansion and contraction result in displacements of points in the components. Due to the imaginary separation, these displacements result from the temperature change, and not from the difference $\beta_m - \beta_p \neq 0$ (see Eqs. (3.100)–(3.102)).

Let the spherical particles are embedded in the infinit matrix. Let $\Delta T = T - T_r \neq 0$ represents the temperature change. Let the condition $\beta_p = \beta_m$ is considered. Due to $\beta_p = \beta_m$, the thermal stresses do not originate in the multi-particle-matrix system, and the infinitesima spherical cap is thus shifted due to the temperature change.

Let $R_{1p} = R_{1p}(T)$ and $R_{1m} = R_{1m}(T)$ represent temperature-dependent functions of radii of these separated spherical particle and the spherical hollows, respectively. Due to $T < T_r$, we get $R_{1q}(T) < R_{1q}(T_r)$ $(q = p, m)$, where $R_{1q}(T_r) = R_{1T_r}$. The same, i.e. $d(T) < d(T_r)$, is also valid for the temperature-dependent function $d = d(T)$ of the inter-particle distance d.

Due to $\beta_q = \beta_q(\varphi, \nu)$ (see Eqs. (3.100)–(3.102)), we get $R_{1q} = R_{1q}(\varphi, \nu, T)$. Consequently, $R_{1p} = R_{1p}(\varphi, \nu, T)$ is a distance from the particle centre to a point on a surface of the separated particle along the axis x_1' which represents the radial direction define by the angles φ, ν. Similarly, $R_{1m} = R_{1m}(\varphi, \nu, T)$ is a distance along the axis x_1' from a centre of the hollow to a point on a surface of the hollow.

If $\beta_p < \beta_m$, then we get $R_{1p}(\varphi, \nu, T) > R_{1m}(\varphi, \nu, T)$. Let the distance R_{1p} is changed to R_1 for each value of the variables $\varphi, \nu \in \langle 0, \pi/2 \rangle$, where $R_1 < R_{1p}$. The change $R_{1p}(\varphi, \nu, T) \rightarrow R_1$ is caused by the radial stress $p_1 = p_1(\varphi, \nu, T)$ which acts, along the axis x_1', on a surface of the separated particle. Due to $\beta_p < \beta_m$, the radial stress p_1 is compressive regarding the surface of the separated particle. Additionally, $[u_{1p}(\varphi, \nu, T)]_{r=R_{1p}} = R_1 - R_{1p}$ represents a **thermal-stress induced radial displacement** on a surface of the separated particle (i.e. for $r = R_{1p}$) at a point define by the coordinates $\varphi, \nu \in \langle 0, \pi/2 \rangle$. The radial displacement $(u_{1p})_{r=R_{1p}}$ along the axis $x_1' = x_r$ (see Fig. 3.1) is induced by the radial stress p_1.

The separated particle can be put into the hollow provided that the distance R_{1m} is also changed to R_1 for each value of the variables $\varphi, \nu \in \langle 0, \pi/2 \rangle$, where $R_1 > R_{1m}$. The change $R_{1m}(\varphi, \nu, T) \rightarrow R_1(\varphi, \nu, T)$ is also caused by the radial stress $p_1 = p_1(\varphi, \nu, T)$ which acts on a surface of the hollow in the matrix. Due to $\beta_p < \beta_m$, the radial stress p_1 is also compressive regarding the surface of the hollow. Additionally, $[u_{1m}(\varphi, \nu, T)]_{r=R_{1m}} = R_1 - R_{1m}$ represents a **thermal-stress induced radial displacement** on a surface of the hollow (i.e. for $r = R_{1m}$) at a point define by the coordinates $\varphi, \nu \in \langle 0, \pi/2 \rangle$. The radial displacement $(u_{1m})_{r=R_{1m}}$ along the axis $x_1' = x_r$ (see Fig. 3.1) is also induced by the radial stress p_1.

After the embedding of the separated particle with the radius R_{1p} in the hollow with the radius R_{1m}, a surface of the particle with the radius R_1 is pushed by a surface of the matrix, where $R_{1p} > R_1 > R_{1m}$. Similarly, a surface of the matrix is pushed by a surface of the particle. The same is also valid in case of the pulling for $\beta_p > \beta_m$. This analysis of the pushing or pulling which considers a concept of the imaginarily separated components is also valid for the multi-particle-envelope-matrix system.

An analysis of a relationship between the radial displacements $(u_{1p})_{r=R_{1p}}, (u_{1q})_{r=R_{iq}}$ and the coefficien $\beta_q = \beta_q(\varphi, \nu)$ $(i=1,2; q=e,m)$ is presented in Section 3.4.4. This analysis results in the determination of the formulae Eqs. (3.109)–(3.111). In case of the multi-particle-matrix system (see Fig. 2.1), the formula Eq. (3.109) is considered for the determination of the radial stresses $p_1 = p_1(\varphi, \nu)$ acting on the particle-matrix boundary. In case of the multi-particle-envelope-matrix system, the formulae (3.110) and (3.111) are considered for the determination of the radial stresses $p_1 = p_1(\varphi, \nu)$ and $p_2 = p_2(\varphi, \nu)$ acting on the particle-envelope and matrix-envelope boundary, respectively.

3.1.3 Cauchy's equations

The Cauchy's equations, representing geometric equations, defin a relationship between displacements and strains of an infinitesima part of a solid continuum. With regard to the infinitesima spherical cap (see Figs. 3.2, 3.3), the radial strain $\varepsilon_{11} = \varepsilon_r$, the tangential strains $\varepsilon_{22} = \varepsilon_\varphi$, $\varepsilon_{33} = \varepsilon_\nu$, and the shear strains $\varepsilon_{12} = \varepsilon_{r\varphi}$, $\varepsilon_{13} = \varepsilon_{r\nu}$ are derived as [1]

$$\varepsilon_{11} = \varepsilon_r = \frac{|1'''1'| - |1''1|}{|1''1|} = \frac{1}{dr}\left[\left(dr + \frac{\partial u_1}{\partial r}dr\right) - dr\right] = \frac{\partial u_1}{\partial r}, \quad (3.1)$$

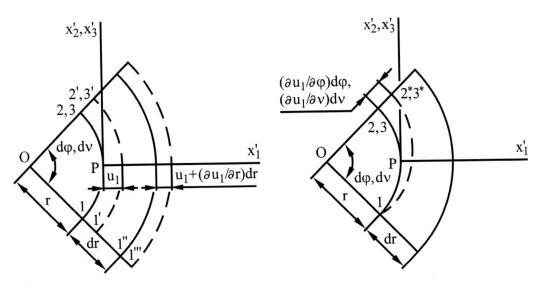

Figure 3.2: The radial displacements u_1 and $u_1 + (\partial u_1/\partial r)\,dr$ of the infinitesima spherical cap at the point P at the radii r and $r + dr$, respectively, in the planes $x'_1 x'_2$ or $x'_1 x'_3$ (see Fig. 3.1), effected by the radial and tangential stresses, $\sigma_{11} = \sigma_r$ and $\sigma_{22} = \sigma_\varphi$ or $\sigma_{33} = \sigma_\nu$ (see Figs. 3.4, 3.5), respectively.

Figure 3.3: The φ, ν-dependent radial displacement u_1 of the infinitesima spherical cap at the point P in the planes $x'_1 x'_2$ and $x'_1 x'_3$ (see Fig. 3.1), effected by the shear stresses $\sigma_{12} = \sigma_{21} = \sigma_{r\varphi} = \sigma_{\varphi r}$ and $\sigma_{13} = \sigma_{31} = \sigma_{r\nu} = \sigma_{\nu r}$ (see Figs. 3.4, 3.5), respectively, where the distances $|22^*| = (\partial u_1/\partial \varphi)\,d\varphi$ and $|33^*| = (\partial u_1/\partial \nu)\,d\nu$.

$$\varepsilon_{22} = \varepsilon_\varphi = \frac{|1'2'| - |12|}{|12|} = \frac{(u_1+r)\,d\varphi - r\,d\varphi}{r\,d\varphi} = \frac{u_1}{r}, \quad (3.2)$$

$$\varepsilon_{33} = \varepsilon_\nu = \frac{|1'3'| - |13|}{|13|} = \frac{(u_1+r)\,d\nu - r\,d\nu}{r\,d\nu} = \frac{u_1}{r}, \quad (3.3)$$

$$\varepsilon_{12} = \varepsilon_{21} = \varepsilon_{r\varphi} = \varepsilon_{\varphi r} = \tan\left[\angle\left(|12|,|12^*|\right)\right] = \frac{1}{r\,d\varphi}\left(\frac{\partial u_1}{\partial \varphi}\,d\varphi\right) = \frac{1}{r}\frac{\partial u_1}{\partial \varphi}, \quad (3.4)$$

$$\varepsilon_{13} = \varepsilon_{31} = \varepsilon_{r\nu} = \varepsilon_{\nu r} = \tan\left[\angle\left(|13|,|13^*|\right)\right] = \frac{1}{r\,d\nu}\left(\frac{\partial u_1}{\partial \nu}\,d\nu\right) = \frac{1}{r}\frac{\partial u_1}{\partial \nu}, \quad (3.5)$$

---

[1] The subscripts p, e and m related to the spherical particle, the spherical envelope and the cell matrix, respectively, are omitted in Sections 3.1.3–3.2.2.

where ε_{ij} $(i,j=1,2,3)$ is strain of the surface of the infinitesima spherical cap with a normal x'_j realized along the axis x'_i, and $\varepsilon_{ij} = \varepsilon_{ji}$ [6, p. 14][2].

With regard to the analysis presented in Sections 3.1.2.1, 3.1.2.2, and accordingly as a consequence of radial loading along a normal to the particle-matrix and particle-envelope, matrix-envelope boundaries, the strain ε_{23} is equal to zero.

3.1.4 Saint-Venant's equations

As a principle valid in Mechanics of Solid Continuum, a system being continuous before deformation is required to exhibit the continuity after deformation as well, and accordingly individual strains are required to be mutually compatible [7, p. 162]. The Saint-Venant's equations, representing equations of the compatibility between individual strains, are derived in the following way. Performing ∂Eq. (3.2) $/\partial r$ and considering Eq. (3.1), we get

$$\varepsilon_{11} - \varepsilon_{22} - r\frac{\partial \varepsilon_{22}}{\partial r} = 0. \tag{3.6}$$

Performing ∂Eq. (3.1) $/\partial \varphi$, ∂Eq. (3.4) $/\partial r$, and ∂Eq. (3.1) $/\partial \nu$, ∂Eq. (3.5) $/\partial r$, we get

$$\frac{\partial \varepsilon_{11}}{\partial \varphi} - \varepsilon_{12} - r\frac{\partial \varepsilon_{12}}{\partial r} = 0, \tag{3.7}$$

$$\frac{\partial \varepsilon_{11}}{\partial \nu} - \varepsilon_{13} - r\frac{\partial \varepsilon_{13}}{\partial r} = 0. \tag{3.8}$$

Performing ∂Eq. (3.2) $/\partial \varphi$ and considering Eq. (3.4), and performing ∂Eq. (3.2) $/\partial \nu$ and considering Eq. (3.5), we get

$$\frac{\partial \varepsilon_{22}}{\partial \varphi} - \varepsilon_{12} = 0, \tag{3.9}$$

$$\frac{\partial \varepsilon_{22}}{\partial \nu} - \varepsilon_{13} = 0. \tag{3.10}$$

The Saint-Venant's equations given by Eqs. (3.6)–(3.10) represents the compatibility equations considered within the mathematical techniques presented in Section 3.2.

3.1.5 Equilibrium equations

As usual in Mechanics of Solid Continuum, stresses acting on an infinitesima part with a suitable shape at an arbitrary point of a solid continuum are required to be derived. Equations of the equilibrium of individual forces acting on the infinitesima spherical cap along the axes x'_1, x'_2, x'_3 (see Figs. 3.4, 3.5) are derived as

---

[2]The points 1, 1′, 1″, 1‴ in Figs. 3.2, 3.3 are in the plane $x'_1 x'_2$ as well as in the plane $x'_1 x'_3$, where $x'_1 = x_r$, $x'_2 = x_\varphi$, $x'_3 = x_\nu$ (see Fig. 3.1). The points 2, 2′, 2* and 3, 3′, 3* in Figs. 3.2, 3.3 along with the infinitesima angles $d\varphi$ and $d\nu$ are in the planes $x'_1 x'_2$ and $x'_1 x'_3$, respectively.

As presented in this book, the conventional subscripts r, φ, ν, $r\varphi$, $r\nu$, $\varphi\nu$ in a connection with the spherical coordinates (r, φ, ν) (see Fig. 3.1) are replaced by the subscripts 11, 22, 33, 12, 13, 23, respectively, due to the mathematical techniques in Section 3.2 (see e.g. Eqs. (3.23)–(3.34))

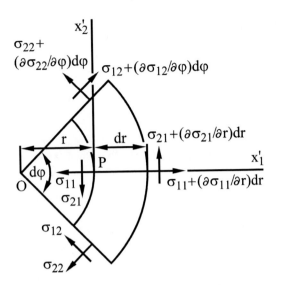 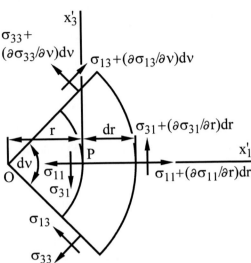

Figure 3.4: The radial, tangential, shear stresses in the plane $x_1' x_2'$, $\sigma_{11} = \sigma_r$, $\sigma_{22} = \sigma_\varphi$, $\sigma_{12} = \sigma_{21} = \sigma_{r\varphi} = \sigma_{\varphi r}$, respectively, acting on surfaces of an infinitesima spherical cap at the point P (see Fig. 3.1), where $x_1' = x_r$, $x_2' = x_\varphi$.

Figure 3.5: The radial, tangential, shear stresses in the plane $x_1' x_3'$, $\sigma_{11} = \sigma_r$, $\sigma_{33} = \sigma_\nu$, $\sigma_{13} = \sigma_{31} = \sigma_{r\nu} = \sigma_{\nu r}$, respectively, acting on surfaces of an infinitesima spherical cap at the point P (see Fig. 3.1), where $x_1' = x_r$, $x_3' = x_\nu$.

$$\left(\sigma_{11} + \frac{\partial \sigma_{11}}{\partial r} dr\right)(r+dr)\, d\varphi\, (r+dr)\, d\nu + \left(\sigma_{12} + \frac{\partial \sigma_{12}}{\partial \varphi} d\varphi\right) \cos\frac{d\varphi}{2} r\, d\nu\, dr$$
$$+ \left(\sigma_{13} + \frac{\partial \sigma_{13}}{\partial \nu} d\nu\right) \cos\frac{d\nu}{2} r\, d\varphi\, dr - \left[\sigma_{11} r\, d\varphi\, r\, d\nu + \left(\sigma_{22} + \frac{\partial \sigma_{22}}{\partial \varphi} d\varphi\right) \sin\frac{d\varphi}{2} r\, d\nu\, dr\right.$$
$$+ \left(\sigma_{33} + \frac{\partial \sigma_{33}}{\partial \nu} d\nu\right) \sin\frac{d\nu}{2} r\, d\varphi\, dr + \sigma_{22} \sin\frac{d\varphi}{2} r\, d\nu\, dr + \sigma_{33} \sin\frac{d\nu}{2} r\, d\varphi\, dr$$
$$\left. + \sigma_{12} \cos\frac{d\varphi}{2} r\, d\nu\, dr + \sigma_{13} \cos\frac{d\nu}{2} r\, d\varphi\, dr \right] = 0, \qquad (3.11)$$

$$\left(\sigma_{22} + \frac{\partial \sigma_{22}}{\partial \varphi} d\varphi\right) \cos\frac{d\varphi}{2} r\, d\nu\, dr + \left(\sigma_{21} + \frac{\partial \sigma_{21}}{\partial r} dr\right)(r+dr)\, d\varphi\, (r+dr)\, d\nu$$
$$+ \left(\sigma_{12} + \frac{\partial \sigma_{12}}{\partial \varphi} d\varphi\right) \sin\frac{d\varphi}{2} r\, d\nu\, dr + \sigma_{12} \sin\frac{d\varphi}{2} r\, d\nu\, dr$$
$$- \left(\sigma_{22} \cos\frac{d\varphi}{2} r\, d\nu\, dr + \sigma_{21} r\, d\varphi\, r\, d\nu\right) = 0, \qquad (3.12)$$

$$\left(\sigma_{33} + \frac{\partial \sigma_{33}}{\partial \nu} d\nu\right) \cos \frac{d\nu}{2} r \, d\varphi \, dr + \left(\sigma_{31} + \frac{\partial \sigma_{31}}{\partial r} dr\right)(r + dr) \, d\varphi \, (r + dr) \, d\nu$$

$$+ \left(\sigma_{13} + \frac{\partial \sigma_{13}}{\partial \nu} d\nu\right) \sin \frac{d\nu}{2} r \, d\varphi \, dr + \sigma_{13} \sin \frac{d\nu}{2} r \, d\varphi \, dr$$

$$- \left(\sigma_{33} \cos \frac{d\nu}{2} r \, d\varphi \, dr + \sigma_{31} r \, d\varphi \, r \, d\nu\right) = 0, \tag{3.13}$$

where the stress σ_{ij} $(i, j = 1,2,3)$ acts along the axis x_i' on the surface of the infinitesima spherical cap with the normal x_j', and $\sigma_{ij} = \sigma_{ji}$ [6, p. 13]. Accordingly, $\sigma_{11} = \sigma_r$ and $\sigma_{22} = \sigma_\varphi$, $\sigma_{33} = \sigma_\nu$ and $\sigma_{12} = \sigma_{21} = \sigma_{r\varphi} = \sigma_{\varphi r}$, $\sigma_{13} = \sigma_{31} = \sigma_{r\nu} = \sigma_{\nu r}$ represent radial and tangential and shear stresses, respectively.

With regard to the analysis presented in Sections 3.1.2.1, 3.1.2.2, and accordingly as a consequence of radial loading along a normal to the particle-matrix and particle-envelope, matrix-envelope boundaries, the stress σ_{23} is equal to zero.

Substituting $\sin d\varphi/2 \approx d\varphi/2$, $\sin d\nu/2 \approx d\nu/2$, $\cos d\varphi/2 = \cos d\nu/2 \approx 1$, $(dr)^2 = (d\varphi)^2 = (d\nu)^2 = 0$, for $d\varphi \approx 0$, $d\nu \approx 0$, $dr \approx 0$ [5, p. 125–126], we get

$$2\sigma_{11} - \sigma_{22} - \sigma_{33} + r \frac{\partial \sigma_{11}}{\partial r} + \frac{\partial \sigma_{12}}{\partial \varphi} + \frac{\partial \sigma_{13}}{\partial \nu} = 0, \tag{3.14}$$

$$\frac{\partial \sigma_{22}}{\partial \varphi} + 3\sigma_{12} + r \frac{\partial \sigma_{12}}{\partial r} = 0, \tag{3.15}$$

$$\frac{\partial \sigma_{33}}{\partial \nu} + 3\sigma_{13} + r \frac{\partial \sigma_{13}}{\partial r} = 0. \tag{3.16}$$

3.1.6 Hooke's law

The Hooke's law which define relationships between stresses and strains of an infinitesima part of an elastic anisotropic solid continuum is derived as [6, p. 30]

$$\varepsilon_{11} = s_{11}' \sigma_{11} + s_{12}' \sigma_{22} + s_{13}' \sigma_{33} + s_{15}' \sigma_{13} + s_{16}' \sigma_{12}, \tag{3.17}$$

$$\varepsilon_{22} = s_{12}' \sigma_{11} + s_{22}' \sigma_{22} + s_{23}' \sigma_{33} + s_{25}' \sigma_{13} + s_{26}' \sigma_{12}, \tag{3.18}$$

$$\varepsilon_{33} = s_{13}' \sigma_{11} + s_{23}' \sigma_{22} + s_{33}' \sigma_{33} + s_{35}' \sigma_{13} + s_{36}' \sigma_{12}, \tag{3.19}$$

$$\varepsilon_{23} = s_{14}' \sigma_{11} + s_{24}' \sigma_{22} + s_{34}' \sigma_{33} + s_{45}' \sigma_{13} + s_{46}' \sigma_{12}, \tag{3.20}$$

$$\varepsilon_{13} = s_{15}' \sigma_{11} + s_{25}' \sigma_{22} + s_{35}' \sigma_{33} + s_{55}' \sigma_{13} + s_{56}' \sigma_{12}, \tag{3.21}$$

$$\varepsilon_{12} = s_{16}' \sigma_{11} + s_{26}' \sigma_{22} + s_{36}' \sigma_{33} + s_{56}' \sigma_{13} + s_{66}' \sigma_{12}. \tag{3.22}$$

The elastic moduli $s_{11}', s_{12}', \ldots, s_{56}', s_{66}'$ in the Cartesian system with the axes x_1', x_2', x_3' derived by $s_{11}, s_{12}, \ldots, s_{56}, s_{66}$ in the Cartesian system $(Ox_1 x_2 x_3)$ are given by Eqs. (9.1)–(9.10).

Thermal stresses in elastic solid continuum 29

Finally, with regard to the subscript $q = p,e,m$, the transformations given by Eq. (9.13) are considered. As mentioned above, the subscripts $q = p$, $q = e$, and $q = m$ are considered for the spherical particle, the spherical envelope, and the matrix, respectively.

Additionally, $\varepsilon_{11} = \varepsilon_r$ or $\sigma_{11} = \sigma_r$; $\varepsilon_{22} = \varepsilon_\varphi$, $\varepsilon_{33} = \varepsilon_\nu$ or $\sigma_{22} = \sigma_\varphi$, $\sigma_{33} = \sigma_\nu$; and $\varepsilon_{12} = \varepsilon_{21} = \varepsilon_{r\varphi}$, $\varepsilon_{13} = \varepsilon_{31} = \varepsilon_{r\nu}$ or $\sigma_{12} = \sigma_{21} = \sigma_{r\varphi} = \sigma_{\varphi r}$, $\sigma_{13} = \sigma_{31} = \sigma_{r\nu} = \sigma_{\nu r}$ represent radial; tangential; and shear strains or stresses related to the Cartesian system $(Px_1'x_2'x_3')$, respectively.

The Hooke's law (see Eqs. (3.17)–(3.18)) is valid for an anisotropic solid continuum which exhibits uniaxial or triaxial anisotropy represented by uniaxial- or triaxial-anisotropic crystal lattices, respectively. Accordingly, different anisotropic components of the anisotropic multi-particle-matrix and multi-particle-envelope-matrix systems can exhibit the same type or the different types of the anisotropy. With regard to the analytical modelling, the only difference concerning the uniaxial and triaxial anisotropy is represented by the elastic modulus $s_{kl} = s_{lk}$ $(k, l = 1, \ldots, 6)$ which is different for the uniaxial-anisotropic and the triaxial-anisotropic crystal lattices [6, p. 17,30], where s_{kl} is included in $s_{ij}' = s_{ji}'$ $(i, j = 1, \ldots, 6)$ (see Eqs. (9.1), (9.14)–(9.34)).

Finally, with regard to the infinitesima spherical cap which exhibits the radial displacement u_1 along the axis x_1' (see Fig. 3.1), we get $\varepsilon_{23} = 0$, $\sigma_{23} = 0$.

3.2   Mathematical techniques

3.2.1   Determination of tangential and shear stresses

With regard to $\varepsilon_{22} = \varepsilon_{33}$, $\varepsilon_{23} = 0$, $\sigma_{23} = 0$ (see Eqs. (3.2), (3.3), (3.18)–(3.20)), the stresses σ_{13}, σ_{12} have the form

$$\sigma_{13} = \sum_{i=1}^{3} c_i \sigma_{ii}, \tag{3.23}$$

$$\sigma_{12} = \sum_{i=1}^{3} c_{3+i} \sigma_{ii}. \tag{3.24}$$

and consequently, using Eqs. (3.23), (3.24), the Hooke's law (see Eqs. (3.17)–(3.22)) is transformed to the forms

$$\varepsilon_{11} = \sum_{i=1}^{3} c_{6+i} \sigma_{ii}, \tag{3.25}$$

$$\varepsilon_{22} = \varepsilon_{33} = \sum_{i=1}^{3} c_{9+i} \sigma_{ii}, \tag{3.26}$$

$$\varepsilon_{13} = \sum_{i=1}^{3} c_{12+i} \sigma_{ii}, \tag{3.27}$$

$$\varepsilon_{12} = \sum_{i=1}^{3} c_{15+i}\sigma_{ii}. \tag{3.28}$$

Substituting $\partial\sigma_{13}/\partial\eta$, $\partial\sigma_{12}/\partial\eta$ ($\eta = \varphi, \nu$), into Eqs. (3.7)–(3.10), (3.15), (3.16), we get

$$\frac{\partial\sigma_{11}}{\partial\varphi} = \sum_{i=1}^{3} c_{18+i}\sigma_{ii} + c_{21+i}r\frac{\partial\sigma_{ii}}{\partial r}, \tag{3.29}$$

$$\frac{\partial\sigma_{22}}{\partial\varphi} = -\sum_{i=1}^{3} c_{3+i}\left(3\sigma_{ii} + r\frac{\partial\sigma_{ii}}{\partial r}\right), \tag{3.30}$$

$$\frac{\partial\sigma_{33}}{\partial\varphi} = \sum_{i=1}^{3} c_{24+i}\sigma_{ii} + c_{27+i}r\frac{\partial\sigma_{ii}}{\partial r}, \tag{3.31}$$

$$\frac{\partial\sigma_{11}}{\partial\nu} = \sum_{i=1}^{3} c_{30+i}\sigma_{ii} + c_{33+i}r\frac{\partial\sigma_{ii}}{\partial r}, \tag{3.32}$$

$$\frac{\partial\sigma_{22}}{\partial\nu} = \sum_{i=1}^{3} c_{36+i}\sigma_{ii} + c_{39+i}r\frac{\partial\sigma_{ii}}{\partial r}, \tag{3.33}$$

$$\frac{\partial\sigma_{33}}{\partial\nu} = -\sum_{i=1}^{3} c_i\left(3\sigma_{ii} + r\frac{\partial\sigma_{ii}}{\partial r}\right). \tag{3.34}$$

The stresses σ_{22}, σ_{33}, and consequently σ_{12}, σ_{13} (see Eqs. (3.23), (3.24)) are derived as a function of σ_{11}, $r\,(\partial\sigma_{11}/\partial r)$, $r^2\,(\partial^2\sigma_{11}/\partial r^2)$ by the following procedure. Performing ∂Eq. (3.7)$/\partial\nu$, ∂Eq. (3.8)$/\partial\varphi$, ∂Eq. (3.9)$/\partial\nu$, ∂Eq. (3.10)$/\partial\varphi$, and assuming $(\partial^2 u_1/\partial\varphi\partial\nu) = (\partial^2 u_1/\partial\nu\partial\varphi)$, $(\partial^2 u_1/\partial r\partial\eta) = (\partial^2 u_1/\partial\eta\partial r)$ ($\eta = \varphi, \nu$) (see Eqs. (3.1)–(3.5)), the Saint-Venant's compatibility equations (3.7)–(3.10) are derived as

$$\frac{\partial}{\partial\varphi}\left(\frac{\partial\varepsilon_{11}}{\partial\nu}\right) - \frac{\partial\varepsilon_{12}}{\partial\nu} - r\frac{\partial}{\partial r}\left(\frac{\partial\varepsilon_{12}}{\partial\nu}\right) = 0, \tag{3.35}$$

$$\frac{\partial}{\partial\nu}\left(\frac{\partial\varepsilon_{11}}{\partial\varphi}\right) - \frac{\partial\varepsilon_{13}}{\partial\varphi} - r\frac{\partial}{\partial r}\left(\frac{\partial\varepsilon_{13}}{\partial\varphi}\right) = 0, \tag{3.36}$$

$$\frac{\partial}{\partial\varphi}\left(\frac{\partial\varepsilon_{22}}{\partial\nu}\right) - \frac{\partial\varepsilon_{12}}{\partial\nu} = 0, \tag{3.37}$$

$$\frac{\partial}{\partial\nu}\left(\frac{\partial\varepsilon_{22}}{\partial\varphi}\right) - \frac{\partial\varepsilon_{13}}{\partial\varphi} = 0. \tag{3.38}$$

Substituting Eqs. (3.25)–(3.28) into Eqs. (3.35)–(3.38), we get

$$\sum_{i=1}^{3} c_{6+i}\frac{\partial}{\partial\nu}\left(\frac{\partial\sigma_{ii}}{\partial\varphi}\right) - c_{12+i}\left[\frac{\partial\sigma_{ii}}{\partial\varphi} + r\frac{\partial}{\partial r}\left(\frac{\partial\sigma_{ii}}{\partial\varphi}\right)\right] = 0, \tag{3.39}$$

$$\sum_{i=1}^{3} c_{6+i} \frac{\partial}{\partial \varphi} \left(\frac{\partial \sigma_{ii}}{\partial \nu} \right) - c_{15+i} \left[\frac{\partial \sigma_{ii}}{\partial \nu} + r \frac{\partial}{\partial r} \left(\frac{\partial \sigma_{ii}}{\partial \nu} \right) \right] = 0, \qquad (3.40)$$

$$\sum_{i=1}^{3} c_{9+i} \frac{\partial}{\partial \nu} \left(\frac{\partial \sigma_{ii}}{\partial \varphi} \right) - c_{12+i} \frac{\partial \sigma_{ii}}{\partial \varphi} = 0. \qquad (3.41)$$

$$\sum_{i=1}^{3} c_{9+i} \frac{\partial}{\partial \varphi} \left(\frac{\partial \sigma_{ii}}{\partial \nu} \right) - c_{15+i} \frac{\partial \sigma_{ii}}{\partial \nu} = 0, \qquad (3.42)$$

Substituting Eqs. (3.23), (3.24) and consequently Eqs. (3.29)–(3.34) into Eq. (3.14), and substituting Eqs. (3.25), (3.26) into Eq. (3.6), the Saint-Venant's compatibility and equilibrium equations (3.6) and (3.14), respectively, are transformed to the forms

$$\sum_{i=1}^{3} (c_{9+i} - c_{6+i}) \sigma_{ii} + c_{9+i} r \frac{\partial \sigma_{ii}}{\partial r} = 0, \qquad (3.43)$$

$$\sum_{i=1}^{3} c_{42+i} \sigma_{ii} + c_{45+i} r \frac{\partial \sigma_{ii}}{\partial r} = 0. \qquad (3.44)$$

Substituting $\partial \sigma_{ii}/\partial \nu$ (see Eqs. (3.32)–(3.34)) and consequently $\partial \sigma_{ii}/\partial \varphi$ ($i = 1 - 3$) (see Eqs. (3.29)–(3.31)) into Eqs. (3.39), (3.41), and substituting $\partial \sigma_{ii}/\partial \varphi$ (see Eqs. (3.29)–(3.31)) and consequently $\partial \sigma_{ii}/\partial \nu$ ($i = 1 - 3$) (see Eqs. (3.32)–(3.34)) into Eqs. (3.40), (3.42), the transformed Saint-Venant's compatibility equations (3.39)–(3.42) are derived as

$$\sum_{i=1}^{3} c_{48+i+9j} \sigma_{ii} + c_{51+i+9j} r \frac{\partial \sigma_{ii}}{\partial r} + c_{54+i+9j} r^2 \frac{\partial^2 \sigma_{ii}}{\partial r^2} = 0, \quad j = 0 - 3. \qquad (3.45)$$

As functions of the radial stress σ_{11} and of the terms $r \left(\partial \sigma_{11}/\partial r \right)$, $r^2 \left(\partial^2 \sigma_{11}/\partial r^2 \right)$, the stresses σ_{ii}, σ_{1i}, and the terms $r \left(\partial \sigma_{ii}/\partial r \right)$, $r^2 \left(\partial^2 \sigma_{ii}/\partial r^2 \right)$ ($i = 2, 3$) can be derived from Eqs. (3.43)–(3.45) in the forms

$$\sigma_{2+i2+i} = (\delta_{1i} - \delta_{0i}) \left(c_{135+i} \sigma_{11} + c_{137+i} r \frac{\partial \sigma_{11}}{\partial r} + c_{139+i} r^2 \frac{\partial^2 \sigma_{11}}{\partial r^2} \right), \quad i = 0, 1, \qquad (3.46)$$

$$r \frac{\partial \sigma_{2+i2+i}}{\partial r} = c_{141+i} \sigma_{11} + c_{143+i} r \frac{\partial \sigma_{11}}{\partial r} + c_{145+i} r^2 \frac{\partial^2 \sigma_{11}}{\partial r^2}, \quad i = 0, 1, \qquad (3.47)$$

$$r^2 \frac{\partial^2 \sigma_{2+i2+i}}{\partial r^2} = (\delta_{0i} - \delta_{1i}) \times$$
$$\left(c_{147+i} \sigma_{11} + c_{149+i} r \frac{\partial \sigma_{11}}{\partial r} + c_{151+i} r^2 \frac{\partial^2 \sigma_{11}}{\partial r^2} \right), \quad i = 0, 1. \qquad (3.48)$$

where δ_{0i}, δ_{1i} are the Kronecker's symbols, and consequently, regarding Eqs. (3.23), (3.24), (3.46), the stresses σ_{12}, σ_{13} are derived as

$$\sigma_{12+i} = c_{153+i}\sigma_{11} + c_{155+i}r\frac{\partial\sigma_{11}}{\partial r} + c_{157+i}r^2\frac{\partial^2\sigma_{11}}{\partial r^2}, \quad i = 0, 1. \tag{3.49}$$

Substituting Eqs. (3.48) into $r\left[\partial\text{Eq. (3.48)}/\partial r\right]$, we get

$$r^3\frac{\partial^3\sigma_{2+i2+i}}{\partial r^3} = (\delta_{0i} - \delta_{1i}) \times$$

$$\left[-2c_{147+i}\sigma_{11} + (c_{147+i} - c_{149+i})\,r\frac{\partial\sigma_{11}}{\partial r} + c_{149+i}r^2\frac{\partial^2\sigma_{11}}{\partial r^2} + c_{151+i}r^3\frac{\partial^3\sigma_{11}}{\partial r^3}\right],$$

$$i = 0, 1. \tag{3.50}$$

Substituting Eqs. (3.50) into $r\left[\partial\text{Eq. (3.50)}/\partial r\right]$, we get

$$r^4\frac{\partial^4\sigma_{2+i2+i}}{\partial r^4} = (\delta_{0i} - \delta_{1i}) \times$$

$$\left[6c_{147+i}\sigma_{11} + 2\left(c_{149+i} - 2c_{147+i}\right)r\frac{\partial\sigma_{11}}{\partial r} + (c_{147+i} - 2c_{149+i})\,r^2\frac{\partial^2\sigma_{11}}{\partial r^2}\right.$$

$$\left. + c_{149+i}r^3\frac{\partial^3\sigma_{11}}{\partial r^3} + c_{151+i}r^4\frac{\partial^4\sigma_{11}}{\partial r^4}\right], \quad i = 0, 1. \tag{3.51}$$

Substituting Eqs. (3.51) into $r\left[\partial\text{Eq. (3.51)}/\partial r\right]$, we get

$$r^5\frac{\partial^5\sigma_{2+i2+i}}{\partial r^5} = (\delta_{0i} - \delta_{1i}) \times$$

$$\left[-24c_{147+i}\sigma_{11} + 6\left(3c_{147+i} - c_{149+i}\right)r\frac{\partial\sigma_{11}}{\partial r} + 2\left(c_{147+i} - 5c_{149+i}\right)r^2\frac{\partial^2\sigma_{11}}{\partial r^2}\right.$$

$$\left. + (c_{147+i} - 3c_{149+i})\,r^3\frac{\partial^3\sigma_{11}}{\partial r^3} + c_{149+i}r^4\frac{\partial^4\sigma_{11}}{\partial r^4} + c_{151+i}r^5\frac{\partial^5\sigma_{11}}{\partial r^5}\right],$$

$$i = 0, 1. \tag{3.52}$$

Substituting Eqs. (3.52) into $r\left[\partial\text{Eq. (3.52)}/\partial r\right]$, we get

$$r^6\frac{\partial^6\sigma_{2+i2+i}}{\partial r^6} = (\delta_{0i} - \delta_{1i}) \times$$

$$\left[120c_{147+i}\sigma_{11} + 24\left(c_{149+i} - 4c_{147+i}\right)r\frac{\partial\sigma_{11}}{\partial r} + 12\left(c_{147+i} + 2c_{149+i}\right)r^2\frac{\partial^2\sigma_{11}}{\partial r^2}\right.$$

$$\left. - 4c_{149+i}r^3\frac{\partial^3\sigma_{11}}{\partial r^3} + (c_{147+i} - 4c_{149+i})\,r^4\frac{\partial^4\sigma_{11}}{\partial r^4} + c_{149+i}r^5\frac{\partial^5\sigma_{11}}{\partial r^5}\right.$$

$$\left. + c_{151+i}r^6\frac{\partial^6\sigma_{11}}{\partial r^6}\right], \quad i = 0, 1. \tag{3.53}$$

3.2.2 Determination of radial stress

With regard to the spherical coordinates (r, φ, ν), the radial stress is assumed to be of the form

$$\sigma_{11} = \sum_{i=1}^{n} C_i r^{\lambda_i}. \tag{3.54}$$

Consequently, the exponent λ_i, representing a real number, is thus derived from a radial stress differential equation of the order $\partial^m \sigma / \partial r^m$ for $n \le m$, where n is a number of boundary conditions for the determination of the integration constant C_i $(i = 1, \ldots, n)$. Applying the mathematical operations, $r(\partial/\partial r)$, $\partial/\partial \varphi$, $\partial/\partial \nu$, to the transformed Saint-Venant's compatibility and equilibrium equations (3.43) and (3.44), respectively, and considering Eqs. (3.29)–(3.34), we get

$$\sum_{i=1}^{3} c_{158+i+3j} r \frac{\partial \sigma_{ii}}{\partial r} + c_{9+i+36j} r^2 \frac{\partial^2 \sigma_{ii}}{\partial r^2} = 0, \quad j = 0, 1, \tag{3.55}$$

$$\sum_{i=1}^{3} c_{164+i+9j} \sigma_{ii} + c_{167+i+9j} r \frac{\partial \sigma_{ii}}{\partial r} + c_{170+i+9j} r^2 \frac{\partial^2 \sigma_{ii}}{\partial r^2} = 0, \quad j = 0 - 3, \tag{3.56}$$

and consequently substituting Eqs. (3.46)–(3.48) into Eqs. (3.55), (3.56), the radial stress differential equations are derived as

$$c_{201+3i} \sigma_{11} + c_{202+3i} r \frac{\partial \sigma_{11}}{\partial r} + c_{203+3i} r^2 \frac{\partial^2 \sigma_{11}}{\partial r^2} = 0, \quad i = 0 - 5. \tag{3.57}$$

Considering $\sigma_{11} = \beta r^\lambda$, Eq. (3.57) is transformed to a parametric algebraic equation of the variable λ and of the parameters φ, ν in the form

$$c_{203+3i} \lambda^2 + (c_{202+3i} - c_{203+3i}) \lambda + c_{201+3i} = 0, \quad i = 0 - 5, \tag{3.58}$$

and then the exponent $\lambda_j^{(i)}$ $(i = 0 - 5; j = 1, 2)$, derived as

$$\lambda_j^{(i)} = \frac{1}{2c_{203+3i}} \times$$
$$\left[c_{203+3i} - c_{202+3i} + (\delta_{1j} - \delta_{2j}) \sqrt{(c_{203+3i} - c_{202+3i})^2 - 4c_{201+3i} c_{203+3i}} \right],$$
$$i = 0 - 5, \quad j = 1, 2, \tag{3.59}$$

represents a root of Eq. (3.58).

If $\lambda_j^{(i)}$ $(i = 0 - 5; j = 1, 2)$, dependent on the parameters φ, ν, is not a real root of or if stresses are not physically acceptable, the following following technique is considered. Applying the mathematical operations $r(\partial/\partial r)$, $\partial/\partial \nu$ to Eqs. (3.55), (3.56), (3.57), and considering Eqs. (3.29)–(3.34), we Eqs. (3.29)–(3.34), we get

$$\sum_{i=1}^{3} c_{218+i+9j} r \frac{\partial \sigma_{ii}}{\partial r} + c_{221+i+9j} r^2 \frac{\partial^2 \sigma_{ii}}{\partial r^2} + c_{224+i+9j} r^3 \frac{\partial^3 \sigma_{ii}}{\partial r^3} = 0, \quad j = 0 - 13, \tag{3.60}$$

$$\sum_{i=1}^{3} c_{344+i+12j}\sigma_{ii} + c_{347+i+12j}r\frac{\partial\sigma_{ii}}{\partial r} + c_{350+i+12j}r^2\frac{\partial^2\sigma_{ii}}{\partial r^2} + c_{354+i+12j}r^3\frac{\partial^3\sigma_{ii}}{\partial r^3} = 0,$$
$$j = 0 - 15. \tag{3.61}$$

Substituting Eqs. (3.46)–(3.48), (3.50) into Eqs. (3.60), (3.61), the radial stress differential equations are derived as

$$c_{537+4i}\sigma_{11} + c_{538+4i}r\frac{\partial\sigma_{11}}{\partial r} + c_{539+4i}r^2\frac{\partial^2\sigma_{11}}{\partial r^2} + c_{540+4i}r^3\frac{\partial^3\sigma_{11}}{\partial r^3} = 0, \quad i = 0 - 29, \tag{3.62}$$

and consequently, considering $\sigma_{11} = \beta r^{\lambda}$, is transformed to the parametric algebraic equation of the variable λ and of the parameters φ, ν in the form

$$c_{540+4i}\lambda^3 + (c_{539+4i} - 3c_{540+4i})\lambda^2$$
$$+ (c_{538+4i} - c_{539+4i} + 2c_{540+4i})\lambda + c_{537+4i} = 0, \quad i = 0 - 29, \tag{3.63}$$

and $\lambda_j^{(i)}$ ($i = 0 - 29; j = 1 - 3$) represents a root of Eq. (3.63).

If $\lambda_j^{(i)}$ ($i = 0 - 29; j = 1 - 3$), dependent on the parameters φ, ν, is not a real root of Eq. (3.63), or if stresses are not physically acceptable, the following technique is considered. Applying the mathematical operations $r(\partial/\partial r)$, $\partial/\partial\varphi$, $\partial/\partial\nu$ to Eqs. (3.60), (3.61), and considering Eqs. (3.29)–(3.34), we get

$$\sum_{i=1}^{3} c_{656+i+12j}r\frac{\partial\sigma_{ii}}{\partial r} + c_{659+i+12j}r^2\frac{\partial^2\sigma_{ii}}{\partial r^2}$$
$$+ c_{662+i+12j}r^3\frac{\partial^3\sigma_{ii}}{\partial r^3} + c_{665+i+12j}r^4\frac{\partial^4\sigma_{ii}}{\partial r^4} = 0, \quad j = 0 - 57, \tag{3.64}$$

$$\sum_{i=1}^{3} c_{1352+i+15j}\sigma_{ii} + c_{1355+i+15j}r\frac{\partial\sigma_{ii}}{\partial r} + c_{1358+i+15j}r^2\frac{\partial^2\sigma_{ii}}{\partial r^2}$$
$$+ c_{1361+i+15j}r^3\frac{\partial^3\sigma_{ii}}{\partial r^3} + c_{1364+i+15j}r^4\frac{\partial^4\sigma_{ii}}{\partial r^4} = 0, \quad j = 0 - 31. \tag{3.65}$$

Substituting Eqs. (3.46)–(3.48), (3.50), (3.51) into Eqs. (3.64), (3.65), the radial stress differential equations are derived as

$$c_{1833+5i}\sigma_{11} + c_{1834+5i}r\frac{\partial\sigma_{11}}{\partial r} + c_{1835+5i}r^2\frac{\partial^2\sigma_{11}}{\partial r^2} + c_{1836+5i}r^3\frac{\partial^3\sigma_{11}}{\partial r^3}$$
$$+ c_{1837+5i}r^4\frac{\partial^4\sigma_{11}}{\partial r^4} = 0, \quad i = 0 - 89, \tag{3.66}$$

and consequently, considering $\sigma_{11} = \beta r^{\lambda}$, is transformed to the parametric algebraic equation of the variable λ and of the parameters φ, ν in the form

$$c_{1837+5i}\lambda^4 + (c_{1836+5i} - 6c_{1837+5i})\lambda^3 + (11c_{1837+5i} - 3c_{1836+5i} + c_{1835+5i})\lambda^2$$
$$+ [2c_{1836+5i} + c_{1834+5i} - (6c_{1837+5i} + c_{1835+5i})]\lambda + c_{1833+5i} = 0,$$
$$i = 0 - 89, \tag{3.67}$$

and $\lambda_j^{(i)}$ $(i = 0 - 89; j = 1 - 4)$ represents a root of Eq. (3.67).

If $\lambda_j^{(i)}$ $(i = 0 - 89; j = 1 - 4)$, dependent on the parameters φ, ν, is not a real root of Eq. (3.67), or if stresses are not physically acceptable, the following technique is considered. Applying the mathematical operations $r\,(\partial/\partial r)$, $\partial/\partial\varphi$, $\partial/\partial\nu$ to Eqs. (3.64), (3.65), and considering Eqs. (3.29)–(3.34), we get

$$\sum_{i=1}^{3} c_{2282+i+15j}\, r\frac{\partial\sigma_{ii}}{\partial r} + c_{2285+i+15j}\, r^2\frac{\partial^2\sigma_{ii}}{\partial r^2} + c_{2288+i+15j}\, r^3\frac{\partial^3\sigma_{ii}}{\partial r^3}$$
$$+ \sum_{i=1}^{3} c_{2291+i+15j}\, r^4\frac{\partial^4\sigma_{ii}}{\partial r^4} + c_{2294+i+15j}\, r^5\frac{\partial^5\sigma_{ii}}{\partial r^5} = 0, \quad j = 0 - 205, \quad (3.68)$$

$$\sum_{i=1}^{3} c_{5372+i+18j}\sigma_{ii} + c_{5375+i+18j}\, r\frac{\partial\sigma_{ii}}{\partial r} + c_{5378+i+18j}\, r^2\frac{\partial^2\sigma_{ii}}{\partial r^2} + c_{5381+i+18j}\, r^3\frac{\partial^3\sigma_{ii}}{\partial r^3}$$
$$+ \sum_{i=1}^{3} c_{5384+i+18j}\, r^4\frac{\partial^4\sigma_{ii}}{\partial r^4} + c_{5387+i+18j}\, r^5\frac{\partial^5\sigma_{ii}}{\partial r^5} = 0, \quad j = 0 - 63. \quad (3.69)$$

Substituting Eqs. (3.46)–(3.48), (3.50)–(3.52) into Eqs. (3.68), (3.69), the radial stress differential equations are derived as

$$c_{6525+6i}\sigma_{11} + c_{6526+6i}\, r\frac{\partial\sigma_{11}}{\partial r} + c_{6527+6i}\, r^2\frac{\partial^2\sigma_{11}}{\partial r^2} + c_{6528+6i}\, r^3\frac{\partial^3\sigma_{11}}{\partial r^3}$$
$$+ c_{6529+6i}\, r^4\frac{\partial^4\sigma_{11}}{\partial r^4} + c_{6530+6i}\, r^5\frac{\partial^5\sigma_{11}}{\partial r^5} = 0, \quad i = 0 - 269, \quad (3.70)$$

and consequently, considering $\sigma_{11} = \beta\, r^\lambda$, is transformed to the parametric algebraic equation of the variable λ and of the parameters φ, ν in the form

$$c_{6530+6i}\lambda^5 + (c_{6529+6i} - 10c_{6530+6i})\,\lambda^4 + (c_{6528+6i} - 6c_{6529+6i} + 35c_{6530+6i})\,\lambda^3$$
$$+ (c_{6527+6i} - 3c_{6528+6i} + 11c_{6529+6i} - 50c_{6530+6i})\,\lambda^2$$
$$+ [c_{6526+6i} - c_{6527+6i} + 2\,(c_{6528+6i} - 3c_{6529+6i} + 12c_{6530+6i})]\,\lambda + c_{6525+6i} = 0,$$
$$i = 0 - 269, \quad (3.71)$$

and $\lambda_j^{(i)}$ $(i = 0 - 269; j = 1 - 5)$ represents a root of Eq. (3.71).

If $\lambda_j^{(i)}$ $(i = 0 - 269; j = 1 - 5)$, dependent on the parameters φ, ν, is not a real root of Eq. (3.71), or if stresses are not physically acceptable, the mathematical technique, applying the mathematical operations $r\,(\partial/\partial r)$, $\partial/\partial\varphi$, $\partial/\partial\nu$ to Eqs. (3.68), (3.69), and consequently leading to the parametric algebraic equation of the order λ^6 and dependent on the parameters φ, ν is considered.

Finally, considering the analyses in Sections 4.1–4.3, the procedures presented above are performed due to the determination of a sufficien number of suitable exponent λ.

3.2.3 Different solutions regarding exponent λ

As analysed in Section 4.1, the exponent $\lambda_{1p} > 0$ (see Eq. (4.7)) included in formulae for the thermal stresses in the spherical particle represents a root of one of Eqs. (3.58), (3.63), (3.67), (3.71), (3.71), considering the Castigliano's theorem (see Section 3.2.5). Consequently, different roots of Eqs. (3.58), (3.63), (3.67), (3.71) fulfillin the condition $\lambda_{1p} > 0$ result in different solutions for the thermal stresses in the spherical particle.

The same concerning the different solutions is also valid for the exponents $\lambda_{1e} \leq -1$ and $\lambda_{1e} \geq -1$ (see Eqs. (4.12), (4.13)) included in formulae for the thermal stresses in the spherical envelope on the conditions $\beta_p \neq \beta_e = \beta_m$ and $\beta_p = \beta_e \neq \beta_m$ (see Section 4.2), respectively.

Additionally, provided that $\beta_p \neq \beta_e \neq \beta_m$, the exponents $\lambda_{1e}, \lambda_{2e}$ both are required to represent roots related to one equation of Eqs. (3.58), (3.63), (3.67), (3.71), where $\lambda_{1e} \neq \lambda_{2e}$.

Similarly, different roots of Eqs. (3.58), (3.63), (3.67), (3.71) fulfillin the conditions $\lambda_{1e} \leq -1$, $\lambda_{1e} \geq -1$, and consequently different sets $\{\lambda_{1e}, \lambda_{2e}\}$ fulfillin the condition $\lambda_{1e} \neq \lambda_{2e}$ result in different solutions for the thermal stresses in the spherical envelope on the conditions $\beta_p \neq \beta_e = \beta_m$, $\beta_p = \beta_e \neq \beta_m$ and $\beta_p \neq \beta_e \neq \beta_m$, respectively.

Similarly, with regard to the mandatory boundary conditions given by Eqs. (4.14), (4.16) and (4.15), (4.16) considered for the cell matrix of the multi-particle-matrix multi-particle-envelope-matrix systems, respectively, the exponents $\lambda_{1m}, \lambda_{2m}$ included in formulae for the thermal stresses in the cell matrix are also required to represent roots related to one equation of Eqs. (3.58), (3.63), (3.67), (3.71), where where $\lambda_{1m} \neq \lambda_{2m}$.

Consequently, different sets $\{\lambda_{1m}, \lambda_{2m}\}$ of the roots $\lambda_{1m}, \lambda_{2m}$ related to one equation of Eqs. (3.58), (3.63), (3.67), (3.71) result result in different solutions for the thermal stresses in the cell matrix. The same concerning the different solutions is also valid provided that the mandatory and additional boundary conditions for the cell matrix are considered, where exponents included in formulae for the thermal stresses in the cell matrix are required to be related to one equation of Eqs. (3.58), (3.63), (3.67), (3.71).

Strictly speaking, let n_q is a number of boundary conditions for the determination of the integration constants $C_{1q}, \ldots, C_{nq}$ related to a component with the subscript $q = p, e, m$. Consequently, the set $\{\lambda_{1q}, \ldots, \lambda_{nq}\}$ is required to be related to one of Eqs. (3.58), (3.63), (3.67), (3.71), where the exponents $\lambda_{1q}, \ldots, \lambda_{nq}$ are required to be real. Finally, if $\lambda_{1q}, \ldots, \lambda_{nq}$ are real, then Equations (3.58); (3.63); (3.67); (3.71) result in $n_q = 1, 2$; $n_q = 1, 2, 3$; $n_q = 1, \ldots, 4$; $n_q = 1, \ldots, 5$, respectively.

Finally, the thermal stresses in a component of the multi-particle-(envelope)-matrix system are determined by mutually different solutions. With regard to the different solutions for a component, the elastic energy W_q of a component ($q = p, e, m$) is required to be determined, and consequently the Castigliano's theorem (see Section 3.2.5) is required to be considered.

Thermal stresses in elastic solid continuum 37

3.2.4 Elastic energy

Induced by elastic stresses acting at an arbitrary point of an elastic solid continuum, the elastic energy density w_q accumulated at the arbitrary point has the form [6, p. 24]

$$w_q = \frac{1}{2}\left(\sum_{i=1}^{3}\sigma_{iiq}\varepsilon_{iiq} + \sum_{i,j=1;\ i\neq j}^{3}\sigma_{ijq}\varepsilon_{ijq}\right), \tag{3.72}$$

and consequently the elastic energy W_q, accumulated in the volume V_q of the elastic solid continuum is derived, using the spherical coordinates (r, φ, ν) (see Fig. 3.1) and considering the φ, ν-dependent elastic energy density, $w_q = w_q(\varphi, \nu)$, in the form

$$W_q = \int_{V_q} w_q\, dV_q = 8\int_0^{\pi/2}\int_0^{\pi/2}\int_{r_1}^{r_2} w_q\, r^2\, dr\, d\varphi\, d\nu, \tag{3.73}$$

where $dV_q = r^2\, dr\, d\varphi\, d\nu$ is the volume of the infinitesima spherical cap for the angles $\varphi, \nu \in \langle 0, \pi/2\rangle$. The integration boundaries r_1, r_2 are as follows:

1. $r_1 = 0, r_2 = R_1$ for the spherical particle of the the multi-particle-matrix and multi-particle-envelope-matrix systems,

2. $r_1 = R_1, r_2 = R_2$ for the spherical envelope,

3. $r_1 = R_1, r_2 = r_c$ for the cell matrix of the multi-particle-matrix system,

4. $r_1 = R_2, r_2 = r_c$ for the cell matrix of the multi-particle-envelope-matrix system.

The integration boundaries define in Items 1, 2 are also valid for the the one-particle-matrix and one-particle-envelope-matrix systems, where

5. $r_1 = R_1, r_2 \to \infty$ for the cell matrix of the one-particle-matrix system,

6. $r_1 = R_2, r_2 \to \infty$ for the cell matrix of the one-particle-envelope-matrix system.

With regard to the elastic energy of the spherical particle and the cell matrix, W_p and W_m, respectively, the elastic energy of the infinit multi-particle-matrix system is represented by the elastic energy of the cell in the form

$$W_c = W_p + W_m. \tag{3.74}$$

Similarly, the elastic energy of the cell of the infinit multi-particle-envelope-matrix system is derived as

$$W_c = W_p + W_e + W_m, \tag{3.75}$$

where W_e is the elastic energy of the spherical envelope.

38 Ladislav Ceniga

3.2.5 Castigliano's theorem

As presented in Section 3.2.3, the thermal stresses in a component of the multi-particle-(envelope)-matrix system are determined by mutually different solutions. Accordingly, the different solutions represent a set of solutions related to the component of the multi-particle-(envelope)-matrix system. With regard to Section 3.2.4, the mutually different solutions result in a set of different values of the elastic energy W_q of the component ($q = p, e, m$) (see Eq. (3.73)).

Consequently, the different values of W_p of the set for the spherical particle (along with the different values of W_e of the set for the spherical envelope) and the different values of W_m of the set for the cell matrix result in different values of the elastic energy W_c of the multi-particle-(envelope)-matrix system (see Eqs. (3.74), (3.75)). Resulting from the tendency of a system to exhibit minimal energy, the Castigliano's theorem is formulated as follows [7, p. 219]:

> Assuming different stress-deformation states which fulfi equilibrium and boundary conditions, the stress-deformation state which exhibits minimal energy of a system is realized.

Accordingly, such a combination of solutions for components of the multi-particle-(envelope)-matrix system is considered to result in minimal thermal-stress induced elastic energy of the cell, W_c (see Eqs. (3.74), (3.75)).

3.3 Mathematical solutions

With regard to Eqs. (3.54), considering Eqs. (3.2), (3.3), (3.23)–(3.28), (3.46), the stresses and strains determined by the mathematical techniques presented in Section 3.2 are derived as

$$\sigma_{11q} = \sum_{i=1}^{n_q} C_{iq} r^{\lambda_{iq}}, \tag{3.76}$$

$$\sigma_{22q} = \sum_{i=1}^{n_q} \xi_{1iq} C_{iq} r^{\lambda_{iq}}, \tag{3.77}$$

$$\sigma_{33q} = \sum_{i=1}^{n_q} \xi_{2iq} C_{iq} r^{\lambda_{iq}}, \tag{3.78}$$

$$\sigma_{12q} = \sum_{i=1}^{n_q} \xi_{3iq} C_{iq} r^{\lambda_{iq}}, \tag{3.79}$$

$$\sigma_{13q} = \sum_{i=1}^{n_q} \xi_{4iq} C_{iq} r^{\lambda_{iq}}, \tag{3.80}$$

$$\varepsilon_{11q} = \sum_{i=1}^{n_q} \xi_{5iq} \, C_{iq} r^{\lambda_{iq}}, \tag{3.81}$$

$$\varepsilon_{22q} = \varepsilon_{33q} = \sum_{i=1}^{n_q} \xi_{6iq} \, C_{iq} r^{\lambda_{iq}}, \tag{3.82}$$

$$\varepsilon_{13q} = \sum_{i=1}^{n_q} \xi_{7iq} \, C_{iq} r^{\lambda_{iq}}, \tag{3.83}$$

$$\varepsilon_{12q} = \sum_{i=1}^{n_q} \xi_{8iq} \, C_{iq} r^{\lambda_{iq}}, \tag{3.84}$$

$$u_{1q} = r \, \varepsilon_{22q} = \sum_{i=1}^{n_q} \xi_{6iq} \, C_{iq} r^{\lambda_{iq}+1}, \tag{3.85}$$

where n_q is a number of boundary conditions for the determination of the integration constants $C_{1q}, \ldots, C_{nq}$ related to the spherical particle ($q = p$) (see Section 4.1), the spherical envelope ($q = e$) (see Section 4.2), and the cell matrix ($q = m$) (see Section 4.3). The set $\{\lambda_{1q}, \ldots, \lambda_{nq}\}$ is required to be related to one of Eqs. (3.58), (3.63), (3.67), (3.71), where the exponents $\lambda_{1q}, \ldots, \lambda_{nq}$ are required to be real. Finally, if $\lambda_{1q}, \ldots, \lambda_{nq}$ are real, then Equations (3.58); (3.63); (3.67); (3.71) result in $n_q = 1, 2$; $n_q = 1, 2, 3$; $n_q = 1, \ldots, 4$; $n_q = 1, \ldots, 5$, respectively.

The coefficient $\xi_{1iq}, \ldots, \xi_{8iq}$ have the forms

$$\xi_{jiq} = (\delta_{2j} - \delta_{1j}) \{c_{146+j} + \lambda_{iq} [c_{148+j} + c_{150+j} (\lambda_{iq} - 1)]\},$$
$$i = 1, \ldots, n_q; \quad j = 1, 2, \tag{3.86}$$

$$\xi_{2+j\,iq} = c_{152+j} + \lambda_{iq} [c_{154+j} + c_{156+j} (\lambda_{iq} - 1)],$$
$$i = 1, \ldots, n_q; \quad j = 1, 2, \tag{3.87}$$

$$\xi_{4+j\,iq} = c_{4+3j} + c_{5+3j} + c_{6+3j}, \quad i = 1, \ldots, n_q; \quad j = 1, \ldots, 4, \tag{3.88}$$

where the transformation concerning the subscript $q = p, e, m$ is presented in Eq. (9.13).

Consequently, with regard to Eqs. (3.74), (3.75), (3.76)–(3.84), considering Items 1–4 in Section 3.2.4, the elastic energy density w_q and the elastic energy W_q in the spherical particle ($q = p$), the spherical envelope ($q = e$) and the cell matrix ($q = m$) are derived as

$$w_q = \frac{1}{2} \sum_{i,j=1}^{n_q} \omega_{ijq} C_{iq} C_{jq} r^{\lambda_{iq}+\lambda_{jq}}, \tag{3.89}$$

$$W_q = \int_0^{\pi/2} \int_0^{\pi/2} \Omega_q \, d\varphi \, d\nu. \tag{3.90}$$

The coefficient ω_{ijq} $(i,j=1,\ldots,n_q)$, Ω_p, Ω_e and $\Omega_m = \Omega_{1+km}$ $(k=0,1)$ have the forms

$$\omega_{ijq} = \xi_{5jq} + \xi_{6jq}\left(\xi_{1iq} + \xi_{2iq}\right) + 2\left(\xi_{3iq}\xi_{8jq} + \xi_{4iq}\xi_{7jq}\right),$$
$$i,j = 1,\ldots,n_q, \tag{3.91}$$

$$\Omega_p = 4 \sum_{i,j=1}^{n_p} \frac{\omega_{ijp}C_{ip}C_{jp}R_1^{\lambda_{ip}+\lambda_{jp}+3}}{\lambda_{ip} + \lambda_{jp} + 3}, \tag{3.92}$$

$$\Omega_e = 4 \sum_{i,j=1}^{n_e} \frac{\omega_{ije}C_{ie}C_{je}R_2^{\lambda_{ie}+\lambda_{je}+3}}{\lambda_{ie} + \lambda_{je} + 3}\left[1 - \left(\frac{R_1}{R_2}\right)^{\lambda_{ie}+\lambda_{je}+3}\right], \tag{3.93}$$

$$\Omega_m = \Omega_{km}$$
$$= 4 \sum_{i,j=1}^{n_m} \frac{\omega_{ijm}C_{im}C_{jm}R_1^{\lambda_{im}+\lambda_{jm}+3}}{\lambda_{im} + \lambda_{jm} + 3}\left[f_c^{\lambda_{im}+\lambda_{jm}+3} - \left(\frac{R_k}{R_1}\right)^{\lambda_{im}+\lambda_{jm}+3}\right],$$
$$k = 1, 2, \tag{3.94}$$

where the subscript $k=1,2$ is as follows:

- $k = 1$ for the multi-particle-matrix system,

- $k = 2$ for the multi-particle-envelope-matrix system.

Considering the determination of the thermal-stress induced strengthening presented in Section 8.4, the stress σ_{iq} resulting in the strain $\varepsilon_{iq} = s_{11q}\sigma_{iq}$, both in a direction of the axis x_i $(i=1,2,3)$ (see Fig. 3.1), along with the elastic energy density w_{iq} induced by the stress σ_{iq}, are required to be derived as

$$\sigma_{iq} = \sum_{j=1}^{3} a_{ji}\sigma_{jjq} + \sum_{j,k=1;\ j\neq k}^{3} a_{ji}\sigma_{jkq}, \quad i = 1,2,3, \tag{3.95}$$

$$w_{iq} = \frac{\varepsilon_i\sigma_{iq}}{2} = \frac{s_{iiq}\sigma_{iq}^2}{2}, \quad i = 1,2,3, \tag{3.96}$$

where $a_{ji}\sigma_{jkq}$ represents a coordinate of the stress σ_{jkq} in a direction represented by the axis x_i, and the coefficien a_{ji} $(i,j,k=1,2,3)$ is given by Eqs. (9.2)–(9.10).

With regard to Eqs. (3.76)–(3.80), (3.95), (3.96), the stress σ_{iq} and the elastic energy density w_{iq} $(i=1,2,3)$ have the forms

$$\sigma_{iq} = \sum_{j=1}^{n_q} \xi_{8+ijq}C_{jq}r^{\lambda_{jq}}, \quad i = 1,2,3, \tag{3.97}$$

$$w_{iq} = \frac{s_{iiq}}{2} \sum_{j,k=1}^{n_q} \xi_{8+ijq}\,\xi_{8+ikq}\,C_{jq}C_{kq}\,r^{\lambda_{jq}+\lambda_{kq}}, \quad i = 1, 2, 3, \tag{3.98}$$

where the coefficien ξ_{8+ijq} is derived as

$$\xi_{8+ijq} = a_{1i} + a_{2i}\xi_{1jq} + a_{3i}\xi_{2jq} + (a_{1i} + a_{2i})\,\xi_{3jq} + (a_{1i} + a_{3i})\,\xi_{4jq},$$
$$i = 1, 2, 3; \quad j = 1, \dots, n_q. \tag{3.99}$$

3.4 Reason of thermal stresses

3.4.1 Coefficient β_p, β_e, β_m

With regard to the multi-particle-matrix and multi-particle-envelope-matrix system, the thermal stresses originate

1. as a consequence of the differences $\alpha'_{1m} - \alpha'_{1p} \neq 0$ and $\alpha'_{1e} - \alpha'_{1p} \neq 0$, $\alpha'_{1m} - \alpha'_{1e} \neq 0$ in the temperature-dependent or temperature-independent thermal expansion coeffi cients α'_{1p}, α'_{1e}, α'_{1m} (see Eqs. (9.2)–(9.4), (9.11)) along the axis x'_1 of the Cartesian system $(Px'_1x'_2x'_3)$ (Fig. 3.1) in the spherical particle, the spherical envelope, the cell matrix, respectively,

2. as a consequence of the differences $\varepsilon_{11tm} - \varepsilon_{11tp} \neq 0$ and $\varepsilon_{11te} - \varepsilon_{11tp} \neq 0$, $\varepsilon_{11tm} - \varepsilon_{11te} \neq 0$ in the radial strains ε_{11tp}, ε_{11te}, ε_{11tm} (see Eq. (3.104)) induced, during a cooling process, by a phase transformation in the spherical particle, the spherical envelope, the cell matrix at the temperature T_{tp}, T_{te}, T_{tm}, respectively,

where α_q and ε_{11tq} are included in the coefficien β_q ($q = p,e,m$) (see Eqs. (3.100)–(3.102)), and $T_{tq} \in \langle T_f, T_r \rangle$ ($q = p,e,m$). The phase-transformation induced radial strain ε_{11tq} ($q = p,e,m$) is determined in Section 3.4.2. The difference $\alpha'_{1q_1} - \alpha'_{1q_2} \neq 0$ ($q_1, q_2 = p,e,m$; $q_1 \neq q_2$) is related to the temperature $T \in \langle T_f, T_r \rangle$, where T_f, T_r are analysed below.

With regard to Item 2, the thermal stresses originate as a consequence of the radial strain ε_{11tq} on the condition of a phase transformation originating at least in one of components of the multi-particle-(envelope)-matrix system. The phase-transformation induced radial strain ε_{11tq} results from a difference in dimensions of mutually transforming cubic crystal lattices of the spherical particle, the spherical envelope, the cell matrix (see Section 3.4.2), thus for $q = p,e,m$, respectively.

In case of the non-zero difference in the thermal expansion coefficients the thermal stresses originate in the temperature interval $\langle T_f, T_r \rangle$, where

3. T_f is fina temperature of a cooling process.

4. T_r is relaxation temperature below that the stress relaxation as a consequence of thermal-activated processes does not occur in components of the multi-particle-(envelope)-matrix system. The relaxation temperature is define approximately by the relationship $T_r = (0.35 - 0.4) \times T_m$ [4, p. 39] and exactly by an experiment.

5. T_m is melting temperature of the multi-particle-(envelope)-matrix system.

   (a) Provided that the particles precipitate from a liquid matrix of multi-particle-matrix system, the melting point T_m represents the minimum of the set $\{T_{mp}, T_{mm}\}$, where T_{mp} and T_{mm} are melting points of the particles and the matrix, respectively. Provided that the particles precipitate from a solid matrix, T_m represents a melting point of the multi-particle-matrix system.

   (b) Similarly, provided that the particles and the envelopes precipitate from a liquid matrix of the multi-particle-envelope-matrix system, the melting point T_m represents the minimum of the set $\{T_{mp}, T_{me}, T_{mm}\}$, where T_{ep} is a melting point of the envelope. Finally, provided that the particles and the envelopes precipitate from a solid matrix, T_m represents a melting point of the multi-particle-envelope-matrix system.

In case of the zero difference in the thermal expansion coefficients and provided that $T_{tq} \in \langle T_f, T_r \rangle$, the thermal stresses originate at the temperature T_t representing maximal temperature from the sets $\{T_{tp}, T_{tm}\}$ and $\{T_{tp}, T_{te}, T_{tm}\}$ related to the multi-particle-matrix and multi-particle-envelope-matrix systems, respectively, where $\langle T_f, T_t \rangle \subset \langle T_f, T_r \rangle$.

Accordingly, the φ, ν-dependent coefficien $\beta_q = \beta_q(\varphi, \nu)$ has the forms

$$\beta_q = \varepsilon_{11tq} + \int_{T_f}^{T_{tq}} \alpha'_{1Iq}\,dT + \int_{T_{tq}}^{T_r} \alpha_{IIq}\,dT, \quad T_{tq} \in \langle T_f, T_r \rangle, \quad T \le T_{tq}, \qquad (3.100)$$

$$\beta_q = \int_{T}^{T_r} \alpha'_{1IIq}\,dT, \quad T_{tq} \in \langle T_f, T_r \rangle, \quad T > T_{tq}, \qquad (3.101)$$

$$\beta_q = \int_{T_f}^{T_r} \alpha'_{1q}\,dT, \quad T_{tq} \notin \langle T_f, T_r \rangle, \qquad (3.102)$$

where α'_{1Iq}, α'_{1IIq} are temperature-dependent or temperature-independent thermal expansion coefficient at the temperature $T \le T_{tq}, T \ge T_{tq}$ provided that $T_{tq} \in \langle T_f, T_r \rangle$, respectively, and α'_{1q} is a temperature-dependent or temperature-independent thermal expansion coefficien on the condition $T_{tq} \notin \langle T_f, T_r \rangle$.

As mentioned in Section 3.1.6, the analytical modelling is valid for an uniaxial- and triaxial-anisotropic components of the multi-particle-(envelope)-matrix system. The uniaxial- and triaxial-anisotropic spherical particle, spherical envelope and cell matrix consists of uniaxial- and triaxial-anisotropic crystal lattices, respectively. The only difference concerning uniaxial and triaxial anisotropy is represented by thermal expansion coeffi cient α_{kq} ($k = 1,2,3$) as well as by the elastic modulus s_{ij} ($i, j = 1, \ldots, 6$) which both result in different α'_{1q}, s'_{ij}, for the uniaxial-/triaxial-anisotropic crystal lattices (see Eqs. (9.1)–(9.34)) [6, p. 30].

Provided that the thermal stresses originate as a consequence of a differences in the thermal expansion coefficient $\alpha'_{1p}, \alpha'_{1e}, \alpha'_{1m}$ and as well as a consequence of a differences in the phase-transformation induced radial strains $\varepsilon_{11tp}, \varepsilon_{11te}, \varepsilon_{11tm}$ (see Items 1, 2, p. 41), contributions of the individual differences to the thermal stresses can be determined by the substitution either $\varepsilon_{11tq} = 0$ or $\alpha'_{1q} = 0$ ($q = p,e,m$), respectively.

Considering $T = T(\tau)$ representing a time-dependent function, the total differential dT has the form [3]

$$dT = \frac{\partial T}{\partial \tau} \, d\tau,\tag{3.103}$$

and the temperature intervals $\langle T_f, T_{tq} \rangle$, $\langle T_{tq}, T_r \rangle$, $\langle T, T_r \rangle$, $\langle T_f, T_r \rangle$ in integrals of Eqs. (3.100)–(3.102) are replaced by the time intervals $\langle \tau_f, \tau_{tq} \rangle$, $\langle \tau_{tq}, \tau_r \rangle$, $\langle \tau, \tau_r \rangle$, $\langle \tau_f, \tau_r \rangle$, where τ_f, τ_{tq}, τ_r, τ are determined numerically from the conditions $T_f = T(\tau_f)$, $T_{tq} = T(\tau_{tq})$, $T_r = T(\tau_r)$, $T = T(\tau)$, respectively.

3.4.2 Phase-transformation induced radial strain

The phase-transformation induced radial strain ε_{11tq} ($q = p,e,m$) in a component of the multi-particle-matrix and multi-particle-envelope-matrix systems is related to the orthorhombic, tetragonal and hexagonal crystal lattices [4, p. 36], where the subscripts $q = p$, $q = e$, $q = m$ are related to a crystal lattice of the spherical particle, the spherical envelope, the cell matrix, respectively.

The orthorhombic crystal lattice (OCL) which represents a rectangular-based prism with mutually perpendicular sides is characterized by the dimensions $a_{01q}, a_{02q}, a_{03q}$ along the axes $x_{01q}, x_{02q}, x_{03q}$, respectively, where $a_{01q} \neq a_{02q} \neq a_{03q}$ and $x_{01q} \perp x_{02q} \perp x_{03q}$. OCL exhibits [4, p. 36]

- a simple modificatio which is characterized by atoms at corner points of OCL,

- a base-centered modification which is, besides atoms at corner points of OCL, characterized by a central atom in each basis of OCL,

- a body-centered modification which is, besides atoms at corner points of OCL, characterized by an atom at the intersection point of diagonals of OCL, i.e. at the geometrical center of OCL,

- a face-centered modification which is, besides atoms at corner points of OCL, characterized by a central atom on each of surfaces of OCL.

The tetragonal crystal lattice (TCL) which represents a square-based prism with mutually perpendicular sides is characterized by the dimensions $a_{T1q}, a_{T2q}, a_{T3q}$ along the axes $x_{T1q}, x_{T2q}, x_{T3q}$, respectively, where $x_{T1q} \perp x_{T2q} \perp x_{T3q}$. TCL is characterized by the same characteristics as OCL, except the condition $a_{T1q} = a_{T2q} \neq a_{T3q}$.

---

[3]The parameter $v_T = \partial T / \partial \tau$ is cooling rate.

The hexagonal crystal lattice (HCL) [4, p. 36] represents a hexagonal-based prism with the height a_{H3q} along the axis x_{H3q} and with bases of a shape of the regular hexagon with the dimension a_{H1q}, where $x_{H1q} \perp x_{H2q} \perp x_{H3q}$. A direction of the axis x_{H1} is identical with a direction of a diagonal of the basis. HCL with atoms at corner points and with a central atom in each basis represents the most known modification

The radial strain ε_{11tq} $(q = p,e,m)$ is thus induced, during a cooling process, by

1. the phase transformation $OCL \rightarrow OCL$ for the orthorhombic crystal lattices with the dimensions $a_{01q}^{(1)}$, $a_{02q}^{(1)}$, $a_{03q}^{(1)}$ and $a_{01q}^{(2)}$, $a_{02q}^{(2)}$, $a_{03q}^{(2)}$, where the dimension $a_{0iq}^{(1)}$ $(i = 1,2,3)$ above the phase transformation temperature is transformed to the dimension $a_{0iq}^{(2)}$ below the phase transformation temperature, i.e. $a_{0iq}^{(1)} \rightarrow a_{0iq}^{(2)}$. Additionally, at least one of the conditions $a_{01q}^{(1)} \neq a_{01q}^{(2)}$, $a_{02q}^{(1)} \neq a_{02q}^{(2)}$, $a_{03q}^{(1)} \neq a_{03q}^{(2)}$ is valid.

2. the phase transformation $OCL \rightarrow TCL$ or $TCL \rightarrow OCL$, where $a_{0iq} \rightarrow a_{Tiq}$ or $a_{Tiq} \rightarrow a_{0iq}$, respectively. Additionally, at least one of the conditions $a_{01q} \neq a_{T1q}$, $a_{02q} \neq a_{T2q}$, $a_{03q} \neq a_{T3q}$ is valid.

3. the phase transformation $OCL \rightarrow HCL$ or $HCL \rightarrow OCL$, i.e. $a_{0i} \rightarrow a_{Hi}$ or $a_{Hiq} \rightarrow a_{0iq}$, respectively.

4. the phase transformation $TCL \rightarrow HCL$ or $HCL \rightarrow TCL$, i.e. $a_{Tiq} \rightarrow a_{Hiq}$ or $a_{Hiq} \rightarrow a_{Tiq}$, respectively.

The phase-transformation induced radial strain ε_{11tq} $(q = p,e,m)$ in a component of the multi-particle-matrix and multi-particle-envelope-matrix systems is given by

$$\varepsilon_{11tq} = \frac{r_c^{(2)} - r_c^{(1)}}{r_c^{(1)}}, \tag{3.104}$$

where the distances $r_c^{(1)}$ and $r_c^{(2)}$ are related to the crystal lattice dimensions at temperature which is above and below the phase transformation temperature as analysed above in Items 1–4, respectively.

The distance r_c $(\equiv r_c^{(1)}, r_c^{(1)})$ for OCL and TCL is given by Eqs. (2.25)–(2.28), (2.34), (2.13), where the cell dimension d_{iex} is replaced by a_{0iq} and a_{Tiq} $(i = 1,2,3)$, respectively.

Finally, the distance r_c $(\equiv r_c^{(1)}, r_c^{(1)})$ for HCL is given by Eqs. (2.25), (2.26), (2.34), (2.13), (2.37), where the cell dimension d_{1ex}, d_{2ex}, d_{3ex} and the angle φ^* included in Eq. (2.37) are replaced by a_{H1q}, $a_{H2q} = 2a_{H1q}\sqrt{3}/3$, a_{H3q} and $\varphi^* = \pi/6$, respectively.

This analysis concerning the distance r_c $(\equiv r_c^{(1)}, r_c^{(1)})$ is valid for $x_i = x_{0iq}$, $x_i = x_{Tiq}$, $x_i = x_{Hiq}$ $(i = 1,2,3)$, where the Cartesian system $(Ox_1x_2x_3)$ is shown in Fig. 3.1.

3.4.3 Analysis of the conditions $\beta_{q1} \neq \beta_{q2}$, $\beta_{q1} = \beta_{q2}$

Section 3.4.3 deals with a determination of intervals of the angles φ, ν which result in the validity of the conditions $\beta_{q1} \neq \beta_{q2}$, $\beta_{q1} = \beta_{q2}$ (see Eqs. (3.100)–(3.102)), where

Thermal stresses in elastic solid continuum 45

- $q_1, q_2 = p, m$ for the multi-particle-matrix system, and $q_1 \neq q_2$,

- either $q_1, q_2 = p, e$, or $q_1, q_2 = m, e$ for the multi-particle-envelope-matrix system, and $q_1 \neq q_2$.

A reason of this determination is a dependence of the coefficien $\beta_q = \beta_q (\varphi, \nu)$ on the angles $\varphi, \nu \in \langle 0, \pi/2 \rangle$ (see Eqs. (3.100)–(3.102), (9.2)–(9.11)). This determination is required to be performed due to the analysis in Section 8.2.4 provided that the analysis of the crack formation in Section 8.2.4 concerns the multi-particle-matrix and multi-particle-envelope-matrix systems which are loaded by the thermal stresses only.

If the model system (see Fig. 2.1) does not exhibit a phase transformation at the temperature $T \in \langle T_f, T_r \rangle$, then the conditions $\beta_{q_1} \neq \beta_{q_2}$, $\beta_{q_1} = \beta_{q_2}$ are transformed to $\alpha_{q_1} (\varphi, \nu) \neq \alpha_{q_2} (\varphi, \nu)$, $\alpha_{q_1} (\varphi, \nu) = \alpha_{q_2} (\varphi, \nu)$, respectively. With regard to Eqs. (9.2)–(9.11), the condition $\alpha_{q_1} (\varphi, \nu) - \alpha_{q_2} (\varphi, \nu) = 0$ is transformed to the form

$$(\alpha_{1q_1} - \alpha_{1q_2}) \cos^2 \varphi \sin^2 \nu + (\alpha_{2q_1} - \alpha_{2q_2}) \sin^2 \varphi \sin^2 \nu + (\alpha_{3q_1} - \alpha_{3q_2}) \cos^2 \nu = 0.$$
(3.105)

Consequently, an analysis of the conditions $\alpha_{q_1} \neq \alpha_{q_2}$, $\alpha_{q_1} = \alpha_{q_2}$ is as follows.

1. If the differences $\alpha_{1q_1} - \alpha_{1q_2}$, $\alpha_{2q_1} - \alpha_{2q_2}$, $\alpha_{3q_1} - \alpha_{3q_2}$ included in Eq. (3.105) exhibits identical signs, i.e. $\alpha_{1q_1} \gtrless \alpha_{1q_2}$, $\alpha_{2q_1} \gtrless \alpha_{2q_2}$, $\alpha_{3q_1} \gtrless \alpha_{3q_2}$, then the condition $\alpha_{q_1} (\varphi, \nu) \neq \alpha_{q_2} (\varphi, \nu)$ is valid for $\varphi, \nu \in \langle 0, \pi/2 \rangle$. Otherwise, the following analysis is considered.

2. If $\alpha_{1q_1} \gtrless \alpha_{1q_2}$, $\alpha_{2q_1} \gtrless \alpha_{2q_2}$, $\alpha_{3q_1} \lessgtr \alpha_{3q_2}$, then the condition $\alpha_{q_1} (\varphi, \nu) = \alpha_{q_2} (\varphi, \nu)$ is valid for $\varphi \in \langle 0, \pi/2 \rangle$ and for $\nu = \nu_{q_1 q_2}$, where the function $\nu_{q_1 q_2} = \nu_{q_1 q_2} (\varphi)$ is given by Eq. (3.106).

3. If $\alpha_{1q_1} \gtrless \alpha_{1q_2}$, $\alpha_{2q_1} \lessgtr \alpha_{2q_2}$, $\alpha_{3q_1} \lessgtr \alpha_{3q_2}$, then the condition $\alpha_{q_1} (\varphi, \nu) = \alpha_{q_2} (\varphi, \nu)$ is valid for $\varphi \in \langle 0, \varphi_{q_1 q_2} \rangle$ and for $\nu = \nu_{q_1 q_2}$, where the angle $\varphi_{q_1 q_2}$ is given by Eq. (3.107).

4. If $\alpha_{1q_1} \gtrless \alpha_{1q_2}$, $\alpha_{2q_1} \lessgtr \alpha_{2q_2}$, $\alpha_{3q_1} \gtrless \alpha_{3q_2}$, then the condition $\alpha_{q_1} (\varphi, \nu) = \alpha_{q_2} (\varphi, \nu)$ is valid for $\varphi \in \langle \varphi_{q_1 q_2}, \pi/2 \rangle$ and for $\nu = \nu_{q_1 q_2}$.

5. If $\alpha_{3q_1} = \alpha_{3q_2}$, then the following analysis is considered.

    (a) If $\alpha_{1q_1} \gtrless \alpha_{1q_2}$, $\alpha_{2q_1} \gtrless \alpha_{2q_2}$, then the condition $\alpha_{q_1} (\varphi, \nu) = \alpha_{q_2} (\varphi, \nu)$ is valid for $\nu = 0$.

    (b) If $\alpha_{1q_1} \gtrless \alpha_{1q_2}$, $\alpha_{2q_1} \lessgtr \alpha_{2q_2}$, then the condition $\alpha_{q_1} (\varphi, \nu) = \alpha_{q_2} (\varphi, \nu)$ is valid for $\varphi = \varphi_{q_1 q_2}$ and for $\nu \in \langle 0, \pi/2 \rangle$.

The function $v_{q_1q_2} = v_{q_1q_2}(\varphi)$ which represents a solution of Eq. (3.105) regarding the variable v has the form

$$v_{q_1q_2} = \arcsin\left(\sqrt{\frac{\alpha_{3q_2} - \alpha_{3q_1}}{(\alpha_{1q_1} - \alpha_{1q_2})\cos^2\varphi + (\alpha_{2q_1} - \alpha_{2q_2})\sin^2\varphi}}\right). \tag{3.106}$$

The angle $\varphi_{q_1q_2}$ which represents a solution of Eq. (3.105) regarding the variable φ for $\alpha_{3q_1} = \alpha_{3q_2}$ is derived as

$$\varphi_{q_1q_2} = \arcsin\left(\sqrt{\frac{\alpha_{1q_2} - \alpha_{1q_1}}{\alpha_{2q_1} - \alpha_{2q_2}}}\right). \tag{3.107}$$

Let the plane x_1x_2 (see Fig. 3.1) in the component which is related to the subscript q_1 is isotropic. In case of this uni-axial anisotropy, we get $\alpha_{1q_1} = \alpha_{2q_1}$.

If $\varepsilon_{11tq_1}(\varphi,v) \neq \varepsilon_{11tq_2}(\varphi,v)$ at $T \in \langle T_f, T_r \rangle$, then the function $v_{q_1q_2} = v_{q_1q_2}(\varphi)$ and the angle $\varphi_{q_1q_2} \in \langle 0, \pi/2 \rangle$ which both result from the condition $\beta_{q_1}(\varphi,v) - \beta_{q_2}(\varphi,v) = 0$ are determined by numerical methods for real two- and three-component materials (see Items 1–4, Section 2.1). In case of $\varepsilon_{11tq_1}(\varphi,v) \neq \varepsilon_{11tq_2}(\varphi,v)$ at $T \in \langle T_f, T_r \rangle$,, the condition $\beta_{q_1}(\varphi,v) - \beta_{q_2}(\varphi,v) = 0$ thus represents a transcendental equation regarding the variables $\varphi, v \in \langle 0, \pi/2 \rangle$.

Let the multi-particle-matrix system (see Fig. 2.1) consists of anisotropic and isotropic components. Let the multi-particle-envelope-matrix system consists of anisotropic and isotropic components. Let the coefficien β_{q_2} is related to the isotropic component. In this case, the coefficien β_{q_2} is not a function of the variables $\varphi, v \in \langle 0, \pi/2 \rangle$. The analysis concerning the condition $\alpha_{q_1}(\varphi,v) - \alpha_{q_2}(\varphi,v) = 0$ is also valid for the condition $\alpha_{q_1}(\varphi,v) - \alpha_{q_2} = 0$. Each of the thermal expansion coefficient $\alpha_{1q_2}, \alpha_{2q_2}, \alpha_{3q_2}$ which are considered within this analysis is only replaced by the thermal expansion coefficien α_{q_2}.

3.4.4 Formulae for determination of the radial stresses p_1, p_2

As presented in Section 3.4.1, the thermal stresses originate at the temperature $T \in \langle T_f, T_r \rangle$, where T_f is fina temperature of a cooling process. The relaxation temperature T_r is analysed in Section 3.4.1.

An analysis of the formulae for determination of the radial stresses p_1, p_2 results from the analysis in Section 3.1.2.

If $T \in \langle T_f, T_r \rangle$, then R_{1p}, R_{iq} $(i = 1,2; q = e,m)$ which are considered in Section 3.1.2 have the forms

$$R_{1p} = R_{1T_r}(1 - \beta_p), \quad R_{iq} = R_{iT_r}(1 - \beta_q), \quad i = 1, 2; \quad q = e, m, \tag{3.108}$$

where R_{1T_r} and t_{T_r} are a radius of the spherical particle and thickness of the spherical envelope both at T_r, respectively, and $R_{2T_r} = R_{1T_r} + t_{T_r}$.

Let the multi-particle-matrix system at $T \in \langle T_f, T_r \rangle$ is then considered. The thermal-stress induced radial displacement $(u_{1q})_{r=R_{1q}}$ is related to the change of $R_{1q} \rightarrow R_1$

$(q = p,m)$, where $(u_{1q})_{r=R_{1q}} = R_1 - R_{1q}$. Strictly speaking, the infinitesima spherical cap is shifted along the axis x_1' from the position $r = R_{1q}$ to the position $r = R_1$. Accordingly, the Cauchy's equations (3.2), (3.3) are thus related to the position $r = R_{1q}$ (see Eq. 3.108). With regard to Eqs. (3.2), (3.108), the difference $(u_{1m})_{r=R_{1m}} - (u_{1p})_{r=R_{1p}}$ results in the formula

$$(1 - \beta_m)(\varepsilon_{22m})_{r=R_{1m}} - (1 - \beta_p)(\varepsilon_{22p})_{r=R_{1p}} = \beta_m - \beta_p, \tag{3.109}$$

which is considered for the determination of the radial stress $p_1 = p_1(\varphi, \nu)$ acting at the particle-matrix boundary (i.e. for $r = R_1$) along the axis x_1' (see Fig. 3.1).

The same analysis is also valid for the multi-particle-system. In case of $\beta_p \neq \beta_e$, the formula

$$(1 - \beta_e)(\varepsilon_{22e})_{r=R_{1e}} - (1 - \beta_p)(\varepsilon_{22p})_{r=R_{1p}} = \beta_e - \beta_p \tag{3.110}$$

is considered for the determination of the radial stress $p_1 = p_1(\varphi, \nu)$ which acts at the particle-envelope boundary.

In case of $\beta_m \neq \beta_e$, the formula

$$(1 - \beta_m)(\varepsilon_{22m})_{r=R_{2m}} - (1 - \beta_e)(\varepsilon_{22e})_{r=R_{2e}} = \beta_m - \beta_e \tag{3.111}$$

is considered for the determination of the radial stress $p_2 = p_2(\varphi, \nu)$ which acts at the matrix-envelope boundary.

Let the multi-particle-matrix system is considered. Let the phase transformation originates at the temperature $T_{tq} \in \langle T_f, T_r \rangle$ $(q = p,m)$ at least in one of components of this system, i.e. $q = p$ and/or $q = m$. If $\alpha_{1np}' \neq \alpha_{1nm}'$ $(n = I, II)$ and $\varepsilon_{11tp} - \varepsilon_{11tm} \neq 0$ (see Eq. (3.100)), then the thermal stresses originate due to the phase transformation at $T_{tq} \in \langle T_f, T_r \rangle$. The same is also valid for the multi-particle-envelope-matrix system.

Finally, with regard to the conditions $\beta_p = \beta_m$ and $\beta_p = \beta_e = \beta_m$, the thermal stresses in the multi-particle-matrix and multi-particle-envelope-matrix systems do not occur, respectively.

3.4.5 Temperature dependence of thermal stresses

Temperature dependencies of the thermal stresses is represented by temperature dependencies of the radial stresses $p_1 = p_1(\varphi, \nu, T)$, $p_2 = p_2(\varphi, \nu, T)$ which are functions of the coefficien β_q $(q = p,e,m)$ (see Eqs. (3.100)–(3.102)) as well as functions of the elastic modulus $s_{ijq}' = s_{jiq}'$ $(i, j = 1, \ldots, 6)$ (see Eqs. (9.14)–(9.34)) included in the tangential strain ε_{22q}, where β_q and ε_{22q} are included in the conditions (3.109), (3.110), (3.111) considered for the determination of p_1, p_2 (see Eqs. (3.100)–(3.102)).

On the conditions $\alpha_{1p}' \neq \alpha_{1m}'$, and $\alpha_{1p}' \neq \alpha_{1e}' \neq \alpha_{1m}'$ or $\alpha_{1p}' \neq \alpha_{1e}' = \alpha_{1m}'$ or $\alpha_{1p}' = \alpha_{1e}' \neq \alpha_{1m}'$ related to the multi-particle-matrix and multi-particle-envelope-matrix systems, respectively, temperature dependencies of the thermal stresses results from temperature intervals in the integrals included in the coefficien β_q, as well as from temperature dependencies of the elastic modulus s_{ijq}' in these temperature intervals. Provided that the thermal expansion coefficient α'_{1Iq}, α'_{1IIq}, α'_{1q} included in β_q are not functions of the temperature T, the integrals in the coefficien β_q are transformed to the forms $\alpha'_{1Iq}(T_{tq} - T_f)$, $\alpha'_{1IIq}(T_r - T_{tq})$, $\alpha'_{1IIq}(T_r - T)$, $\alpha'_{1q}(T_r - T_f)$.

On the conditions $\alpha'_{1p} \neq \alpha'_{1m}$ and $\alpha'_{1p} = \alpha'_{1e} = \alpha'_{1m}$, provided that $\varepsilon_{11tm} - \varepsilon_{11tp} \neq 0$, and $\varepsilon_{11te} - \varepsilon_{11tp} \neq 0$ or $\varepsilon_{11tm} - \varepsilon_{11te} \neq 0$ (see Eq. (3.104)), respectively, a phase transformation originating at least in one of components of the multi-particle-(envelope)-matrix system is a reason of the thermal stresses originating at the temperature $T_{tq} \in \langle T_f, T_r \rangle$. Accordingly, temperature dependencies of the thermal stresses results only from temperature dependencies of the elastic modulus s'_{ijq} in the temperature interval $\langle T_f, T_t \rangle$. T_t represents maximal temperature from the sets $\{T_{tp}, T_{tm}\}$ and $\{T_{tp}, T_{te}, T_{tm}\}$ related to the multi-particle-matrix and multi-particle-envelope-matrix systems, respectively, where $\langle T_f, T_t \rangle \subset \langle T_f, T_r \rangle$.

Accordingly, provided that $s_{11q} \neq f(T)$, $s_{12q} \neq f(T)$, $s_{44q} \neq f(T)$, the thermal stresses are temperature-independent.

3.4.6 Temperature range

The analytical model of the thermal stresses in the multi-particle-matrix and multi-particle-envelope-matrix systems presented in this book is related to thermal-stress induced strains which are elastic, i.e. the stress-strain relationships are define by the Hooke's law. Accordingly, the validity of the analytical model is required to consider yield stresses of components of the multi-particle-matrix and multi-particle-envelope-matrix systems as well as a maximal stress of the thermal-stress fiel which is induced in the cell due to $\beta_m \neq \beta_p$, and $\beta_p \neq \beta_e = \beta_m$ (see Sections 5.2.1) or $\beta_p \neq \beta_e \neq \beta_m$ (see Sections 5.2.2) or $\beta_p = \beta_e \neq \beta_m$ (see Sections 5.2.3), respectively.

Let the multi-particle-matrix system is considered. The radial stress $p_1 = p_1(\varphi, \nu, T)$ acting at the particle-matrix boundary represents a maximal stress of the thermal-stress fiel which is induced in the cell due to $\beta_m \neq \beta_p$. Let $|p_{1max}| = |p_{1max}(T)|$ is a maximal value of the φ, ν-dependent function $|p_1| = |p_1(\varphi, \nu, R_1, v, T)|$, where $\varphi, \nu \in \langle 0, \pi/2 \rangle$. Let σ_{ys} representing the minimum of the set $\{\sigma_{ysp}, \sigma_{ysm}\}$ is a yield stress in compression or tension provided that $p_{1max} > 0$ or $p_{1max} < 0$, where $\sigma_{ysp} > 0$ and $\sigma_{ysm} > 0$ represent yield stresses in compression or tension for the spherical particle and the cell matrix, respectively. Consequently, the analytical model of the thermal stresses is valid to the critical fina temperature of a cooling process, T_{ys}, which is determined by the condition $|p_{1max}(T)| = \sigma_{ys}$.

The same analysis is also valid for the multi-particle-envelope-matrix system. In this case, T_{ys} is determined by the condition $|p_{max}(T)| = \sigma_{ys}$, where $|p_{max}|$ and σ_{ys} are the maximum and the minimum of the sets $\{|p_{1max}|, |p_{2max}|\}$ and $\{\sigma_{ysp}, \sigma_{yse}, \sigma_{ysm}\}$, respectively; $|p_{imax}| = |p_{imax}(T)|$ ($i = 1,2$) is a maximal value of the φ, ν-dependent function $|p_i| = |p_i(\varphi, \nu, R_1, v, T)|$ for $\varphi, \nu \in \langle 0, \pi/2 \rangle$; and σ_{yse} is a yield stress of the spherical envelope.

The analytical model of the thermal stresses presented in this book is valid for the temperature $T \in \langle T_{ys}, T_r \rangle$, where $\langle T_f, T_r \rangle \subset \langle T_{ys}, T_r \rangle$.

Chapter 4

Boundary conditions

Chapter 4 deals with the determination of boundary conditions for components of the multi-particle-matrix and multi-particle-envelope-matrix systems (see Fig. 2.1). In case of the multi-particle-envelope-matrix system, the conditions $\beta_p \neq \beta_e = \beta_m$, $\beta_p \neq \beta_e \neq \beta_m$, $\beta_p = \beta_e \neq \beta_m$ are considered. In case of the cell matrix (see Section 4.3), mandatory and additionally boundary conditions are determined.

With regard to the analysis in Section 3.1.2, in case of the multi-particle-matrix system, R_1 represents a radius of the embedded spherical particle at the temperature $T \in \langle T_f, T_r \rangle$, where T_f is a fina temperature of a cooling process, and the relaxation temperature T_r is determined in Section 3.4.1. In case of the multi-particle-envelope-matrix system, R_1, R_2 and $t = R_2 - R_1$ represents radii and thickness of the embedded spherical envelope at the temperature $T \in \langle T_f, T_r \rangle$, respectively.

4.1 Spherical particle

The thermal stresses σ_{iip}, σ_{12+jp} ($i = 1,2,3$; $j = 0,1$), and the radial displacement u_{1p} are required to fulfi the boundary conditions determined as

$$(u_{1p})_{r=0} = 0, \tag{4.1}$$

$$(\sigma_{iip})_{r \to 0} \nrightarrow \pm\infty, \quad i = 1, 2, 3, \tag{4.2}$$

$$(\sigma_{12+ip})_{r \to 0} \nrightarrow \pm\infty, \quad i = 0, 1, \tag{4.3}$$

$$(\sigma_{11p})_{r=R_1} = -p_1, \tag{4.4}$$

where (4.1) and (4.2)–(4.4) represent geometric and stress boundary conditions, respectively.

The boundary conditions (4.1)–(4.4) are related to the multi-particle-matrix and multi-particle-envelope-matrix systems on the conditions $\beta_p \neq \beta_m$ and $\beta_p \neq \beta_e = \beta_m$, $\beta_p \neq \beta_e \neq \beta_m$ (see Eqs. (3.100)–(3.102)).

The radial stress p_1 acting at the particle-matrix boundary is determined by the condition (3.109).

If either $\beta_p \neq \beta_e = \beta_m$ or $\beta_p \neq \beta_e \neq \beta_m$, then the radial stress p_1 acting at the particle-envelope boundary is determined by either the condition (3.110) or by the conditions (3.110), (3.111), respectively.

If $\beta_p = \beta_e \neq \beta_m$, then the boundary condition (4.4) is replaced by

$$(\varepsilon_{22p})_{r=R_1} = (\varepsilon_{22e})_{r=R_1} = \frac{(u_{1e})_{r=R_1}}{R_1} = -p_2 \varrho_{1e}^{pe}, \tag{4.5}$$

In this case, the radial stress p_2 acting at the matrix-envelope boundary is determined by the condition (3.111), and the coefficien ϱ_{1e}^{pe} results from the determination of $(\varepsilon_{22e})_{r=R_1}$. Equations (4.1)–(4.4) result in one integration constant only, and then we get

$$n_p = 1, \tag{4.6}$$

where the integration constant $C_{1p} \neq 0$ is determined by Eq. (4.4). Consequently, with regard to Eqs. (3.76)–(3.80), (4.2), (4.3), the exponent λ_{1p} is required to fulfi the following condition

$$\lambda_{1p} > 0. \tag{4.7}$$

Additionally, the absolute value $|u_{1p}|$ is required to represent an increasing function of $r \in \langle 0, R_1 \rangle$. This increasing function exhibits a maximal value at either the particle-matrix or particle-envelope boundary, thus for $r = R_1$. Finally, considering Eq. (4.7), the increasing course of the dependence $|u_{1p}| - r$ is fulfille due to $\partial |u_{1p}| / \partial r \propto (\lambda_{1p} + 1) r^{\lambda_{1p}} \leq 0$.

4.2 Spherical envelope

If $\beta_p \neq \beta_e \neq \beta_m$ (see Eqs. (3.100)–(3.102)), then the following boundary conditions

$$(\sigma_{11e})_{r=R_1} = -p_1, \tag{4.8}$$

$$(\sigma_{11e})_{r=R_2} = -p_2, \tag{4.9}$$

are considered for the determination of the integration constants $C_{1e} \neq 0$ and $C_{2e} \neq 0$. The real exponents λ_{1e} and λ_{2e} are not required to be define by proper conditions similar to that given by Eq. (4.7). Accordingly, we get

$$n_e = 2. \tag{4.10}$$

If $\beta_p \neq \beta_e = \beta_m$, then the boundary condition (4.8) is considered for the determination of the integration constant $C_{1e} \neq 0$. Accordingly, we get

$$n_e = 1. \tag{4.11}$$

The absolute value $|u_{1e}|$ exhibits a maximal value at the particle-envelope boundary, thus for $r = R_1$. Additionally, $|u_{1e}|$ is required to represent a decreasing function of the variable $r \in \langle R_1, R_2 \rangle$ or, at the very most, a constant function regarding the variable r. Consequently, with regard to Eq. (3.85), the exponent λ_{1e} is required to fulfi the following condition derived as

$$\lambda_{1e} \leq -1. \tag{4.12}$$

If $\beta_p = \beta_e \neq \beta_m$, then the boundary condition (4.9) along with Eq. (4.11) are considered for the determination of the integration constant $C_{1e} \neq 0$. The absolute value $|u_{1e}|$ exhibits a maximal value at the matrix-envelope boundary, thus for $r = R_2 = R_1 + t$. Accordingly, $|u_{1e}|$ is required to represent an increasing function of the variable $r \in \langle R_1, R_2 \rangle$ or, at the very most, a constant function regarding the variable r. Finally, the exponent λ_{1e} is required to fulfi the following condition derived as

$$\lambda_{1e} \geq -1. \tag{4.13}$$

4.3 Cell matrix

The thermal stresses and corresponding quantities (see Chapter 8) in the cell matrix are derived regarding standard boundary conditions (see Eqs. (4.14)–(4.16)) which are mandatory, without or with consideration of additional boundary conditions. The additional boundary conditions analysed in Section 4.3.2 (see Eqs. (4.19), (4.20)) might be considered as a possible state of deformation of the matrix at cell boundaries. The condition (4.19) assumes the minimum or the maximum of the $u_{1m} - r$ dependence on the cell surface, thus for $r = r_c$, regarding $u_{1m} > 0$ or $u_{1m} < 0$, respectively. Finally, the condition (4.19) assumes the minimum of the $w_m - r$ dependence on the cell surface regarding $w_m \propto \sigma_m^2 > 0$ (see Eq. (3.72)).

The mandatory boundary conditions

1. for the multi-particle-matrix system are given by Eqs. (4.14), (4.16),

2. for the multi-particle-envelope-matrix system,

    (a) provided that $\beta_p \neq \beta_e \neq \beta_m$ or $\beta_p = \beta_e \neq \beta_m$ (see Eqs. (3.100)–(3.102)), are given by Eqs. (4.15), (4.16).

    (b) provided that $\beta_p \neq \beta_e = \beta_m$, are given by Eqs. (4.16), (4.17).

Accordingly, the thermal stresses in the cell matrix are derived regarding the following boundary conditions:

3. the mandatory boundary conditions (4.14), (4.16) or (4.15), (4.16).

4. the mandatory boundary conditions (4.14), (4.16) or (4.15), (4.16) along with the additional boundary condition (4.19).

5. the mandatory boundary conditions (4.14), (4.16) or (4.15), (4.16) along with the additional boundary condition (4.20).

6. the mandatory boundary conditions (4.14), (4.16) or (4.15), (4.16) along with the additional boundary conditions (4.19) and (4.20).

4.3.1 Mandatory boundary conditions

The thermal radial stress σ_{11m} and the radial displacement u_{1m} are required to fulfi the mandatory boundary conditions determined as

$$(\sigma_{11m})_{r=R_1} = -p_1, \tag{4.14}$$

$$(\sigma_{11m})_{r=R_2} = -p_2, \tag{4.15}$$

$$(u_{1m})_{r=r_c} = 0. \tag{4.16}$$

The mandatory boundary conditions (4.14), (4.16) and (4.15), (4.16) are considered for the multi-particle-matrix and multi-particle-envelope-matrix systems provided that $\beta_p \neq \beta_m$, and either $\beta_p \neq \beta_e \neq \beta_m$ or $\beta_p = \beta_e \neq \beta_m$, respectively.

The radial stress p_1 acting at the particle-matrix boundary is determined by the condition (3.109).

If $\beta_p \neq \beta_e \neq \beta_m$, then the radial stresses p_1 and p_2 acting at the particle-envelope and matrix-envelope boundaries, respectively, are both determined by the conditions (3.110), (3.111).

If $\beta_p = \beta_e \neq \beta_m$, then p_2 is determined by the condition (3.111). Finally, the distance r_c is given by Eqs. (2.25)–(2.41).

If $\beta_p \neq \beta_e = \beta_m$, then the boundary condition (4.15) is replaced by

$$(\varepsilon_{22m})_{r=R_2} = (\varepsilon_{22e})_{r=R_2} = \frac{(u_{1e})_{r=R_2}}{R_2} = -p_1\,\varrho_{2e}^{me}, \tag{4.17}$$

The radial stress p_2 acting at the matrix-envelope boundary is determined by the condition (3.111), and the coefficien ϱ_{2e}^{me} is given by Eq. (3.93). Considering the analysis concerning the boundary conditions for the cell matrix, we get

$$n_m = 2. \tag{4.18}$$

In case of all conditions mentioned above, i.e. $\beta_p \neq \beta_m$, $\beta_p \neq \beta_e \neq \beta_m$, $\beta_p = \beta_e \neq \beta_m$, and $\beta_p \neq \beta_e = \beta_m$, the absolute value $|u_{1m}|$ represents a decreasing function of $r \in \langle R_{1+i}, r_c \rangle$ $(i=0,1)$ due to $(|u_{1m}|)_{r=R_{1+i}} \neq 0$ and $(u_{1m})_{r=r_c} = 0$ (see Eq. (4.16)). The subscripts $i = 0$ and $i = 1$ are related to the multi-particle-matrix and multi-particle-envelope-matrix system, respectively. Accordingly, $|u_{1m}|$ exhibits a maximal value at the particle-matrix or matrix-envelope boundary, then for $r = R_1$ or $r = R_2$, respectively. Finally, the decreasing course of the dependence $|u_{1m}| - r$ is ensured by the integration constants C_{1m}, C_{2m}.

4.3.2 Additional boundary conditions

The point C ($\equiv C_1, C_3$) on the cell surface (see Figs. 2.2, 2.3) is related to two neighbouring identical cells with the Cartesian systems $(O_1 x_1' x_2' x_3')$, $(O_2 x_1' x_2' x_3')$ and with the distance $r_c = |O_1 C| = |O_2 C|$ as length of the abscissae $O_1 C$, $O_2 C$ in a radial direction determined by the angles φ, ν and represented by the axis x_1'. With regard to the imaginary division of the infinit matrix, the surface of the cell which represents the part of the multi-particle-(envelope)-matrix system related to one spherical particle is not a physical boundary. Consequently, a smooth course of the r-dependent functions $u_{1m} = u_{1m}(r)$ and $w_m = w_m(r)$ is assumed to be exhibited on the cell surface. Consequently, the dependencies $u_{1m}^{(1)} - r$ and $u_{1m}^{(2)} - r$ related to the Cartesian systems $(O_1 x_1' x_2' x_3')$ and $(O_2 x_1' x_2' x_3')$ are required not to mutually create a singular point at the point C, as also required for the dependencies $w_m^{(1)} - r$ and $w_m^{(2)} - r$, respectively.

To fulfi the non-singularity assumption, the functions $u_{1m}^{(i)}$, $w_m^{(i)}$ ($i = 1,2$) are required to be extremal on the cell surface, thus for $r = r_c$. Accordingly, with regard to the condition of the extremum, the functions $f_u = u_{1m}^{(1)} + u_{2m}^{(1)}$, $f_w = w_m^{(1)} + w_m^{(1)}$ are smooth regarding the variable r and for $r = r_c$. Finally, resulting from decreasing courses of the r-dependent functions $|u_{1m}|$, $w_m \propto \sigma_m^2 > 0$ (see Eq. (3.84)), the dependencies $|u_{1m}| - r$, $w_m - r$ exhibit minima on the cell surface. Considering the boundary condition (4.16), provided that $u_{1m} < 0$, the dependence $u_{1m} - r$ exhibits the maximum on the cell surface.

Accordingly, in addition to the standard boundary conditions given by Eqs. (4.14)–(4.17) which are mandatory, the additional boundary conditions for the cell matrix are derived as

$$(\varepsilon_{11m})_{r=r_c} = \left(\frac{\partial u_{1m}}{\partial r}\right)_{r=r_c} = 0, \tag{4.19}$$

$$\left(\frac{\partial w_m}{\partial r}\right)_{r=r_c} = 0. \tag{4.20}$$

4.4 One-particle-(envelope)-matrix system

Boundary conditions for the spherical particle, for the spherical envelope and for the infinite matrix of the one-particle-(envelope)-matrix system are identical to those presented in Sections 4.1–4.3, except for Eq. (4.16) replacing by Eqs. (4.21) or (4.22) derived as

$$(u_{1m})_{r\to\infty} = 0, \tag{4.21}$$

$$(\sigma_{iim})_{r\to\infty} = 0, \quad (\sigma_{12+jm})_{r\to\infty} = 0, \quad i = 1,2,3; \quad j = 0,1. \tag{4.22}$$

Accordingly, disregarding the conditions concerning the coefficient β_p, β_e, β_m (see Eqs. (3.100)–(3.102)), but with regard to Eqs. (4.14), (4.15) and due to $(u_{1m})_{r=R_{1+i}} \neq 0$ ($i = 0,1$), the absolute values $|u_{1m}|$, $|\sigma_{jjm}|$, $|\sigma_{12+km}|$ ($j = 1,2,3; k = 0,1$) are required to be decreasing functions of the variable $r \in \langle R_1, \infty)$ and $r \in \langle R_2, \infty)$ related to the one-particle-matrix and one-particle-envelope-matrix systems, respectively.

The boundary conditions (4.21) and (4.22) defin courses of the dependencies $u_{1m} - r$ and $\sigma_{11m} - r$. Consequently, Eqs. (4.14) or (4.15) or (4.17) determine one integration constant only. With regard to (4.21) or (4.22), the additional boundary conditions (4.19), (4.20) for $r \to \infty$ are fulfille due to $\lambda_{1m} < 0$, and another integration constant is not required to be considered. Additionally, considering Items 5, 6, Section 3.2.4, we get $W_m \propto [1/(2\lambda_{1m} + 3)] \lim_{r_2 \to \infty} \left(r_2^{2\lambda_{1m}+3} - r_1^{2\lambda_{1m}+3} \right) \nrightarrow \pm\infty$, and then $2\lambda_{1m} + 3 < 0$. Accordingly, with regard to Eqs. (3.76)–(3.80), (3.85), the number n_m and the exponent λ_{1m} have the forms

$$n_m = 1, \tag{4.23}$$

$$\lambda_{1m} < -\frac{3}{2}. \tag{4.24}$$

4.5 Supplement

Provided that $\beta_p = \beta_e \neq \beta_m$ and $\beta_p \neq \beta_e = \beta_m$ (see Sections 4.1, 4.3.1), the boundary conditions (4.5) and (4.17) for the spherical particle and the cell matrix are related to the radial displacement u_{1e} in the spherical envelope determined by the boundary conditions (4.8) and (4.9), respectively. A detailed analysis of reasons for the determination of the boundary conditions (4.5) and (4.17) by the radial displacement u_{1e} instead of u_{1p} and u_{1m} is presented in Sections 4.5.1 and 4.5.2 concerning the conditions $\beta_p = \beta_e \neq \beta_m$ and $\beta_p \neq \beta_e = \beta_m$, respectively.

With regard to the one-particle-(envelope)-matrix system, the boundary condition (4.16) mentioned in Sections 4.5.1, 4.5.2 is meant to be replaced by (4.21), and additionally the boundary condition (4.19) is not considered.

4.5.1 Condition $\beta_p = \beta_e \neq \beta_m$

With regard to Eq. (4.5), the integration constant C_{1p} for $n_p = 1$ included in the solution $u_{1p} = \xi_{61p} C_{1p} r^{\lambda_{1p}+1}$ (see Eq. (3.88), $q = p$) for the spherical particle is derived by the derived by the coefficien ϱ_{1e}^{pe} related to the spherical envelope. Conversely, assuming two boundary conditions for the spherical envelope given by Eq. (4.9) and determined as

$$(\varepsilon_{22e})_{r=R_1} = (\varepsilon_{22p})_{r=R_1} = \frac{(u_{1p})_{r=R_1}}{R_1} = -p_2\, \varrho_{1p}^{pe}, \tag{4.25}$$

the integration constants C_{1e}, C_{2e} for $n_e = 2$ included in the solution $u_{1e} = \sum_{i=1}^{n_q} \xi_{61e} C_{1e} r^{\lambda_{1e}+1} + \xi_{62e} C_{2e} r^{\lambda_{2e}+1}$ are derived by the coefficien ϱ_{1m}^{pe} related to the spherical particle. Consequently, the radial displacement u_{1p} in the spherical particle could be determined using the boundary condition (4.1), transforming to the form

$$\xi_{61p} C_{1p} \left(r^{\lambda_{1p}+1} \right)_{r=0} = 0. \tag{4.26}$$

Boundary conditions   55

Accordingly, due to the fact that no further boundary condition exists to be considered for the determination of C_{lp}, ϱ^{pe}_{1p}, neither the integration constant C_{lp} nor the coefficien ϱ^{pe}_{1p} can be determined. Finally, the boundary condition (4.5) for $\epsilon_{22p}(u_{1p})$ is required to be related to $\epsilon_{22e}(u_{1e})$ determined by the boundary condition (4.8).

4.5.2   Condition $\beta_p \neq \beta_e = \beta_m$

With regard to Eqs. (4.16), (4.17), the integration constants C_{1m}, C_{2m} included in the solution $u_{1m} = \xi_{61m} C_{1m} r^{\lambda_{1m}+1} + \xi_{62m} C_{2m} r^{\lambda_{2m}+1}$ (see Eqs. (3.85), (3.88), $q = m$) for the cell matrix are derived by the coefficien ϱ^{me}_{2e} related to the spherical envelope. Conversely, assuming two boundary conditions for the spherical envelope given by Eq. (4.8) and determined as

$$(\varepsilon_{22e})_{r=R_2} = (\varepsilon_{22m})_{r=R_2} = \frac{(u_{1m})_{r=R_2}}{R_2} = -p_1 \varrho^{me}_{2m}, \tag{4.27}$$

the integration constants C_{1e}, C_{2e} included in the solution $u_{1e} = \xi_{61e} C_{1e} r^{\lambda_{1e}+1} + \xi_{62e} C_{2e} r^{\lambda_{2e}+1}$ for the spherical envelope are derived by the coefficien ϱ^{me}_{2m} related to the cell matrix. Consequently, the radial displacement u_{1m} in the cell matrix could be determined using one boundary condition only, given by Eq. (4.16), transforming to the form

$$\xi_{61m} C_{1m} \left(r^{\lambda_{1m}+1} \right)_{r=r_c} = 0. \tag{4.28}$$

Accordingly, due to the fact that no further boundary condition exists to be considered for the determination of C_{lm}, ϱ^{me}_{2m}, neither the integration constant C_{lm} nor the coefficien ϱ^{me}_{2m} can be determined. Finally, the boundary condition (4.17) for $\epsilon_{22m}(u_{1m})$ is required to be related to $\epsilon_{22e}(u_{1e})$ which is determined by the boundary condition (4.9).

Similarly, provided that the radial displacement u_{1m} could be determined using two boundary conditions given by Eqs. (4.16), (4.19), and with regard to Eqs. (3.81), (3.85), we get

$$\left(\xi_{61m} C_{1m} r^{\lambda_{1m}+1} + \xi_{62m} C_{2m} r^{\lambda_{2m}+1} \right)_{r=r_c} = 0, \tag{4.29}$$

$$\left(\xi_{51m} C_{1m} r^{\lambda_{1m}} + \xi_{52m} C_{2m} r^{\lambda_{2m}} \right)_{r=r_c} = 0. \tag{4.30}$$

The integration constants C_{1m}, C_{2m}, determined by a system of the linear algebraic equations (4.29), (4.30) are equal to zero, and then $u_{1m} = \xi_{61m} C_{1m} r^{\lambda_{1m}+1} + \xi_{62m} C_{2m} r^{\lambda_{2m}+1} = 0$ (see Eq. (3.85)) what is physically unacceptable.

Chapter 5

Thermal stresses in multi-particle-(envelope)-matrix system I

Chapter 5 deals with the thermal stresses in the multi-particle-(envelope)-matrix system which consists of anisotropic components.

With regard to the analysis in Section 3.1.2, in case of the multi-particle-matrix system, R_1 represents a radius of the embedded spherical particle at the temperature $T \in \langle T_f, T_r \rangle$, where T_f is a fina temperature of a cooling process, and the relaxation temperature T_r is determined in Section 3.4.1. In case of the multi-particle-envelope-matrix system, R_1, R_2 and $t = R_2 - R_1$ represents radii and thickness of the embedded spherical envelope at the temperature $T \in \langle T_f, T_r \rangle$, respectively.

5.1 Multi-particle-matrix system

5.1.1 Thermal stresses in spherical particle

With regard to the analyses in Sections 3.2.3–3.2.5, 4.1, the integration constant C_{1p} is derived as

$$C_{1p} = -\frac{p_1}{R_1^{\lambda_{1p}}}, \tag{5.1}$$

where the radial stress p_1 acting at the particle-matrix boundary is given by Eq. (5.61).

Equations (3.76)–(3.84), (3.89), (3.90), (3.92), (3.97), (3.98) ($q = p$) are thus transformed to the forms

$$\sigma_{11p} = -p_1 \left(\frac{r}{R_1}\right)^{\lambda_{1p}}, \tag{5.2}$$

$$\sigma_{22p} = -p_1 \, \xi_{11p} \left(\frac{r}{R_1}\right)^{\lambda_{1p}}, \tag{5.3}$$

$$\sigma_{33p} = -p_1\,\xi_{21p}\left(\frac{r}{R_1}\right)^{\lambda_{1p}}, \tag{5.4}$$

$$\sigma_{12p} = -p_1\,\xi_{31p}\left(\frac{r}{R_1}\right)^{\lambda_{1p}}, \tag{5.5}$$

$$\sigma_{13p} = -p_1\,\xi_{41p}\left(\frac{r}{R_1}\right)^{\lambda_{1p}}, \tag{5.6}$$

$$\varepsilon_{11p} = -p_1\,\xi_{51p}\left(\frac{r}{R_1}\right)^{\lambda_{1p}}, \tag{5.7}$$

$$\varepsilon_{22p} = \varepsilon_{33p} = -p_1\,\xi_{61p}\left(\frac{r}{R_1}\right)^{\lambda_{1p}}, \tag{5.8}$$

$$\varepsilon_{13p} = -p_1\,\xi_{71p}\left(\frac{r}{R_1}\right)^{\lambda_{1p}}, \tag{5.9}$$

$$\varepsilon_{12p} = -p_1\,\xi_{81p}\left(\frac{r}{R_1}\right)^{\lambda_{1p}}, \tag{5.10}$$

$$w_p = \frac{p_1^2\,\omega_{11p}}{2}\left(\frac{r}{R_1}\right)^{2\lambda_{1p}}, \tag{5.11}$$

$$W_p = 4R_1^3 \int_0^{\pi/2}\int_0^{\pi/2} \frac{p_1^2\,\omega_{11p}}{2\lambda_{1p}+3}\,d\varphi\,d\nu, \tag{5.12}$$

$$\sigma_{ip} = -p_1\,\xi_{8+i1p}\left(\frac{r}{R_1}\right)^{\lambda_{1p}}, \quad i = 1,2,3, \tag{5.13}$$

$$w_{ip} = \frac{s_{iip}\,p_1^2\,\xi_{8+i1p}^2}{2}\left(\frac{r}{R_1}\right)^{2\lambda_{1p}}, \quad i = 1,2,3, \tag{5.14}$$

where the coefficient $\xi_{11p}, \ldots, \xi_{81p}, \omega_{11p}, \xi_{8+i\,1p}$ $(i = 1,2,3)$ are given by Eqs. (3.86)–(3.88) $(q = p)$, (3.91), (3.100), respectively.

5.1.2 Thermal stresses in cell matrix

As presented in Section 4.3, the thermal stresses in the cell matrix are determined by the four combinations of boundary conditions (see Sections 4.3.1, 4.3.2) define in Items 3–6 (see p. 51). With regard to the analyses in Sections 3.2.3–3.2.5, 4.3, the integration constant C_{im} $(i = 1, \ldots, n_m)$ is derived as

$$C_{im} = -\frac{p_1\,\kappa_{i1m}^{(x)}}{R_1^{\lambda_{im}}}, \quad i = 1, \ldots, n_q \tag{5.15}$$

where n_m is a number of boundary conditions for the cell matrix, and the radial stress p_1 acting at the particle-matrix boundary is given by Eq. (5.61).

Provided that the boundary conditions for the cell matrix which are

- define in Item 3, Section 4.3 are considered, the coefficien $\kappa_{i1m}^{(x)}$ is replaced by $\kappa_{i1m}^{(1)}$ for $i = 1,2$, $n_q = 2$.

- define in Item 4, Section 4.3 are considered, the coefficien $\kappa_{i1m}^{(x)}$ is replaced by $\kappa_{i1m}^{(2)}$ for $i = 1,2,3$, $n_q = 3$.

- define in Item 5, Section 4.3 are considered, the coefficien $\kappa_{i1m}^{(x)}$ is replaced by $\kappa_{i1m}^{(3)}$ for $i = 1,2,3$, $n_q = 3$.

- define in Item 5, Section 4.3 are considered, the coefficien $\kappa_{i1m}^{(x)}$ is replaced by $\kappa_{i1m}^{(4)}$ for $i = 1,\ldots,4$, $n_q = 4$.

Consequently, the coefficient $\kappa_{i1m}^{(1)}$, $\kappa_{i1m}^{(2)}$, $\kappa_{i1m}^{(3)}$, $\kappa_{i1m}^{(4)}$ have the forms

$$\kappa_{ijm}^{(1)} = \frac{\xi_{63-im}}{\xi_{63-im} - \xi_{60+im}\left(\dfrac{R_1 f_c}{R_j}\right)^{\lambda_{im}-\lambda_{3-im}}}, \quad i,j = 1,2, \tag{5.16}$$

$$\kappa_{ijm}^{(2)} = \frac{\psi_{im}}{\displaystyle\sum_{k=1}^{3} (-1)^{k+1}\,\psi_{km}\left(\dfrac{R_1 f_c}{R_j}\right)^{\lambda_{im}-\lambda_{km}}}, \quad i = 1,2,3; \quad j = 1,2, \tag{5.17}$$

$$\kappa_{ijm}^{(3)} = \kappa_{ijm}^{(1)} - \kappa_{3jm}^{(3)}\left[\kappa_{ijm}^{(1)} - \xi_{11+ijm}\right], \quad i,j = 1,2, \tag{5.18}$$

$$\kappa_{3jm}^{(3)} = \frac{-\xi_{15\,jm} \pm \sqrt{\xi_{15\,jm}^2 - 4\,\xi_{14\,jm}\,\xi_{16\,jm}}}{2\,\xi_{14\,jm}}, \quad j = 1,2, \tag{5.19}$$

$$\kappa_{ijm}^{(4)} = \xi_{16+ijm} - \xi_{19+ijm}\,\kappa_{4jm}^{(4)}, \quad i = 1,2,3; \quad j = 1,2, \tag{5.20}$$

$$\kappa_{4jm}^{(4)} = \frac{-\xi_{24\,jm} \pm \sqrt{\xi_{24\,jm}^2 - 4\,\xi_{23\,jm}\,\xi_{25\,jm}}}{2\,\xi_{23\,jm}}, \quad j = 1,2, \tag{5.21}$$

where the conditions $\xi_{15\,jm}^2 - 4\,\xi_{14\,jm}\,\xi_{16\,jm} \geq 0$, $\xi_{24\,jm}^2 - 4\,\xi_{23\,jm}\,\xi_{25\,jm} \geq 0$ (Eqs. (5.19), (5.21)) are required to be fulfille for $\varphi, \nu \in \langle 0, \pi/2\rangle$ due to the function $f_c = f_c(\varphi, \nu)$ (see Eqs. (2.18), (2.25)–(2.28), (2.30)–(2.33), (2.36)) included in the coefficient ξ_{13+ijm}, ξ_{22+ijm} $(i=1,2,3;\ j=1,2)$ (see Eqs. (5.33)–(5.35), (5.42)–(5.44)). Otherwise, the additional boundary condition given by Eq. (4.20) can not be considered. Additionally, the Castigliano's theorem (see Section 3.2.5) is required to be considered due to the two solutions for each of the coefficient $\kappa_{3jm}^{(3)}$, $\kappa_{4jm}^{(4)}$ $(j=1,2)$ (see Eqs. (5.19), (5.21)).

The coefficient ψ_{im} $(i=1,\ldots,9)$; ξ_{11+ijm} $(i=1,\ldots,14;\ j=1,2)$ are derived as

$$\psi_{1m} = \xi_{53m}\,\xi_{62m} - \xi_{52m}\,\xi_{63m}, \tag{5.22}$$

$$\psi_{2m} = \xi_{51m}\,\xi_{63m} - \xi_{53m}\,\xi_{61m}, \tag{5.23}$$

$$\psi_{3m} = \xi_{52m}\,\xi_{61m} - \xi_{51m}\,\xi_{62m}, \tag{5.24}$$

$$\psi_{4m} = \xi_{54m}\,\xi_{63m} - \xi_{53m}\,\xi_{64m}, \tag{5.25}$$

$$\psi_{5m} = \xi_{52m}\,\xi_{64m} - \xi_{54m}\,\xi_{62m}, \tag{5.26}$$

$$\psi_{6m} = \xi_{53m}\,\xi_{64m} - \xi_{54m}\,\xi_{63m}, \tag{5.27}$$

$$\psi_{7m} = \xi_{54m}\,\xi_{61m} - \xi_{51m}\,\xi_{64m}, \tag{5.28}$$

$$\psi_{8m} = \xi_{54m}\,\xi_{62m} - \xi_{52m}\,\xi_{64m}, \tag{5.29}$$

$$\psi_{9m} = \xi_{51m}\,\xi_{64m} - \xi_{54m}\,\xi_{61m}, \tag{5.30}$$

$$
\begin{aligned}
\psi_{9+im} &= \psi_{1m}\left(1 + \delta_{1i}\,\xi_{61m}\right)\left(\frac{R_1 f_c}{R_2}\right)^{\lambda_{2m}+\lambda_{3m}} \\
&+ \psi_{2m}\left(1 + \delta_{1i}\,\xi_{62m}\right)\left(\frac{R_1 f_c}{R_2}\right)^{\lambda_{1m}+\lambda_{3m}} + \psi_{3m}\left(1 + \delta_{1i}\,\xi_{63m}\right)\left(\frac{R_1 f_c}{R_2}\right)^{\lambda_{1m}+\lambda_{2m}}, \\
i &= 1,2,
\end{aligned}
\tag{5.31}
$$

$$
\begin{aligned}
\xi_{11+ijm} &= \frac{\xi_{63m}}{\xi_{63-im} - \xi_{60+im}\left(\frac{R_1 f_c}{R_j}\right)^{\lambda_{im}-\lambda_{3-im}}}\left(\frac{R_1 f_c}{R_j}\right)^{\lambda_{3m}-\lambda_{3-im}}, \\
i,j &= 1,2,
\end{aligned}
\tag{5.32}
$$

$$
\begin{aligned}
\xi_{14jm} &= 2\lambda_{1m}\,\omega_{11m}\left[\kappa_{1jm}^{(1)} - \xi_{12jm}\right]^2\left(\frac{R_1 f_c}{R_j}\right)^{2\lambda_{1m}} \\
&+ 2\lambda_{2m}\,\omega_{22m}\left[\kappa_{2jm}^{(1)} - \xi_{13jm}\right]^2\left(\frac{R_1 f_c}{R_j}\right)^{2\lambda_{2m}} + 2\lambda_{3m}\,\omega_{33m}\left(\frac{R_1 f_c}{R_j}\right)^{2\lambda_{3m}} \\
&+ (\lambda_{1m}+\lambda_{2m})(\omega_{12m}+\omega_{21m})\left[\kappa_{1jm}^{(1)} - \xi_{12jm}\right]\left[\kappa_{2jm}^{(1)} - \xi_{13jm}\right]\left(\frac{R_1 f_c}{R_j}\right)^{\lambda_{1m}+\lambda_{2m}} \\
&- (\lambda_{1m}+\lambda_{3m})(\omega_{13m}+\omega_{31m})\left(\frac{R_1 f_c}{R_j}\right)^{\lambda_{1m}+\lambda_{3m}} \\
&- (\lambda_{2m}+\lambda_{3m})(\omega_{23m}+\omega_{32m})\left[\kappa_{2jm}^{(1)} - \xi_{13jm}\right]\left(\frac{R_1 f_c}{R_j}\right)^{\lambda_{2m}+\lambda_{3m}}, \\
j &= 1,2,
\end{aligned}
\tag{5.33}
$$

$$\xi_{15\,jm} = -4\lambda_{1m}\,\omega_{11m}\,\kappa_{1jm}^{(1)}\left[\kappa_{1jm}^{(1)} - \xi_{12\,jm}\right]\left(\frac{R_1 f_c}{R_j}\right)^{2\lambda_{1m}}$$

$$- 4\lambda_{2m}\,\omega_{22m}\,\kappa_{2jm}^{(1)}\left[\kappa_{2jm}^{(1)} - \xi_{13\,jm}\right]\left(\frac{R_1 f_c}{R_j}\right)^{2\lambda_{2m}}$$

$$- \left(\lambda_{1m} + \lambda_{2m}\right)\left(\omega_{12m} + \omega_{21m}\right)$$

$$\times \left\{\kappa_{1jm}^{(1)}\left[\kappa_{2jm}^{(1)} - \xi_{13\,jm}\right] + \kappa_{2jm}^{(1)}\left[\kappa_{1jm}^{(1)} - \xi_{12\,jm}\right]\right\}\left(\frac{R_1 f_c}{R_j}\right)^{\lambda_{1m}+\lambda_{2m}}$$

$$+ \kappa_{1jm}^{(1)}\left(\lambda_{1m} + \lambda_{3m}\right)\left(\omega_{13m} + \omega_{31m}\right)\left(\frac{R_1 f_c}{R_j}\right)^{\lambda_{1m}+\lambda_{3m}}$$

$$+ \kappa_{2jm}^{(1)}\left(\lambda_{2m} + \lambda_{3m}\right)\left(\omega_{23m} + \omega_{32m}\right)\left(\frac{R_1 f_c}{R_j}\right)^{\lambda_{2m}+\lambda_{3m}}, \quad j = 1, 2, \qquad (5.34)$$

$$\xi_{16\,jm} = 2\lambda_{1m}\,\omega_{11m}\left[\kappa_{1jm}^{(1)}\right]^2\left(\frac{R_1 f_c}{R_j}\right)^{2\lambda_{1m}} + 2\lambda_{2m}\,\omega_{22m}\left[\kappa_{2jm}^{(1)}\right]^2\left(\frac{R_1 f_c}{R_j}\right)^{2\lambda_{2m}}$$

$$+ \kappa_{1jm}^{(1)}\,\kappa_{2jm}^{(1)}\left(\lambda_{1m} + \lambda_{2m}\right)\left(\omega_{12m} + \omega_{21m}\right)\left(\frac{R_1 f_c}{R_j}\right)^{\lambda_{1m}+\lambda_{2m}}, \quad j = 1, 2. \quad (5.35)$$

$$\xi_{17\,jm} = \frac{\psi_{1m}}{\psi_{10m}}\left(\frac{R_1 f_c}{R_j}\right)^{\lambda_{2m}+\lambda_{3m}}, \quad j = 1, 2, \qquad (5.36)$$

$$\xi_{18\,jm} = \frac{\psi_{2m}}{\psi_{10m}}\left(\frac{R_1 f_c}{R_j}\right)^{\lambda_{1m}+\lambda_{3m}}, \quad j = 1, 2, \qquad (5.37)$$

$$\xi_{19\,jm} = \frac{\psi_{3m}}{\psi_{10m}}\left(\frac{R_1 f_c}{R_j}\right)^{\lambda_{1m}+\lambda_{2m}}, \quad j = 1, 2, \qquad (5.38)$$

$$\xi_{20\,jm} =$$
$$\frac{1}{\psi_{10m}}\left[\psi_{4m}\left(\frac{R_1 f_c}{R_j}\right)^{\lambda_{3m}+\lambda_{4m}} + \psi_{5m}\left(\frac{R_1 f_c}{R_j}\right)^{\lambda_{2m}+\lambda_{4m}} + \psi_{1m}\left(\frac{R_1 f_c}{R_j}\right)^{\lambda_{2m}+\lambda_{3m}}\right],$$
$$j = 1, 2, \qquad (5.39)$$

$$\xi_{21\,jm} =$$
$$\frac{1}{\psi_{10m}}\left[\psi_{6m}\left(\frac{R_1 f_c}{R_j}\right)^{\lambda_{3m}+\lambda_{4m}} + \psi_{7m}\left(\frac{R_1 f_c}{R_j}\right)^{\lambda_{1m}+\lambda_{4m}} + \psi_{2m}\left(\frac{R_1 f_c}{R_j}\right)^{\lambda_{1m}+\lambda_{3m}}\right],$$
$$j = 1, 2, \qquad (5.40)$$

$$\xi_{22\,jm} =$$

$$\frac{1}{\psi_{10m}} \left[\psi_{8m} \left(\frac{R_1 f_c}{R_j} \right)^{\lambda_{2m}+\lambda_{4m}} + \psi_{9m} \left(\frac{R_1 f_c}{R_j} \right)^{\lambda_{1m}+\lambda_{4m}} + \psi_{3m} \left(\frac{R_1 f_c}{R_j} \right)^{\lambda_{1m}+\lambda_{2m}} \right],$$
$$j = 1, 2, \tag{5.41}$$

$$\begin{aligned}
\xi_{23\,jm} &= 2\lambda_{1m}\,\omega_{11m}\,\xi_{20\,jm}^2 \left(\frac{R_1 f_c}{R_j} \right)^{2\lambda_{1m}} + 2\lambda_{2m}\,\omega_{22m}\,\xi_{21\,jm}^2 \left(\frac{R_1 f_c}{R_j} \right)^{2\lambda_{2m}} \\
&\quad + 2\lambda_{3m}\,\omega_{33m}\,\xi_{22\,jm}^2 \left(\frac{R_1 f_c}{R_j} \right)^{2\lambda_{3m}} + 2\lambda_{4m}\,\omega_{44m} \left(\frac{R_1 f_c}{R_j} \right)^{2\lambda_{4m}} \\
&\quad + \xi_{20\,jm}\,\xi_{21\,jm} \left(\lambda_{1m} + \lambda_{2m} \right) \left(\omega_{12m} + \omega_{21m} \right) \left(\frac{R_1 f_c}{R_j} \right)^{\lambda_{1m}+\lambda_{2m}} \\
&\quad + \xi_{20\,jm}\,\xi_{22\,jm} \left(\lambda_{1m} + \lambda_{3m} \right) \left(\omega_{13m} + \omega_{31m} \right) \left(\frac{R_1 f_c}{R_j} \right)^{\lambda_{1m}+\lambda_{3m}} \\
&\quad + \xi_{21\,jm}\,\xi_{22\,jm} \left(\lambda_{2m} + \lambda_{3m} \right) \left(\omega_{23m} + \omega_{32m} \right) \left(\frac{R_1 f_c}{R_j} \right)^{\lambda_{2m}+\lambda_{3m}} \\
&\quad - \xi_{20\,jm} \left(\lambda_{1m} + \lambda_{4m} \right) \left(\omega_{14m} + \omega_{41m} \right) \left(\frac{R_1 f_c}{R_j} \right)^{\lambda_{1m}+\lambda_{4m}} \\
&\quad - \xi_{21\,jm} \left(\lambda_{2m} + \lambda_{4m} \right) \left(\omega_{24m} + \omega_{42m} \right) \left(\frac{R_1 f_c}{R_j} \right)^{\lambda_{2m}+\lambda_{4m}} \\
&\quad - \xi_{22\,jm} \left(\lambda_{3m} + \lambda_{4m} \right) \left(\omega_{34m} + \omega_{43m} \right) \left(\frac{R_1 f_c}{R_j} \right)^{\lambda_{3m}+\lambda_{4m}}, \quad j = 1, 2, \tag{5.42}
\end{aligned}$$

$$\begin{aligned}
\xi_{24\,jm} &= -4\lambda_{1m}\,\omega_{11m}\,\xi_{17\,jm}\,\xi_{20\,jm} \left(\frac{R_1 f_c}{R_j} \right)^{2\lambda_{1m}} \\
&\quad - 4\lambda_{2m}\,\omega_{22m}\,\xi_{18\,jm}\,\xi_{21\,jm} \left(\frac{R_1 f_c}{R_j} \right)^{2\lambda_{2m}} \\
&\quad - 4\lambda_{3m}\,\omega_{33m}\,\xi_{19\,jm}\,\xi_{22\,jm} \left(\frac{R_1 f_c}{R_j} \right)^{2\lambda_{3m}} \\
&\quad - \left(\lambda_{1m} + \lambda_{2m} \right) \left(\omega_{12m} + \omega_{21m} \right) \left(\xi_{17\,jm}\,\xi_{21\,jm} + \xi_{18\,jm}\,\xi_{20\,jm} \right) \left(\frac{R_1 f_c}{R_j} \right)^{\lambda_{1m}+\lambda_{2m}} \\
&\quad - \left(\lambda_{1m} + \lambda_{3m} \right) \left(\omega_{13m} + \omega_{31m} \right) \left(\xi_{17\,jm}\,\xi_{22\,jm} + \xi_{19\,jm}\,\xi_{20\,jm} \right) \left(\frac{R_1 f_c}{R_j} \right)^{\lambda_{1m}+\lambda_{3m}} \\
&\quad - \left(\lambda_{2m} + \lambda_{3m} \right) \left(\omega_{23m} + \omega_{32m} \right) \left(\xi_{18\,jm}\,\xi_{22\,jm} + \xi_{19\,jm}\,\xi_{21\,jm} \right) \left(\frac{R_1 f_c}{R_j} \right)^{\lambda_{2m}+\lambda_{3m}} \\
&\quad + \xi_{17\,jm} \left(\lambda_{1m} + \lambda_{4m} \right) \left(\omega_{14m} + \omega_{41m} \right) \left(\frac{R_1 f_c}{R_j} \right)^{\lambda_{1m}+\lambda_{4m}}
\end{aligned}$$

$$+ \xi_{18jm} \left(\lambda_{2m} + \lambda_{4m} \right) \left(\omega_{24m} + \omega_{42m} \right) \left(\frac{R_1 f_c}{R_j} \right)^{\lambda_{2m}+\lambda_{4m}}$$

$$+ \xi_{19jm} \left(\lambda_{3m} + \lambda_{4m} \right) \left(\omega_{34m} + \omega_{43m} \right) \left(\frac{R_1 f_c}{R_j} \right)^{\lambda_{3m}+\lambda_{4m}} , \quad j = 1, 2, \tag{5.43}$$

$$\xi_{25jm} = 2\lambda_{1m} \omega_{11m} \xi_{17jm}^2 \left(\frac{R_1 f_c}{R_j} \right)^{2\lambda_{1m}} 2\lambda_{2m} \omega_{22m} \xi_{18jm}^2 \left(\frac{R_1 f_c}{R_j} \right)^{2\lambda_{2m}}$$

$$+ 2\lambda_{3m} \omega_{33m} \xi_{19jm}^2 \left(\frac{R_1 f_c}{R_j} \right)^{2\lambda_{3m}}$$

$$+ \xi_{17jm} \xi_{18jm} \left(\lambda_{1m} + \lambda_{2m} \right) \left(\omega_{12m} + \omega_{21m} \right) \left(\frac{R_1 f_c}{R_j} \right)^{\lambda_{1m}+\lambda_{2m}}$$

$$+ \xi_{17jm} \xi_{19jm} \left(\lambda_{1m} + \lambda_{3m} \right) \left(\omega_{13m} + \omega_{31m} \right) \left(\frac{R_1 f_c}{R_j} \right)^{\lambda_{1m}+\lambda_{3m}}$$

$$+ \xi_{18jm} \xi_{19jm} \left(\lambda_{2m} + \lambda_{3m} \right) \left(\omega_{23m} + \omega_{32m} \right) \left(\frac{R_1 f_c}{R_j} \right)^{\lambda_{2m}+\lambda_{3m}} ,$$

$$j = 1, 2, \tag{5.44}$$

where the coefficient ξ_{5im}, ξ_{6jm} ($i, j = 1,2,3$) included in Eqs. (5.22)–(5.30) are given by Eq. (3.88) for $q = m$.

Equations (3.76)–(3.84), (3.89), (3.90), (3.94), (3.97), (3.98) ($q = m$) along with $\left(\varepsilon_{22m} \right)_{r=R_{1m}}$ (see Eq. (3.109)) are thus transformed to the forms

$$\sigma_{11m} = -p_1 \sum_{i=1}^{n_m} \kappa_{i1m}^{(x)} \left(\frac{r}{R_1} \right)^{\lambda_{im}} , \tag{5.45}$$

$$\sigma_{22m} = -p_1 \sum_{i=1}^{n_m} \xi_{1im} \kappa_{i1m}^{(x)} \left(\frac{r}{R_1} \right)^{\lambda_{im}} , \tag{5.46}$$

$$\sigma_{33m} = -p_1 \sum_{i=1}^{n_m} \xi_{2im} \kappa_{i1m}^{(x)} \left(\frac{r}{R_1} \right)^{\lambda_{im}} , \tag{5.47}$$

$$\sigma_{12m} = -p_1 \sum_{i=1}^{n_m} \xi_{3im} \kappa_{i1m}^{(x)} \left(\frac{r}{R_1} \right)^{\lambda_{im}} , \tag{5.48}$$

$$\sigma_{13m} = -p_1 \sum_{i=1}^{n_m} \xi_{4im} \kappa_{i1m}^{(x)} \left(\frac{r}{R_1} \right)^{\lambda_{im}} , \tag{5.49}$$

$$\varepsilon_{11m} = -p_1 \sum_{i=1}^{n_m} \xi_{5im} \kappa_{i1m}^{(x)} \left(\frac{r}{R_1} \right)^{\lambda_{im}} , \tag{5.50}$$

$$\varepsilon_{22m} = \varepsilon_{33m} = -p_1 \sum_{i=1}^{n_m} \xi_{6im}\, \kappa_{i1m}^{(x)} \left(\frac{r}{R_1}\right)^{\lambda_{im}}, \tag{5.51}$$

$$\varepsilon_{13m} = -p_1 \sum_{i=1}^{n_m} \xi_{7im}\, \kappa_{i1m}^{(x)} \left(\frac{r}{R_1}\right)^{\lambda_{im}}, \tag{5.52}$$

$$\varepsilon_{12m} = -p_1 \sum_{i=1}^{n_m} \xi_{8im}\, \kappa_{i1m}^{(x)} \left(\frac{r}{R_1}\right)^{\lambda_{im}}, \tag{5.53}$$

$$w_m = \frac{p_1^2}{2} \sum_{i,j=1}^{n_m} \omega_{ijm}\, \kappa_{i1m}^{(x)} \kappa_{j1m}^{(x)} \left(\frac{r}{R_1}\right)^{\lambda_{im}+\lambda_{jm}}, \tag{5.54}$$

$$W_m = 4R_1^3 \int_0^{\pi/2} \int_0^{\pi/2} p_1^2\, \Omega_{1m}\, d\varphi\, d\nu, \tag{5.55}$$

$$\Omega_{1m} = \sum_{i,j=1}^{n_m} \frac{\omega_{ijm}\, \kappa_{i1m}^{(x)} \kappa_{j1m}^{(x)}}{\lambda_{im} + \lambda_{jm} + 3} \left(f_c^{\lambda_{im}+\lambda_{jm}+3} - 1\right), \tag{5.56}$$

$$\sigma_{im} = -p_1 \sum_{j=1}^{n_m} \xi_{8+i\,jm}\, \kappa_{j1m}^{(x)} \left(\frac{r}{R_1}\right)^{\lambda_{jm}}, \quad i = 1,2,3, \tag{5.57}$$

$$w_{im} = \frac{s_{iiq}\, p_1^2}{2} \sum_{j,k=1}^{n_m} \xi_{8+i\,jm}\, \xi_{8+i\,km}\, \kappa_{j1m}^{(x)} \kappa_{k1m}^{(x)} \left(\frac{r}{R_1}\right)^{\lambda_{jm}+\lambda_{km}},$$
$$i = 1,2,3, \tag{5.58}$$

$$(\varepsilon_{22m})_{r=R_{1m}} = -p_1 \varrho_{1m}, \tag{5.59}$$

where the coefficient $\xi_{1im}, \ldots, \xi_{8im}, \omega_{ijm}, \xi_{8+k\,im}$ $(i,j=1,\ldots,n_m;\ k=1,2,3)$ are given by Eqs. (3.86)–(3.88) $(q=m)$, (3.91), (3.100), respectively. The coefficien ϱ_{1m} is derived as

$$\varrho_{im} = (1 - \beta_m) \sum_{j=1}^{n_m} \xi_{6jm}\, \kappa_{jim}^{(x)} \left(\frac{R_{im}}{R_1}\right)^{\lambda_{jm}}, \quad i = 1,2, \tag{5.60}$$

where the radius R_{im} $(i=1,2)$ is given by Eq. (3.108).

5.1.3 Radial stress p_1

With regard to Eqs. (3.109), (5.8), (5.59), the radial stress p_1 acting at the particle-matrix boundary, considered in Sections 5.1.1, 5.1.2 has the form

$$p_1 = \frac{\beta_m - \beta_p}{\xi_{61p}\left(1 - \beta_p\right)\left(\frac{R_{1p}}{R_1}\right)^{\lambda_{1p}} - \varrho_{1m}}, \tag{5.61}$$

where the radius R_{1p} and the coefficient $\xi_{61p},\, \varrho_{1m}$ are given by Eqs. (3.108) and (3.88) $(q=p)$, (5.60), respectively.

5.2 Multi-particle-envelope-matrix system

5.2.1 Condition $\beta_p \neq \beta_e = \beta_m$

5.2.1.1 Thermal stresses in spherical particle

The integration constant C_{1p} for the spherical particle is given by Eq. (5.1). Accordingly, formulae in Section 5.1.1 are considered for the spherical particle on the condition $\beta_p \neq \beta_e = \beta_m$. The radial stress p_1 included in these formulae is given by Eq. (5.106).

5.2.1.2 Thermal stresses in spherical envelope

With regard to the analyses in Sections 3.2.3–3.2.5, 4.2, the integration constant C_{1e} is derived as

$$C_{1e} = -\frac{p_1}{R_1^{\lambda_{1e}}}, \tag{5.62}$$

where the radial stress p_1 acting at the particle-envelope boundary is given by Eq. (5.106).
Equations (3.76)–(3.84), (3.89), (3.90), (3.92), (3.97), (3.98) ($q = e$) along with $(\varepsilon_{22e})_{r=R_{1e}}$ (see Eq. (3.110)) are thus transformed to the forms

$$\sigma_{11e} = -p_1 \left(\frac{r}{R_1}\right)^{\lambda_{1e}}, \tag{5.63}$$

$$\sigma_{22e} = -p_1 \, \xi_{11e} \left(\frac{r}{R_1}\right)^{\lambda_{1e}}, \tag{5.64}$$

$$\sigma_{33e} = -p_1 \, \xi_{21e} \left(\frac{r}{R_1}\right)^{\lambda_{1e}}, \tag{5.65}$$

$$\sigma_{12e} = -p_1 \, \xi_{31e} \left(\frac{r}{R_1}\right)^{\lambda_{1e}}, \tag{5.66}$$

$$\sigma_{13e} = -p_1 \, \xi_{41e} \left(\frac{r}{R_1}\right)^{\lambda_{1e}}, \tag{5.67}$$

$$\varepsilon_{11e} = -p_1 \, \xi_{51e} \left(\frac{r}{R_1}\right)^{\lambda_{1e}}, \tag{5.68}$$

$$\varepsilon_{22e} = \varepsilon_{33e} = -p_1 \, \xi_{61e} \left(\frac{r}{R_1}\right)^{\lambda_{1e}}, \tag{5.69}$$

$$\varepsilon_{13e} = -p_1 \, \xi_{71e} \left(\frac{r}{R_1}\right)^{\lambda_{1e}}, \tag{5.70}$$

$$\varepsilon_{12e} = -p_1 \, \xi_{81e} \left(\frac{r}{R_1}\right)^{\lambda_{1e}}, \tag{5.71}$$

$$w_e = \frac{p_1^2 \omega_{11e}}{2} \left(\frac{r}{R_1} \right)^{2\lambda_{1e}}, \tag{5.72}$$

$$W_e = 4R_1^3 \int_0^{\pi/2} \int_0^{\pi/2} \frac{p_1^2 \omega_{11e}}{2\lambda_{1e} + 3} \left[\left(\frac{R_2}{R_1} \right)^{2\lambda_{1e}+3} - 1 \right] d\varphi \, d\nu, \tag{5.73}$$

$$\sigma_{ie} = -p_1 \xi_{8+i1e} \left(\frac{r}{R_1} \right)^{\lambda_{1e}}, \quad i = 1, 2, 3, \tag{5.74}$$

$$w_{ie} = \frac{s_{iie} \, p_1^2 \, \xi_{8+i1e}^2}{2} \left(\frac{r}{R_1} \right)^{2\lambda_{1e}}, \quad i = 1, 2, 3, \tag{5.75}$$

$$(\varepsilon_{22e})_{r=R_{1e}} = -p_1 \varrho_{1e}^{me}, \tag{5.76}$$

where the coefficient $\xi_{11e}, \ldots, \xi_{81e}, \omega_{11e}, \xi_{8+i\,1e}$ $(i=1,2,3)$ are given by Eqs. (3.86)–(3.88) $(q=e)$, (3.91), (3.100), respectively. The coefficien ϱ_{1m}^{me} is derived as

$$\varrho_{ie}^{me} = \xi_{61e} (1 - \beta_e) \left(\frac{R_{ie}}{R_1} \right)^{\lambda_{1e}}, \quad i = 1, 2, \tag{5.77}$$

where the radius R_{ie} $(i=1,2)$ is given by Eq. (3.108).

5.2.1.3 Thermal stresses in cell matrix

As presented in Section 4.3, the thermal stresses in the cell matrix are determined by the four combinations of boundary conditions (see Sections 4.3.1, 4.3.2) define in Items 3–6 (see p. 51).

With regard to the analyses in Sections 3.2.3–3.2.5, 4.3, the integration constant C_{im} $(i=1,\ldots,n_m)$ is derived as

$$C_{im} = -\frac{p_1 \varrho_{2e}^{me} \kappa_{i2m}^{(x)}}{R_2^{\lambda_{im}}}, \quad i = 1, \ldots, n_q, \tag{5.78}$$

where the coefficien ϱ_{2e}^{me} and the radial stress p_1 acting at the particle-envelope boundary are given by Eqs. (5.77) and (5.106), respectively.

Considering the analysis in Section 5.1.2, the coefficient $\kappa_{i2m}^{(1)}$ $(i=1,2)$, $\kappa_{i2m}^{(2)}$, $\kappa_{i2m}^{(3)}$ $(i=1,2,3)$ have the forms

$$\kappa_{i2m}^{(1)} = \frac{1}{\xi_{60+im} \left[1 - \left(\frac{R_1 f_c}{R_2} \right)^{\lambda_{im} - \lambda_{3-im}} \right]}, \quad i = 1, 2, \tag{5.79}$$

$$\kappa_{i2m}^{(2)} = \frac{\psi_{11+im}}{\sum\limits_{k=1}^{3} (-1)^{k+1} \xi_{60+km} \, \psi_{11+km} \left(\frac{R_1 f_c}{R_2} \right)^{\lambda_{im} - \lambda_{km}}}, \quad i = 1, 2, 3, \tag{5.80}$$

$$\kappa_{i2m}^{(3)} = \kappa_{i2m}^{(1)} - \kappa_{32m}^{(3)} \left[\kappa_{i2m}^{(1)} - \xi_{11+i\,2m} \right], \quad i = 1, 2, \tag{5.81}$$

$$\kappa_{32m}^{(3)} = \frac{-\xi_{15\,2m} \pm \sqrt{\xi_{15\,2m}^2 - 4\,\xi_{14\,2m}\,\xi_{16\,2m}}}{2\,\xi_{14\,2m}}, \tag{5.82}$$

where the coefficient ψ_{11+im}, $\xi_{13+i\,2m}$ ($i=1,2,3$) are given by Eq. (5.22)–(5.24), (5.33)–(5.35), respectively. The coefficien $\xi_{11+i\,2m}$ ($i=1,2$) included in Eq. (5.22)–(5.24) is derived as

$$\xi_{11+i\,2m} = \frac{\xi_{63m}}{\xi_{60+im} \left[1 - \left(\dfrac{R_1 f_c}{R_2} \right)^{\lambda_{im} - \lambda_{3-im}} \right]} \left(\frac{R_1 f_c}{R_2} \right)^{\lambda_{3m} - \lambda_{3-im}}, \quad i = 1, 2. \tag{5.83}$$

Similarly, the coefficient $\kappa_{i2m}^{(4)}$ ($i=1,\ldots,4$) have the forms

$$\kappa_{i2m}^{(4)} = \xi_{16+i\,2m} - \xi_{19+i\,2m}\,\kappa_{42m}^{(4)}, \quad i = 1, 2, 3, \tag{5.84}$$

$$\kappa_{42m}^{(4)} = \frac{-\xi_{24\,2m} \pm \sqrt{\xi_{24\,2m}^2 - 4\,\xi_{23\,2m}\,\xi_{25\,2m}}}{2\,\xi_{23\,2m}}, \quad j = 1, 2, \tag{5.85}$$

where the coefficien $\xi_{22+i\,2m}$ ($i=1,2,3$) is given by Eq. (5.42)–(5.44). The coefficien $\xi_{16+i\,2m}$ ($i=1,\ldots,6$) included in Eq. (5.42)–(5.44) is derived as

$$\xi_{17\,2m} = \frac{\psi_{1m}}{\psi_{11m}} \left(\frac{R_1 f_c}{R_2} \right)^{\lambda_{2m} + \lambda_{3m}}, \tag{5.86}$$

$$\xi_{18\,2m} = \frac{\psi_{2m}}{\psi_{11m}} \left(\frac{R_1 f_c}{R_2} \right)^{\lambda_{1m} + \lambda_{3m}}, \tag{5.87}$$

$$\xi_{19\,2m} = \frac{\psi_{3m}}{\psi_{11m}} \left(\frac{R_1 f_c}{R_2} \right)^{\lambda_{1m} + \lambda_{2m}}, \tag{5.88}$$

$$\xi_{20\,2m} = \frac{1}{\psi_{11m}} \left[\psi_{4m}\,\xi_{62m} \left(\frac{R_1 f_c}{R_2} \right)^{\lambda_{3m} + \lambda_{4m}} + \psi_{5m}\,\xi_{63m} \left(\frac{R_1 f_c}{R_2} \right)^{\lambda_{2m} + \lambda_{4m}} \right.$$
$$\left. + \psi_{1m}\,\xi_{64m} \left(\frac{R_1 f_c}{R_2} \right)^{\lambda_{2m} + \lambda_{3m}} \right], \tag{5.89}$$

$$\xi_{21\,2m} = \frac{1}{\psi_{11m}} \left[\psi_{6m}\,\xi_{61m} \left(\frac{R_1 f_c}{R_2} \right)^{\lambda_{3m} + \lambda_{4m}} + \psi_{7m}\,\xi_{63m} \left(\frac{R_1 f_c}{R_2} \right)^{\lambda_{1m} + \lambda_{4m}} \right.$$
$$\left. + \psi_{2m}\,\xi_{64m} \left(\frac{R_1 f_c}{R_2} \right)^{\lambda_{1m} + \lambda_{3m}} \right], \tag{5.90}$$

$$\xi_{22\,2m} = \frac{1}{\psi_{11m}} \left[\psi_{8m}\, \xi_{61m} \left(\frac{R_1 f_c}{R_2} \right)^{\lambda_{2m}+\lambda_{4m}} + \psi_{9m}\, \xi_{62m} \left(\frac{R_1 f_c}{R_2} \right)^{\lambda_{1m}+\lambda_{4m}} \right.$$
$$\left. + \psi_{3m}\, \xi_{64m} \left(\frac{R_1 f_c}{R_2} \right)^{\lambda_{1m}+\lambda_{2m}} \right], \tag{5.91}$$

where the coefficien ψ_{im} $(i = 1, \ldots, 9)$ is given by Eqs. (5.22)–(5.31).

Equations (3.76)–(3.84), (3.89), (3.90), (3.94), (3.97), (3.98) $(q = m)$ are thus transformed to the forms

$$\sigma_{11m} = -p_1\, \varrho_{2e}^{me} \sum_{i=1}^{n_m} \kappa_{i2m}^{(x)} \left(\frac{r}{R_2} \right)^{\lambda_{im}}, \tag{5.92}$$

$$\sigma_{22m} = -p_1\varrho_{2e}^{me} \sum_{i=1}^{n_m} \xi_{1im}\, \kappa_{i2m}^{(x)} \left(\frac{r}{R_2} \right)^{\lambda_{im}}, \tag{5.93}$$

$$\sigma_{33m} = -p_1\varrho_{2e}^{me} \sum_{i=1}^{n_m} \xi_{2im}\, \kappa_{i2m}^{(x)} \left(\frac{r}{R_2} \right)^{\lambda_{im}}, \tag{5.94}$$

$$\sigma_{12m} = -p_1\varrho_{2e}^{me} \sum_{i=1}^{n_m} \xi_{3im}\, \kappa_{i2m}^{(x)} \left(\frac{r}{R_2} \right)^{\lambda_{im}}, \tag{5.95}$$

$$\sigma_{13m} = -p_1\varrho_{2e}^{me} \sum_{i=1}^{n_m} \xi_{4im}\, \kappa_{i2m}^{(x)} \left(\frac{r}{R_2} \right)^{\lambda_{im}}, \tag{5.96}$$

$$\varepsilon_{11m} = -p_1\varrho_{2e}^{me} \sum_{i=1}^{n_m} \xi_{5im}\, \kappa_{i2m}^{(x)} \left(\frac{r}{R_2} \right)^{\lambda_{im}}, \tag{5.97}$$

$$\varepsilon_{22m} = \varepsilon_{33m} = -p_1\varrho_{2e}^{me} \sum_{i=1}^{n_m} \xi_{6im}\, \kappa_{i2m}^{(x)} \left(\frac{r}{R_2} \right)^{\lambda_{im}}, \tag{5.98}$$

$$\varepsilon_{13m} = -p_1\varrho_{2e}^{me} \sum_{i=1}^{n_m} \xi_{7im}\, \kappa_{i2m}^{(x)} \left(\frac{r}{R_2} \right)^{\lambda_{im}}, \tag{5.99}$$

$$\varepsilon_{12m} = -p_1\varrho_{2e}^{me} \sum_{i=1}^{n_m} \xi_{8im}\, \kappa_{i2m}^{(x)} \left(\frac{r}{R_2} \right)^{\lambda_{im}}, \tag{5.100}$$

$$w_m = \frac{(p_1\,\varrho_{2e}^{me})^2}{2} \sum_{i,j=1}^{n_m} \omega_{ijm}\, \kappa_{i2m}^{(x)}\, \kappa_{j2m}^{(x)} \left(\frac{r}{R_2} \right)^{\lambda_{im}+\lambda_{jm}}, \tag{5.101}$$

$$W_m = 4\,(\varrho_{2e}^{me})^2\, R_2^3 \int_0^{\pi/2} \int_0^{\pi/2} p_1^2\, \Omega_{2m}\, d\varphi\, d\nu, \tag{5.102}$$

Thermal stresses in multi-particle-(envelope)-matrix system I   69

$$\Omega_{2m} = \sum_{i,j=1}^{n_m} \frac{\omega_{ijm} \, \kappa_{i2m}^{(x)} \, \kappa_{j2m}^{(x)}}{\lambda_{im} + \lambda_{jm} + 3} \left[\left(\frac{R_1 f_c}{R_2} \right)^{\lambda_{im}+\lambda_{jm}+3} - 1 \right], \tag{5.103}$$

$$\sigma_{im} = -p_1 \varrho_{2e}^{me} \sum_{j=1}^{n_m} \xi_{8+i\,jm} \, \kappa_{j2m}^{(x)} \left(\frac{r}{R_2} \right)^{\lambda_{jm}}, \quad i = 1,2,3, \tag{5.104}$$

$$w_{im} = \frac{s_{iiq} \, (p_1 \varrho_{2e}^{me})^2}{2} \sum_{j,k=1}^{n_m} \xi_{8+i\,jm} \, \xi_{8+i\,km} \, \kappa_{j2m}^{(x)} \, \kappa_{k2m}^{(x)} \left(\frac{r}{R_2} \right)^{\lambda_{jm}+\lambda_{km}},$$

$$i = 1,2,3, \tag{5.105}$$

where the coefficient $\xi_{1im}, \dots, \xi_{8im}, \omega_{ijm}, \xi_{8+k\,im}$ $(i,j=1,\dots,n_m;\ k=1,2,3)$ are given by Eqs. (3.86)–(3.88) $(q=m)$, (3.91), (3.100), respectively.

5.2.1.4   Radial stress p_1

With regard to Eqs. (3.110), (5.8), (5.76), the radial stress p_1 acting at the particle-envelope boundary, considered in Sections 5.2.1.1–5.2.1.3 has the form

$$p_1 = \frac{\beta_e - \beta_p}{\xi_{61p} \, (1 - \beta_p) \left(\frac{R_{1p}}{R_1} \right)^{\lambda_{1p}} - \varrho_{1e}^{me}}, \tag{5.106}$$

where the radius R_{1p} and the coefficient $\xi_{61p}, \varrho_{1e}^{me}$ are given by Eqs. (3.108) and (3.88) $(q=p)$, (5.77), respectively.

5.2.2   Condition $\beta_p \neq \beta_e \neq \beta_m$

5.2.2.1   Thermal stresses in spherical particle

The integration constant C_{1p} for the spherical particle is given by Eq. (5.1). Accordingly, formulae in Section 5.1.1 are considered for the spherical particle on the condition $\alpha'_{1p} \neq \alpha'_{1e} = \alpha'_{1m}$. The radial stress p_1 included in these formulae is given by Eq. (5.140).

5.2.2.2   Thermal stresses in spherical envelope

With regard to the analyses in Sections 3.2.3–3.2.5, 4.2, the integration constant C_{ie} $(i=1,2)$ is derived as

$$C_{ie} = -\sum_{j=1}^{2} \frac{p_j \, \kappa_{ije}}{R_j^{\lambda_{ie}}}, \quad i = 1,2. \tag{5.107}$$

where the coefficien κ_{ije} $(i,j=1,2)$ has the form

$$\kappa_{ije} = \frac{1}{1 - \left(\frac{R_i}{R_{3-i}} \right)^{\lambda_{3-je}-\lambda_{je}}}, \quad i,j = 1,2. \tag{5.108}$$

The radial stresses p_1 and p_2 acting at the particle-envelope and matrix-envelope boundaries are given by Eqs. (5.140) and (5.141), respectively.

Equations (3.76)–(3.84), (3.89), (3.90), (3.92), (3.97), (3.98) ($q = e$) along with $(\varepsilon_{22e})_{r=R_{ie}}$ ($i = 1,2$) (see Eqs. (3.110), (3.111)) are thus transformed to the forms

$$\sigma_{11e} = -\sum_{i,j=1}^{2} p_j \kappa_{ije} \left(\frac{r}{R_j}\right)^{\lambda_{ie}}, \tag{5.109}$$

$$\sigma_{22e} = -\sum_{i,j=1}^{2} p_j \, \xi_{1ie} \, \kappa_{ije} \left(\frac{r}{R_j}\right)^{\lambda_{ie}}, \tag{5.110}$$

$$\sigma_{33e} = -\sum_{i,j=1}^{2} p_j \, \xi_{2ie} \, \kappa_{ije} \left(\frac{r}{R_j}\right)^{\lambda_{ie}}, \tag{5.111}$$

$$\sigma_{12e} = -\sum_{i,j=1}^{2} p_j \, \xi_{3ie} \, \kappa_{ije} \left(\frac{r}{R_j}\right)^{\lambda_{ie}}, \tag{5.112}$$

$$\sigma_{13e} = -\sum_{i,j=1}^{2} p_j \, \xi_{4ie} \, \kappa_{ije} \left(\frac{r}{R_j}\right)^{\lambda_{ie}}, \tag{5.113}$$

$$\varepsilon_{11e} = -\sum_{i,j=1}^{2} p_j \, \xi_{5ie} \, \kappa_{ije} \left(\frac{r}{R_j}\right)^{\lambda_{ie}}, \tag{5.114}$$

$$\varepsilon_{22e} = \varepsilon_{33e} = -\sum_{i,j=1}^{2} p_j \, \xi_{6ie} \, \kappa_{ije} \left(\frac{r}{R_j}\right)^{\lambda_{ie}}, \tag{5.115}$$

$$\varepsilon_{13e} = -\sum_{i,j=1}^{2} p_j \, \xi_{7ie} \, \kappa_{ije} \left(\frac{r}{R_j}\right)^{\lambda_{ie}}, \tag{5.116}$$

$$\varepsilon_{12e} = -\sum_{i,j=1}^{2} p_j \, \xi_{8ie} \, \kappa_{ije} \left(\frac{r}{R_j}\right)^{\lambda_{ie}}, \tag{5.117}$$

$$w_e = \frac{1}{2} \sum_{i,j,k,l=1}^{2} p_k \, p_l \, \omega_{ije} \, \kappa_{ike} \, \kappa_{jle} \left(\frac{r}{R_k}\right)^{\lambda_{ie}} \left(\frac{r}{R_l}\right)^{\lambda_{je}}, \tag{5.118}$$

$$W_e = 4R_2^3 \int_0^{\pi/2} \int_0^{\pi/2} \Omega_e \, d\varphi \, d\nu. \tag{5.119}$$

$$\Omega_e = \sum_{i,j,k,l=1}^{2} \frac{p_k \, p_l \, \omega_{ije} \, \kappa_{ike} \, \kappa_{jle}}{\lambda_{ie} + \lambda_{je} + 3} \left(\frac{R_2}{R_k}\right)^{\lambda_{ie}} \left(\frac{R_2}{R_l}\right)^{\lambda_{je}} \left[1 - \left(\frac{R_1}{R_2}\right)^{\lambda_{ie} + \lambda_{je} + 3}\right], \tag{5.120}$$

Thermal stresses in multi-particle-(envelope)-matrix system I   71

$$\sigma_{ie} = - \sum_{j,k=1}^{2} p_k\, \xi_{8+ije}\, \kappa_{jkm} \left(\frac{r}{R_k}\right)^{\lambda_{je}}, \quad i = 1, 2, 3, \tag{5.121}$$

$$w_{ie} = \frac{s_{iie}}{2} \sum_{j,k,l,s=1}^{2} p_l\, p_s\, \xi_{8+ije}\, \xi_{8+ike}\, \kappa_{jlm}\, \kappa_{ksm} \left(\frac{r}{R_l}\right)^{\lambda_{je}} \left(\frac{r}{R_s}\right)^{\lambda_{ke}},$$
$$i = 1, 2, 3, \tag{5.122}$$

$$(\varepsilon_{22e})_{r=R_{ie}} = - \sum_{j=1}^{2} p_j\, \varrho_{ije}, \quad i = 1, 2, \tag{5.123}$$

where the coefficient $\xi_{1ie}, \ldots, \xi_{8ie}$, ω_{ije}, $\xi_{8+k\,1e}$ $(i, j = 1, 2;\ k = 1, 2, 3)$ are given by Eqs. (3.86)–(3.88) $(q = e)$, (3.91), (3.100), respectively. The coefficien ϱ_{ije} is derived as

$$\varrho_{ije} = (1 - \beta_e) \sum_{k=1}^{2} \xi_{6ke}\, \kappa_{kje} \left(\frac{R_{ie}}{R_j}\right)^{\lambda_{ke}}, \quad i, j = 1, 2, \tag{5.124}$$

where the radius R_{ie} $(i = 1, 2)$ is given by Eq. (3.108).

5.2.2.3 Thermal stresses in cell matrix

As presented in Section 4.3, the thermal stresses in the cell matrix are determined by the four combinations of boundary conditions (see Sections 4.3.1, 4.3.2) define in Items 3–6 (see p. 51).

With regard to the analyses in Sections 3.2.3–3.2.5, 4.3, the integration constant C_{im} $(i = 1, \ldots, n_m)$ is derived as

$$C_{im} = - \frac{p_2\, \kappa_{i2m}^{(x)}}{R_2^{\lambda_{im}}}, \quad i = 1, \ldots, n_q, \tag{5.125}$$

where the radial stress p_2 acting at the matrix-envelope boundary is given by Eq. (5.141).

Considering the analysis in Section 5.1.2, the coefficient $\kappa_{i2m}^{(1)}$ $(i = 1, 2)$, $\kappa_{i2m}^{(2)}$, $\kappa_{i2m}^{(3)}$ $(i = 1, 2, 3)$, $\kappa_{i2m}^{(4)}$ $(i = 1, \ldots, 4)$ are given by Eqs. (5.16)–(5.44).

Equations (3.76)–(3.84), (3.89), (3.90), (3.94), (3.97), (3.98) $(q = m)$ along with $(\varepsilon_{22m})_{r=R_{2m}}$ (see Eq. (3.111)) are thus transformed to the forms

$$\sigma_{11m} = -p_2 \sum_{i=1}^{n_m} \kappa_{i2m}^{(x)} \left(\frac{r}{R_2}\right)^{\lambda_{im}}, \tag{5.126}$$

$$\sigma_{22m} = -p_2 \sum_{i=1}^{n_m} \xi_{1im}\, \kappa_{i2m}^{(x)} \left(\frac{r}{R_2}\right)^{\lambda_{im}}, \tag{5.127}$$

$$\sigma_{33m} = -p_2 \sum_{i=1}^{n_m} \xi_{2im} \, \kappa_{i2m}^{(x)} \left(\frac{r}{R_2} \right)^{\lambda_{im}} , \tag{5.128}$$

$$\sigma_{12m} = -p_2 \sum_{i=1}^{n_m} \xi_{3im} \, \kappa_{i2m}^{(x)} \left(\frac{r}{R_2} \right)^{\lambda_{im}} , \tag{5.129}$$

$$\sigma_{13m} = -p_2 \sum_{i=1}^{n_m} \xi_{4im} \, \kappa_{i2m}^{(x)} \left(\frac{r}{R_2} \right)^{\lambda_{im}} , \tag{5.130}$$

$$\varepsilon_{11m} = -p_2 \sum_{i=1}^{n_m} \xi_{5im} \, \kappa_{i2m}^{(x)} \left(\frac{r}{R_2} \right)^{\lambda_{im}} , \tag{5.131}$$

$$\varepsilon_{22m} = \varepsilon_{33m} = -p_2 \sum_{i=1}^{n_m} \xi_{6im} \, \kappa_{i2m}^{(x)} \left(\frac{r}{R_2} \right)^{\lambda_{im}} , \tag{5.132}$$

$$\varepsilon_{13m} = -p_2 \sum_{i=1}^{n_m} \xi_{7im} \, \kappa_{i2m}^{(x)} \left(\frac{r}{R_2} \right)^{\lambda_{im}} , \tag{5.133}$$

$$\varepsilon_{12m} = -p_2 \sum_{i=1}^{n_m} \xi_{8im} \, \kappa_{i2m}^{(x)} \left(\frac{r}{R_2} \right)^{\lambda_{im}} , \tag{5.134}$$

$$w_m = \frac{p_2^2}{2} \sum_{i,j=1}^{n_m} \omega_{ijm} \, \kappa_{i2m}^{(x)} \, \kappa_{j2m}^{(x)} \left(\frac{r}{R_2} \right)^{\lambda_{im}+\lambda_{jm}} , \tag{5.135}$$

$$W_m = 4R_2^3 \int_0^{\pi/2} \int_0^{\pi/2} p_2^2 \, \Omega_{2m} \, d\varphi \, d\nu, \tag{5.136}$$

$$\sigma_{im} = -p_2 \sum_{j=1}^{n_m} \xi_{8+ijm} \, \kappa_{j2m}^{(x)} \left(\frac{r}{R_2} \right)^{\lambda_{jm}} , \quad i = 1, 2, 3, \tag{5.137}$$

$$w_{im} = \frac{s_{iiq} \, p_2^2}{2} \sum_{j,k=1}^{n_m} \xi_{8+ijm} \, \xi_{8+ikm} \, \kappa_{j2m}^{(x)} \, \kappa_{k2m}^{(x)} \left(\frac{r}{R_2} \right)^{\lambda_{jm}+\lambda_{km}} ,$$
$$i = 1, 2, 3, \tag{5.138}$$

$$\left(\varepsilon_{22m} \right)_{r=R_{2m}} = -p_2 \varrho_{2m}, \tag{5.139}$$

where the coefficient $\xi_{1im}, \dots, \xi_{8im}, \omega_{ijm}, \xi_{8+kim}$ $(i, j = 1, \dots, n_m; k = 1,2,3)$, Ω_{2m}, ϱ_{2m} are given by Eqs. (3.86)–(3.88) $(q = m)$, (3.91), (3.100), (5.103), (5.60), respectively.

5.2.2.4 Radial stresses p_1, p_2

With regard to Eqs. (3.110), (3.111), (5.8), (5.123), (5.139), the radial stresses p_1 and p_2 acting at the particle-envelope and matrix-envelope boundaries, respectively, considered in Sections 5.2.2.1–5.2.2.3 have the forms

$$p_1 = \frac{\varrho_{12e}\left(\beta_m - \beta_e\right) - \left(\varrho_{2m} - \varrho_{22e}\right)\left(\beta_e - \beta_p\right)}{\varrho_{12e}\varrho_{21e} - \left(\varrho_{2m} - \varrho_{22e}\right)\left[\xi_{61p}\left(1 - \beta_p\right)\left(\frac{R_{1p}}{R_1}\right)^{\lambda_{1p}} - \varrho_{11e}\right]}, \tag{5.140}$$

$$p_2 = \frac{\left[\xi_{61p}\left(1 - \beta_p\right)\left(\frac{R_{1p}}{R_1}\right)^{\lambda_{1p}} - \varrho_{11e}\right]\left(\beta_m - \beta_e\right) - \varrho_{21e}\left(\beta_e - \beta_p\right)}{\varrho_{12e}\varrho_{21e} - \left(\varrho_{2m} - \varrho_{22e}\right)\left[\xi_{61p}\left(1 - \beta_p\right)\left(\frac{R_{1p}}{R_1}\right)^{\lambda_{1p}} - \varrho_{11e}\right]}, \tag{5.141}$$

where the radius R_{1p} and the coefficient ξ_{61p}, ϱ_{2m}, ϱ_{ije} $(i,j=1,2)$ are given by Eqs. (3.108) and (3.88) $(q=p)$, (5.60), (5.124), respectively.

5.2.3 Condition $\beta_p = \beta_e \neq \beta_m$

5.2.3.1 Thermal stresses in spherical particle

With regard to the analyses in Sections 3.2.3–3.2.5, 4.1, the integration constant C_{1p} is derived as

$$C_{1p} = -\frac{p_2\,\varrho_{1e}^{pe}}{\xi_{61p}R_1^{\lambda_{1p}}}, \tag{5.142}$$

where the coefficient ξ_{61p}, ϱ_{1e}^{pe} and the radial stress p_2 acting at the matrix-envelope boundary are given by Eqs. (3.88) $(q=p)$, (5.171) and (5.172), respectively.

Equations (3.76)–(3.84), (3.89), (3.90), (3.92), (3.97), (3.98) $(q=p)$ are thus transformed to the forms

$$\sigma_{11p} = -\frac{p_2\,\varrho_{1e}^{pe}}{\xi_{61p}}\left(\frac{r}{R_1}\right)^{\lambda_{1p}}, \tag{5.143}$$

$$\sigma_{22p} = -\frac{p_2\,\varrho_{1e}^{pe}\,\xi_{11p}}{\xi_{61p}}\left(\frac{r}{R_1}\right)^{\lambda_{1p}}, \tag{5.144}$$

$$\sigma_{33p} = -\frac{p_2\,\varrho_{1e}^{pe}\,\xi_{21p}}{\xi_{61p}}\left(\frac{r}{R_1}\right)^{\lambda_{1p}}, \tag{5.145}$$

$$\sigma_{12p} = -\frac{p_2\,\varrho_{1e}^{pe}\,\xi_{31p}}{\xi_{61p}}\left(\frac{r}{R_1}\right)^{\lambda_{1p}}, \tag{5.146}$$

$$\sigma_{13p} = -\frac{p_2\,\varrho_{1e}^{pe}\,\xi_{41p}}{\xi_{61p}}\left(\frac{r}{R_1}\right)^{\lambda_{1p}}, \tag{5.147}$$

$$\varepsilon_{11p} = -\frac{p_2\,\varrho_{1e}^{pe}\,\xi_{51p}}{\xi_{61p}}\left(\frac{r}{R_1}\right)^{\lambda_{1p}}, \tag{5.148}$$

$$\varepsilon_{22p} = \varepsilon_{33p} = -p_2 \, \varrho_{1e}^{pe} \left(\frac{r}{R_1} \right)^{\lambda_{1p}}, \tag{5.149}$$

$$\varepsilon_{13p} = -\frac{p_2 \, \varrho_{1e}^{pe} \, \xi_{71p}}{\xi_{61p}} \left(\frac{r}{R_1} \right)^{\lambda_{1p}}, \tag{5.150}$$

$$\varepsilon_{12p} = -\frac{p_2 \, \varrho_{1e}^{pe} \, \xi_{81p}}{\xi_{61p}} \left(\frac{r}{R_1} \right)^{\lambda_{1p}}, \tag{5.151}$$

$$w_p = \frac{\omega_{11p}}{2} \left(\frac{p_2 \, \varrho_{1e}^{pe}}{\xi_{61p}} \right)^2 \left(\frac{r}{R_1} \right)^{2\lambda_{1p}}, \tag{5.152}$$

$$W_p = 4R_1^3 \int_0^{\pi/2} \int_0^{\pi/2} \frac{\omega_{11p}}{2\lambda_{1p} + 3} \left(\frac{p_2 \, \varrho_{1e}^{pe}}{\xi_{61p}} \right)^2 d\varphi \, d\nu, \tag{5.153}$$

$$\sigma_{ip} = -\frac{p_2 \, \varrho_{1e}^{pe} \, \xi_{8+i1p}}{\xi_{61p}} \left(\frac{r}{R_1} \right)^{\lambda_{1p}}, \quad i = 1, 2, 3, \tag{5.154}$$

$$w_{ip} = \frac{s_{iip} \, \xi_{8+i1p}^2}{2} \left(\frac{p_2 \, \varrho_{1e}^{pe}}{\xi_{61p}} \right)^2 \left(\frac{r}{R_1} \right)^{2\lambda_{1p}}, \quad i = 1, 2, 3, \tag{5.155}$$

where the coefficient $\xi_{11p}, \ldots, \xi_{81p}, \omega_{11p}, \xi_{8+i\,1p}$ ($i = 1, 2, 3$) are given by Eqs. (3.86)–(3.88) ($q = p$), (3.91), (3.100), respectively.

5.2.3.2 Thermal stresses in spherical envelope

With regard to the analyses in Sections 3.2.3–3.2.5, 4.2, the integration constant C_{1e} is derived as

$$C_{1e} = -\frac{p_2}{R_2^{\lambda_{1e}}}, \tag{5.156}$$

where the radial stress p_2 acting at the matrix-envelope boundary is given by Eq. (5.172).

Equations (3.76)–(3.84), (3.89), (3.90), (3.92), (3.97), (3.98) ($q = e$) along with $(\varepsilon_{22e})_{r=R_{2e}}$ (see Eq. (3.111)) are thus transformed to the forms

$$\sigma_{11e} = -p_2 \left(\frac{r}{R_2} \right)^{\lambda_{1e}}, \tag{5.157}$$

$$\sigma_{22e} = -p_2 \, \xi_{11e} \left(\frac{r}{R_2} \right)^{\lambda_{1e}}, \tag{5.158}$$

$$\sigma_{33e} = -p_2 \, \xi_{21e} \left(\frac{r}{R_2} \right)^{\lambda_{1e}}, \tag{5.159}$$

$$\sigma_{12e} = -p_2 \, \xi_{31e} \left(\frac{r}{R_2} \right)^{\lambda_{1e}}, \tag{5.160}$$

Thermal stresses in multi-particle-(envelope)-matrix system I 75

$$\sigma_{13e} = -p_2 \, \xi_{41e} \left(\frac{r}{R_2} \right)^{\lambda_{1e}}, \tag{5.161}$$

$$\varepsilon_{11e} = -p_2 \, \xi_{51e} \left(\frac{r}{R_2} \right)^{\lambda_{1e}}, \tag{5.162}$$

$$\varepsilon_{22e} = \varepsilon_{33e} = -p_2 \, \xi_{61e} \left(\frac{r}{R_2} \right)^{\lambda_{1e}}, \tag{5.163}$$

$$\varepsilon_{13e} = -p_2 \, \xi_{71e} \left(\frac{r}{R_2} \right)^{\lambda_{1e}}, \tag{5.164}$$

$$\varepsilon_{12e} = -p_2 \, \xi_{81e} \left(\frac{r}{R_2} \right)^{\lambda_{1e}}, \tag{5.165}$$

$$w_e = \frac{p_2^2 \, \omega_{11e}}{2} \left(\frac{r}{R_2} \right)^{2\lambda_{1e}}, \tag{5.166}$$

$$W_e = 4R_2^3 \int\limits_0^{\pi/2} \int\limits_0^{\pi/2} \frac{p_2^2 \, \omega_{11e}}{2\lambda_{1e} + 3} \left[1 - \left(\frac{R_1}{R_2} \right)^{2\lambda_{1e}+3} \right] d\varphi \, d\nu, \tag{5.167}$$

$$\sigma_{ie} = -p_2 \, \xi_{8+i1e} \left(\frac{r}{R_2} \right)^{\lambda_{1e}}, \quad i = 1, 2, 3, \tag{5.168}$$

$$w_{ie} = \frac{s_{iie} \, p_2^2 \, \xi_{8+i1e}^2}{2} \left(\frac{r}{R_2} \right)^{2\lambda_{1e}}, \quad i = 1, 2, 3, \tag{5.169}$$

$$(\varepsilon_{22e})_{r=R_{2e}} = -p_2 \varrho_{2e}^{pe}, \tag{5.170}$$

where the coefficient $\xi_{11e}, \dots, \xi_{81e}, \omega_{11e}, \xi_{8+i1e}$ $(i = 1,2,3)$ are given by Eqs. (3.86)–(3.88) $(q = e)$, (3.91), (3.100), respectively. The coefficien ϱ_{2e}^{pe} is derived as

$$\varrho_{ie}^{pe} = \xi_{61e} \, (1 - \beta_e) \left(\frac{R_{ie}}{R_{2e}} \right)^{\lambda_{1e}}, \quad i = 1, 2, \tag{5.171}$$

where the radius R_{ie} $(i = 1,2)$ is given by Eq. (3.108).

5.2.3.3 Thermal stresses in cell matrix

The integration constant C_{im} $(i = 1, \dots, n_m)$ for the cell matrix is given by Eq. (5.125). Accordingly, formulae in Section 5.2.2.3 are considered for the cell matrix on the condition $\beta_p = \beta_e \neq \beta_m$. The radial stress p_2 included in these formulae is given by Eq. (5.172).

5.2.3.4 Radial stress p_2

With regard to Eqs. (3.111), (5.139), (5.170), the radial stress p_2 acting at the matrix-envelope boundary, considered in Sections 5.2.3.1–5.2.1.3 has the form

$$p_2 = \frac{\beta_m - \beta_e}{\varrho_{2e}^{pe} - \varrho_{2m}},$$

(5.172)

where the coefficient ϱ_{2e}^{pe}, ϱ_{2m} are given by Eqs. (5.171), (5.60), respectively.

Chapter 6

Thermal stresses in one-particle-(envelope)-matrix system

Chapter 6 deals with the thermal stresses in the multi-particle-(envelope)-matrix system which consists of anisotropic components.

With regard to the analysis in Section 3.1.2, in case of the multi-particle-matrix system, R_1 represents a radius of the embedded spherical particle at the temperature $T \in \langle T_f, T_r \rangle$, where T_f is a fina temperature of a cooling process, and the relaxation temperature T_r is determined in Section 3.4.1. In case of the multi-particle-envelope-matrix system, R_1, R_2 and $t = R_2 - R_1$ represents radii and thickness of the embedded spherical envelope at the temperature $T \in \langle T_f, T_r \rangle$, respectively.

6.1 Thermal stresses in spherical particle and envelope

Formulae for quantities given by Eqs. (3.76)–(3.99) are identical with those presented in Sections 5.1.1, 5.2.1.1, 5.2.1.2, 5.2.2.1, 5.2.2.2, 5.2.3.1, 5.2.3.2. The coefficien ϱ_{im} ($i = 1,2$) included in formulae for the radial stresses p_1 and p_2 (see Eqs. (5.61), (5.140), (5.141), (5.172)) acting at the particle-matrix, particle-envelope and matrix-envelope boundaries, respectively, is replaced by ξ_{61m} given by Eq. (3.88).

6.2 Thermal stresses in infinit matrix

6.2.1 Conditions $\beta_m \neq \beta_p$, $\beta_p \neq \beta_e \neq \beta_p$, $\beta_p = \beta_e \neq \beta_p$

The subscript $i = 1,2$ in Section 6.2.1 is related to

- the multi-particle-matrix system on the condition $\beta_m \neq \beta_p$, and then $i = 1$,

78 Ladislav Ceniga

- the multi-particle-envelope-matrix system on the conditions $\beta_p \neq \beta_e \neq \beta_p$, $\beta_p\beta_e \neq \beta_p$, and then $i = 2$.

With regard to the analyses in Section 3.2.3–3.2.3, 4.4, the integration constant C_{im} ($i = 1,2$) is derived as

$$C_{im} = -\frac{p_i}{R_i^{\lambda_{1m}}}, \quad i = 1, 2, \tag{6.1}$$

where the radial stresses p_1 and p_2 acting at the particle-matrix and matrix-envelope boundaries are given by Eqs. (5.61), (5.140), (5.141), (5.172). The coefficien ϱ_{im} ($i = 1,2$) included Eqs. (5.61), (5.140), (5.141), (5.172) is replaced by ξ_{61m} given by Eq. (3.88).

Equations (3.76)–(3.84), (3.89), (3.90), (3.97), (3.98) ($q = m$) are thus transformed to the forms

$$\sigma_{11m} = -p_i \left(\frac{r}{R_i}\right)^{\lambda_{1m}}, \tag{6.2}$$

$$\sigma_{22m} = -p_i \, \xi_{11m} \left(\frac{r}{R_i}\right)^{\lambda_{1m}}, \tag{6.3}$$

$$\sigma_{33m} = -p_i \, \xi_{21m} \left(\frac{r}{R_i}\right)^{\lambda_{1m}}, \tag{6.4}$$

$$\sigma_{12m} = -p_i \, \xi_{31m} \left(\frac{r}{R_i}\right)^{\lambda_{1m}}, \tag{6.5}$$

$$\sigma_{13m} = -p_i \, \xi_{41m} \left(\frac{r}{R_i}\right)^{\lambda_{1m}}, \tag{6.6}$$

$$\varepsilon_{11m} = -p_i \, \xi_{51m} \left(\frac{r}{R_i}\right)^{\lambda_{1m}}, \tag{6.7}$$

$$\varepsilon_{22m} = \varepsilon_{33m} = -p_i \, \xi_{61m} \left(\frac{r}{R_i}\right)^{\lambda_{1m}}, \tag{6.8}$$

$$\varepsilon_{13m} = -p_i \, \xi_{71m} \left(\frac{r}{R_i}\right)^{\lambda_{1m}}, \tag{6.9}$$

$$\varepsilon_{12m} = -p_i \, \xi_{81m} \left(\frac{r}{R_i}\right)^{\lambda_{1m}}, \tag{6.10}$$

$$w_m = \frac{p_i^2 \, \omega_{11m}}{2} \left(\frac{r}{R_i}\right)^{2\lambda_{1m}}, \tag{6.11}$$

$$W_m = 4R_i^3 \int_0^{\pi/2} \int_0^{\pi/2} \frac{p_i^2 \, \omega_{11m}}{|2\lambda_{1m} + 3|} \, d\varphi \, d\nu, \tag{6.12}$$

Thermal stresses in one-particle-(envelope)-matrix system   79

$$\sigma_{jp} = -p_1\,\xi_{8+j1p}\left(\frac{r}{R_i}\right)^{\lambda_{1m}}, \quad i=1,2;\quad j=1,2,3, \tag{6.13}$$

$$w_{jm} = \frac{s_{jjm}\,p_i^2\,\xi_{8+j1m}^2}{2}\left(\frac{r}{R_i}\right)^{2\lambda_{1m}}, \quad i=1,2;\quad j=1,2,3, \tag{6.14}$$

where the coefficient $\xi_{11m},\ldots,\xi_{81m},\omega_{11m},\xi_{8+j\,1m}$ $(j=1,2,3)$ are given by Eqs. (3.86)–(3.88) $(q=m)$, (3.91), (3.100), respectively.

6.2.2 Condition $\beta_p \neq \beta_e = \beta_p$

With regard to the analyses in Sections 3.2.3–3.2.5, 4.4, the integration constant C_{1p} is derived as

$$C_{1m} = -\frac{p_1\,\varrho_{2e}^{me}}{\xi_{61m}R_1^{\lambda_{1m}}}, \tag{6.15}$$

where the coefficient ξ_{61m}, ϱ_{2e}^{me} and the radial stress p_1 acting at the particle-envelope boundary are given by Eqs. (3.88) $(q=p)$, (5.77) and (5.106), respectively.

Equations (3.76)–(3.84), (3.89), (3.90), (3.97), (3.98) $(q=m)$ are thus transformed to the forms

$$\sigma_{11m} = -\frac{p_1\,\varrho_{2e}^{me}}{\xi_{61m}}\left(\frac{r}{R_2}\right)^{\lambda_{1m}}, \tag{6.16}$$

$$\sigma_{22m} = -\frac{p_1\,\varrho_{2e}^{me}\,\xi_{11m}}{\xi_{61m}}\left(\frac{r}{R_2}\right)^{\lambda_{1m}}, \tag{6.17}$$

$$\sigma_{33m} = -\frac{p_1\,\varrho_{2e}^{me}\,\xi_{21m}}{\xi_{61m}}\left(\frac{r}{R_2}\right)^{\lambda_{1m}}, \tag{6.18}$$

$$\sigma_{12m} = -\frac{p_1\,\varrho_{2e}^{me}\,\xi_{31m}}{\xi_{61m}}\left(\frac{r}{R_2}\right)^{\lambda_{1m}}, \tag{6.19}$$

$$\sigma_{13m} = -\frac{p_1\,\varrho_{2e}^{me}\,\xi_{41m}}{\xi_{61m}}\left(\frac{r}{R_2}\right)^{\lambda_{1m}}, \tag{6.20}$$

$$\varepsilon_{11m} = -\frac{p_1\,\varrho_{2e}^{me}\,\xi_{51m}}{\xi_{61m}}\left(\frac{r}{R_2}\right)^{\lambda_{1m}}, \tag{6.21}$$

$$\varepsilon_{22m} = \varepsilon_{33m} = -p_1\,\varrho_{2e}^{me}\left(\frac{r}{R_2}\right)^{\lambda_{1m}}, \tag{6.22}$$

$$\varepsilon_{13m} = -\frac{p_1\,\varrho_{2e}^{me}\,\xi_{71m}}{\xi_{61m}}\left(\frac{r}{R_2}\right)^{\lambda_{1m}}, \tag{6.23}$$

$$\varepsilon_{12m} = -\frac{p_1\,\varrho_{2e}^{me}\,\xi_{81m}}{\xi_{61m}}\left(\frac{r}{R_2}\right)^{\lambda_{1m}}, \tag{6.24}$$

$$w_m = \frac{\omega_{11m}}{2} \left(\frac{p_1 \, \varrho_{2e}^{me}}{\xi_{61m}} \right)^2 \left(\frac{r}{R_2} \right)^{2\lambda_{1m}}, \tag{6.25}$$

$$W_m = 4R_2^3 \int\limits_0^{\pi/2} \int\limits_0^{\pi/2} \frac{\omega_{11m}}{|2\lambda_{1m} + 3|} \left(\frac{p_1 \, \varrho_{2e}^{me}}{\xi_{61m}} \right)^2 d\varphi \, d\nu, \tag{6.26}$$

$$\sigma_{im} = - \frac{p_1 \, \varrho_{2e}^{me} \, \xi_{8+i1m}}{\xi_{61m}} \left(\frac{r}{R_2} \right)^{\lambda_{1m}}, \quad i = 1, 2, 3, \tag{6.27}$$

$$w_{im} = \frac{s_{iim} \, \xi_{8+i1m}^2}{2} \left(\frac{p_1 \, \varrho_{2e}^{me}}{\xi_{61m}} \right)^2 \left(\frac{r}{R_2} \right)^{2\lambda_{1m}}, \quad i = 1, 2, 3, \tag{6.28}$$

where the coefficient $\xi_{11m}, \ldots, \xi_{81m}, \omega_{11m}, \xi_{8+i1m}$ $(i = 1,2,3)$ are given by Eqs. (3.86)–(3.88) $(q = m)$, (3.91), (3.100), respectively.

Chapter 7

Thermal stresses in multi-particle-(envelope)-matrix system II

Chapter 7 deals with the thermal stresses in the multi-particle-(envelope)-matrix system which consists of isotropic and anisotropic components, where formulae for the thermal stresses are determined by the following procedure.

In case of the multi-particle-matrix system which consists of isotropic spherical particles and an anisotropic infinit matrix, formulae for the thermal stresses acting in the isotropic and anisotropic components, derived in Volumes I, II [9, 10] and Volume III of this trilogy, respectively, contain the radial stress p_1 acting at the particle-matrix boundary (see Eq. (5.61)).

If the spherical particle and the cell matrix are isotropic and anisotropic, the coefficient ϱ_{1p} and ϱ_{1m} are taken from Volumes I, II and III, respectively. Additionally, the thermal expansion coefficient $\alpha'_{1Ip} = \alpha'_{1Ip}(\varphi, \nu)$, $\alpha'_{1IIp} = \alpha'_{1IIp}(\varphi, \nu)$, $\alpha'_{1p} = \alpha'_{1p}(\varphi, \nu)$ included in the coefficien β_p (see Eqs. (3.100)–(3.102)), which are functions of $\varphi, \nu \in \langle 0, \pi/2 \rangle$ (see Fig. 3.1), are replaced by $\alpha_{Ip} \neq f(\varphi, \nu)$, $\alpha_{IIp} \neq f(\varphi, \nu)$, $\alpha_p \neq f(\varphi, \nu)$, respectively.

Contrarily, if the spherical particle and the cell matrix are anisotropic and isotropic, the coefficient ϱ_{1p} and ϱ_{1m} are taken from Volumes III and I, II, respectively. Similarly, the thermal expansion coefficient $\alpha'_{1Im} = \alpha'_{1Im}(\varphi, \nu)$, $\alpha'_{1IIm} = \alpha'_{1IIm}(\varphi, \nu)$, $\alpha'_{1m} = \alpha'_{1m}(\varphi, \nu)$ included in the coefficien β_m (see Eqs. (3.100)–(3.102)) are replaced by $\alpha_{Im} \neq f(\varphi, \nu)$, $\alpha_{IIm} \neq f(\varphi, \nu)$, $\alpha_m \neq f(\varphi, \nu)$, respectively.

The same procedure is also considered for the multi-particle-envelope-matrix system. Accordingly, the coefficien ϱ included in formulae for p_1, p_2 (see Eqs. (5.106), (5.140), (5.141), (5.172)) which is related to isotropic and anisotropic components are taken from Volumes I, II and III, respectively. Anisotropic and isotropic components are characterized by $\alpha'_{1Iq_1} = \alpha'_{1Iq_1}(\varphi, \nu)$, $\alpha'_{1IIq_1} = \alpha'_{1IIq_1}(\varphi, \nu)$, $\alpha'_{1q_1} = \alpha'_{1q_1}(\varphi, \nu)$ and $\alpha_{Iq_2} \neq f(\varphi, \nu)$, $\alpha_{IIq_2} \neq f(\varphi, \nu)$, $\alpha_{q_2} \neq f(\varphi, \nu)$ ($q_1, q_2 = p,e,m$; $q_1 \neq q_2$) (see Eqs. (3.100)–(3.102)), respectively.

Finally, if the coefficien ϱ related to an isotropic or anisotropic component of the multi-particle-(envelope)-matrix system is determined by different mathematical techniques, the thermal-stress induced elastic energy W_c of the cell (see Eqs. (3.74), (3.75)) and consequently the Castigliano's theorem (see Section 3.2.5) are required to be considered.

Chapter 8

Related phenomena

8.1 Introduction

The related phenomena presented in Chapter 8 include

1. analytical fracture mechanics (see Section 8.2),

2. analytical model of energy barrier (see Section 8.3),

3. analytical model of strengthening (see Section 8.4),

4. analytical-computational and analytical-experimental-computational methods of life-time prediction (see Section 8.5).

The analytical modelling in Items 1–4 considers the multi-particle-matrix and multi-particle-envelope-matrix systems which consist of cubic cells with the dimension d which is equal to the inter-particle distance (see Fig. 2.1a).

8.1.1 Analytical fracture mechanics

The analytical modelling of the fracture mechanics presented in Section 8.2 is based on the comparison of energy accumulated in the cubic cell with energy for the creation of a crack surface. This comparison results from a general analysis (see Section 8.2.1) which is applied to a solid continuum of a general shape (see Fig. 8.1a). The solid continuum is assumed to be acted by stress-deformation field which induce energy density given by Eq. (8.1). The stress-deformation field acting in the solid continuum are of a general character, i.e. a reason of the stress-deformation field are not only the thermal stresses.

The comparison considers infinitesima volume of the solid continuum along with an infinitesima crack surface (see Fig. 8.1c), and thus results in a formula for a function (see Eq. (8.9)) which describes the crack shape in a plane (see the plane $x_{ij}x_k$, 8.1a,b) which is perpendicular to the crack formation plane (see the plane x_ix_j, Fig. 8.1a,b). Consequently, this formula (see Eq. (8.9)) results in the crack condition given by Eq. (8.10) which includes fracture toughness and a curve integral of the energy density (see Eq. (8.3)) along an abscissa in the solid continuum.

This general analysis in Section 8.2.1 is applied to the cubic cell (see Section 8.2.2) as expressed by Eqs. (8.12), (8.13). The subscript $q = p,e,m$ in the fracture toughness and the curve integral, both included in Eqs. (8.12), (8.13), is related to such component of the multi-particle-matrix and multi-particle-envelope-matrix systems which fulfil the crack formation conditions presented in Section 8.2.2.1.

The curve integral as a function of a position in the cubic cell includes the parameters R_1, v and R_1, t, v of the multi-particle-matrix and multi-particle-envelope-matrix systems which represent microstructural parameters of the two- and three-component materials define in Items 1–4, Section 2.1, respectively. Consequently, the crack initiation condition (8.17) related to such position in which the curve integral is maximal is considered for the determination of a critical particle radius. The critical particle radius is a reason of the initiation of the crack with infinitesima length in the crack propagation plane (see Section 8.2.2.2.A.1). Additionally, as presented in Section 8.2.2.2.A.3, the critical radius thus define a limit state regarding the crack initiation.

The curve integral is assumed to exhibit different courses which result in different positions of the crack initiation as well as in different courses of the crack propagation (see Figs. 8.3–8.7) as analysed in Section 8.2.2.2.A.2.

As presented in Section 8.2.2.3, the stress-deformation field along with the energy density induced by the stress-deformation field are modifie (changed) due to the crack propagation. On the one hand, the analyses in Section 8.2.2.2 concerning the crack initiation and propagation in the cubic cell of the multi-particle-matrix and multi-particle-envelope-matrix systems exhibit a general character regarding a reason of the stress-deformation field as well as regarding the curve integral of the energy density. On the other hand, as presented in Section 8.2.2.3, if a modificatio (a change) of the stress-deformation field during the crack propagation along with a modificatio of the curve integral can not be analytically and/or computationally determined, then the determination of the function which describes the crack shape in a plane perpendicular to the crack propagation plane is valid for a ceramic component only. In general, a ceramic component exhibits a high-speed crack propagation during which the modificatio can be assumed to be neglected.

The curve integral is determined for the multi-particle-matrix (see Sections 8.2.3.1, 8.2.4.1) and multi-particle-envelope-matrix systems (see Sections 8.2.3.2, 8.2.4.2) regarding positions in the spherical particle, the spherical envelope and the cell matrix. Additionally, this determination considers the conditions $w_q \neq 0$ (see Section 8.2.3) and $w_q = 0$ (see Section 8.2.4), where $w_q = w_q(x_1, x_2, x_3, R_1, v)$ or $w_q = w_q(x_1, x_2, x_3, R_1, t, v)$ is energy density induced by the stress-deformation field which is a function of position given by the coordinates (x_1, x_2, x_3) in the spherical particle $(q = p)$, the spherical envelope $(q = e)$ and the cell matrix $(q = m)$.

The condition $w_q(x_1, x_2, x_3, R_1, v) = 0$ or $w_q(x_1, x_2, x_3, R_1, t, v) = 0$ considered for the multi-particle-matrix or multi-particle-envelope-matrix system, respectively, results in

Related phenomena 85

such dependence between the variables x_1, x_2, x_3 which define two volumes of the cubic cell. The two volumes are characterized by mutually different stress conditions (see Items 1, 2, Section 8.2.4.1; Items 1, 2, Section 8.2.4.2). With regard to the condition $w_q(x_1, x_2, x_3, R_1, v) = 0$ or $w_q(x_1, x_2, x_3, R_1, t, v) = 0$, the curve integral is determined by two method. The method 1 (see Sections 8.2.4.1.A, 8.2.4.2.A) is based on the crack formation conditions presented in Section 8.2.2.1. The method 2 (see Sections 8.2.4.1.B, 8.2.4.2.B) results from a principle that release of energy of a system considers 'minimal resistance' of the system, i.e. the energy is released through 'minimal resistance' of the system.

As an example, let the multi-particle-matrix system is considered. Let two cracks are expected to be formed regarding these crack formation conditions. Let one of them is expected to be formed in the plane $x_1 x_2$ in the spherical particle for $R_1 > R_{1pc-1}^{(12)}$. $R_{1pc-1}^{(12)}$ is a critical particle radius for the crack formation in the plane $x_1 x_2$ in the spherical particle. Let the other crack is expected to be formed in the plane $x_1 x_3$ in the cell matrix for $R_1 > R_{1mc-1}^{(13)}$. $R_{1mc-1}^{(13)}$ is a critical particle radius for the crack formation in the plane $x_1 x_3$ in the cell matrix. $R_{1pc-1}^{(12)}$ and $R_{1mc-1}^{(13)}$ are determined by the condition (8.17) for $q = p$ and $q = m$, respectively, where the curve integral included in Eq. (8.17) is determined by the method 1.

Let the critical particle radius for the crack formation in the plane $x_1 x_2$ in the cell matrix, $R_{1mc-2}^{(12)}$, is determined by the condition (8.17) for $q = m$, where the curve integral included in Eq. (8.17) is determined by the method 2. Let the conditions $R_{1mc-2}^{(12)} < R_{1pc-1}^{(12)}$ and $R_{1mc-2}^{(12)} < R_{1mc-1}^{(13)}$ are valid. Consequently, in spite of the crack formation conditions presented in Section 8.2.2.1, only one crack is formed in the plane $x_1 x_2$ in the cell matrix instead of the two cracks which are expected to be formed in the planes $x_1 x_2$ and $x_1 x_3$ in the spherical particle and the cell matrix, respectively. Accordingly, the critical particle radius $R_{1mc-2}^{(12)}$ represents the 'minimal resistance' through which energy of the cubic cell is released.

Transformations of material parameters (elastic moduli, thermal expansion coefficients regarding the crack formation in the planes $x_1 x_2$, $x_2 x_3$, $x_1 x_3$ of the Cartesian system $(Ox_1 x_2 x_3)$ are presented in Section 8.2.5.

Finally, the curve integrals, the critical particle radii, and the functions describing a shape of cracks presented in Section 8.2 are determined for real two- and three-component materials by numerical methods.

8.1.2 Analytical model of energy barrier

As presented in Section 8.3.1, the determination of the energy barrier considers a solid continuum with a general shape (see Fig. 8.37). The solid continuum is also assumed to be acted by stress-deformation field which induce energy density given by Eq. (8.1). The stress-deformation field acting in the solid continuum are of a general character, i.e. a reason of the stress-deformation field are not only the thermal stresses.

In general, energy barriers influenc motion of dislocations, magnetic domain walls (see Fig. 8.38), etc. [11, p. 63]. The energy barrier along the axis x_i ($i = 1,2,3$) of the Cartesian system $(Ox_1x_2x_3)$ (see Fig. 8.37) represents a surface integral of the energy density $w = w(x_1, x_2, x_3)$ (see Eqs. (8.92)–(8.94)). The integration surface $S_i = S_i(x_i)$ which is perpendicular to the axis x_i represents a cross-section of the solid continuum.

The surface integrals within the cubic cells of the multi-particle-matrix and multi-particle-envelope-matrix systems along with integration boundaries are presented in Sections 8.3.2 and 8.3.3 (see Eqs. (8.98)–(8.101)), respectively.

Finally, the surface integrals presented in Section 8.3 are determined for real two- and three-component materials by numerical methods.

8.1.3 Analytical model of strengthening

As presented in Section 8.4.1, the analytical model of the micro-/macro-strengthening along the axis x_i ($i = 1,2,3$) of the Cartesian system $(Ox_1x_2x_3)$ (see Fig. 8.37) is based on the surface integral $W_i = W_i(x_1, x_2, x_3)$ (see Eqs. (8.103)) of the energy density $w_i = w_i(x_1, x_2, x_3)$ over the surface $S_i = S_i(x_i)$. In contrast to the energy barrier, the energy density $w_i = w_i(x_1, x_2, x_3)$ is induced by the stress $\sigma_i = \sigma_i(x_1, x_2, x_3)$ of the stress-deformation field which acts along the axis x_i. Additionally, $W_i = W_i(x_1, x_2, x_3)$ thus represents energy which is accumulated on the surface area $S_i = S_i(x_i)$ and is induced by the stress $\sigma_i = \sigma_i(x_1, x_2, x_3)$.

The micro-strengthening $\sigma_{si} = \sigma_{si}(x_i)$ (see Eq. (8.104)) along the axis x_i represents a stress which acts along the axis x_i and is constant at each point of the surface $S_i = S_i(x_i)$. Additionally, $W_{\sigma_{si}} = W_{\sigma_{si}}(x_1, x_2, x_3)$ given by Eq. (8.102) represents energy which is also accumulated on the surface area $S_i = S_i(x_i)$ and is induced by the stress $\sigma_{si} = \sigma_{si}(x_i)$.

The determination of micro-strengthening $\sigma_{si} = \sigma_{si}(x_i)$ is thus based on the comparison of $W_i = W_i(x_1, x_2, x_3)$ with $W_{\sigma_{si}} = W_{\sigma_{si}}(x_1, x_2, x_3)$, where the macro-strengthening $\overline{\sigma_{si}}$ represents a mean value of $\sigma_{si} = \sigma_{si}(x_i)$ regarding the interval $x_i \in \langle x_{1i}, x_{2i} \rangle$.

Analyses of signs of the micro-/macro-strengthening regarding compressive or tensile mechanical loading are presented in Items 1–6, Section 8.4.1.

The micro-/macro-strengthening within the cubic cells of the multi-particle-matrix and multi-particle-envelope-matrix systems along with integration boundaries is presented in Sections 8.4.2.1 and 8.4.2.2 (see Eqs. (8.106)–(8.124)), respectively.

Finally, the surface integrals, and the micro-/macro-strengthening presented in Section 8.4 are determined for real two- and three-component materials by numerical methods.

8.1.4 Analytical-computational and analytical-experimental-computational methods of lifetime prediction

The methods of the lifetime prediction are applicable to the three-component material define in Item 3, Section 2.1, which consists of grains with and without a continuous component on a grain surface. The grains with the continuous component and the grains without

Related phenomena

the continuous component are identical or different from microstructural point of view, and thus exhibit identical or different thermal expansion coefficients As analysed in Section 2.1

With regard to analytical modelling of the thermal stresses, this three-component material is replaced by the multi-particle-envelope-matrix system as analysed in Section 2.1. Strictly speaking, the grain with the continuous component on the grain surface corresponds to the spherical particle with the radius R_1. The grain without the continuous component on the grain surface corresponds to the cell matrix. The continuous component corresponds to the spherical envelope with the radii $R_1 < R_2$, i.e. with the thickness $t = R_2 - R_1 > 0$.

Finally, the $R_1 - t$ dependences presented in Sections 8.1.4.1, 8.5 are determined for a real two- and three-component materials by numerical methods.

8.1.4.1   Resistive and contributory effects of thermal stresses

As presented in Section 8.5.1, the methods of the lifetime prediction are based on the conditions $\overline{p_i}(R_1, t, v) = 0$ and $|\overline{p_i}(R_1, t, v)| = |\overline{\sigma_{ri}}(R_1, t, v)|$ ($i = 1,2$) which are required to be valid for each value of the parameter $v \in (0, v_{max})$ (see Eq. (2.7)), where $\overline{p_1} = \overline{p_1}(R_1, t, v)$ and $\overline{p_2} = \overline{p_2}(R_1, t, v)$ (see Eq. (8.125)) are mean values of the φ, ν-dependent radials stresses $p_1 = p_1(\varphi, \nu, R_1, t, v)$ and $p_2 = p_2(\varphi, \nu, R_1, t, v)$ acting at the particle-envelope and matrix-envelope boundaries, respectively.

The radial stresses $\overline{\sigma_{r1}} = \overline{\sigma_{r1}}(R_1, t, v)$ and $\overline{\sigma_{r2}} = \overline{\sigma_{r2}}(R_1, t, v)$ acting at the particle-envelope and matrix-envelope boundaries thus represent mean values of the φ, ν-dependent radial stresses $\sigma_{r1} = \sigma_{r1}(\varphi, \nu, R_1, t, v)$ and $\sigma_{r2} = \sigma_{r2}(\varphi, \nu, R_1, t, v)$ acting also at the particle-envelope and matrix-envelope boundaries, respectively.

The radial stress $\sigma_{ri} = \sigma_{ri}(\varphi, \nu, R_1, t, v)$ ($i = 1,2$) represents a 'response' of the mechanical loading $\sigma_{1\,mech}, \sigma_{2\,mech}, \sigma_{3\,mech}$ acting along the axes x_1, x_2, x_3 (see Fig. 3.1), respectively, where $\sigma_{ri} = \sigma_{ri}(\varphi, \nu, R_1, t, v)$ can be determined by e.g. the Eshelby's model. The Eshelby's model and its development [1]– [3] defin the disturbance of an applied stress-fiel in a solid continuum, where the applied stress-fiel is disturbed due to the presence of inclusions in a solid continuum.

Consequently, the conditions $\overline{p_i}(R_1, t, v) = 0$ and $|\overline{p_i}(R_1, t, v)| = |\overline{\sigma_{ri}}(R_1, t, v)|$ result in the dependences $R_{1Bc} = f_i(t_{Bc})$ and $R_{1Ac} = f_i(t_{Ac}, \overline{\sigma_{ri}})$ ($i = 1,2$) between the critical microstructural parameters R_{1Bc}, t_{Bc} and R_{1Ac}, t_{Ac} which represent critical values of the variables R_1, t of the functions $\overline{p_i} = \overline{p_i}(R_1, t, v), \overline{\sigma_{ri}} = \overline{\sigma_{ri}}(R_1, t, v)$, respectively.

The dependences $R_{1Ac} = f_i(t_{Ac}, \overline{\sigma_{ri}})$, $R_{1Bc} = f_i(t_{Bc})$ ($i = 1,2$) defin the areas A, B, C shown in Fig. 8.39. The areas A, B and C are characterized by such coordinates (R_1, t) for which the thermal stress exhibit resistive, quasi-resistive and contributory effects regarding the mechanical loading.

8.1.4.2   Analytical-computational method

As presented in Section 8.5.2, the analytical-computational method of lifetime prediction results from the function $F(R_1, t, \tau, T) = 0$ which is determined by a computational simulation of the time-temperature-dependent development of the microstructure, where the grain radius R_1, the thickness t and the time τ represent variables of the function $F(R_1, t, \tau, T) = 0$ at the exploitation temperature $T \,(\equiv T_e)$ which is a parameter of the function $F(R_1, t, \tau, T) = 0$, where $T < T_r$ (see Items 4, 5, Section 3.4.1). With regard to Fig. 8.40, intersections of $R_{1Ac} = f_i(t_{Ac}, \overline{\sigma_{ri}})$ and $R_{1Bc} = f_i(t_{Bc})$ with the function $F_p(R_1, t, T) = 0$ defin the time $\tau_{Ac} = \tau_{Ac}(\overline{\sigma_r}) \,(\equiv \tau_{Aci} = \tau_{Aci}(\overline{\sigma_{ri}}))$ and $\tau_{Bc} \,(\equiv \tau_{Bci})$ $(i = 1,2)$, where $F_p(R_1, t, T) = 0$ represents a projection of $F(R_1, t, \tau, T) = 0$ into the plane $R_1 - t$.

Consequently, the intervals $\tau < \tau_{Ac}(\overline{\sigma_r})$, $\tau \in \langle \tau_{Ac}(\overline{\sigma_r}), \tau_{Bc} \rangle$, $\tau > \tau_{Bc}$ define the time τ which is related to the resistive, quasi-resistive, contributory effects of the thermal stresses against the mechanical loading, respectively. The intervals $\tau < \tau_{Ac}(\overline{\sigma_r})$, $\tau \in \langle \tau_{Ac}(\overline{\sigma_r}), \tau_{Bc} \rangle$, $\tau > \tau_{Bc}$ determine non-critical, quasi-critical, critical time periods of the exploitation, respectively.

Finally, τ_{Acmin} and τ_{Bcmin} which represent minimal values of the sets $\{\tau_{Ac1}(\overline{\sigma_{r1}}), \tau_{Ac2}(\overline{\sigma_{r2}})\}$ and $\{\tau_{Bc1}, \tau_{Bc2}\}$ are considered to represent the lifetime regarding the 'resistance–quasi-resistance' and 'quasi-resistance–contribution' transformations (see Fig. 8.39), respectively.

8.1.4.3 Analytical-computational-experimental method

As presented in Section 8.5.3, the analytical-computational method of lifetime prediction (see Eqs. (8.126)–(8.131)) represents a connection of the analytical dependences $R_{1Ac} = f_i(t_{Ac}, \overline{\sigma_{ri}})$, $R_{1Bc} = f_i(t_{Bc})$ $(i = 1,2)$

- with the function $\tau = f(t)$ which is determined by a computational simulation, and with the function $R_1 = f(\tau)$ which is determined by an experiment.

- or with the function $\tau = f(t)$ which is determined by an experiment, and with the function $R_1 = f(\tau)$ which is determined by a computational simulation.

8.2 Analytical fracture mechanics

8.2.1 General analysis

Analytical fracture mechanics presented in Section 8.2 is based on the following analysis concerning a crack formation in a solid continuum with the volume V and with a general shape as shown in Figs. 8.1a and 8.1b without and with a crack propagating in the plane $x_i x_j$, respectively.

The shaded area in Fig. 8.1a represents cuts of the solid continuum in the planes $x_i x_j$, $x_i x_k$, $x_{ij} x_k$. The curves *1, 2, 3, 4* on a surface of the solid continuum for $x_k > 0$ are outlines of the cuts in the planes $x_i x_k$, $x_i x_j$, $x_j x_k$, $x_{ij} x_k$, respectively. The plane $x_{ij} x_k$ with a general position determined by the angle $\varphi = \angle(x_i, x_{ij}) \in \langle 0, 2\pi \rangle$ is perpendicular to the plane $x_i x_j$ of the crack formation, where $x_{ij} \subset x_i x_j$, and then $x_{ij} \equiv x_i$, $x_{ij} \equiv x_j$

Related phenomena

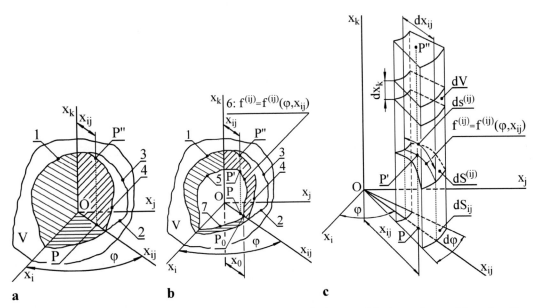

Figure 8.1: (a) The solid continuum with a general shape with the volume V and without a crack, and (b) with a crack in the plane $x_i x_j$. The shaded area represents cuts of the solid continuum in the planes $x_i x_j$, $x_i x_k$, $x_{ij} x_k$. The curves 1, 2, 3, 4 on a surface of the solid continuum for $x_k > 0$ are outlines of the cuts in the planes $x_i x_k$, $x_i x_j$, $x_j x_k$, $x_{ij} x_k$, respectively. The plane $x_{ij} x_k$ with a general position determined by the angle $\varphi = \angle(x_i, x_{ij}) \in \langle 0, 2\pi \rangle$ is perpendicular to $x_i x_j$ of the crack formation, where $x_{ij} \subset x_i x_j$, and then $x_{ij} \equiv x_i$, $x_{ij} \equiv x_j$ for $\varphi = 0$, $\varphi = \pi/2$, respectively. P and P'' both with the coordinate x_{ij} related to the axis x_{ij} are arbitrary points on the axis x_{ij} and the curve 4, respectively. The curves 5 and 6 (b) represent the crack shape in the planes $x_i x_k$ and $x_{ij} x_k$ which are perpendicular to $x_i x_j$ of the crack formation, respectively. The curve 7 (b) determines a position of the crack tip in the plane $x_i x_j$. The function $f^{(ij)} = f^{(ij)}(x_{ij}, \varphi, x_k)$ related to the curve 6 (b) describes the crack shape in the plane $x_{ij} x_k$ with the crack tip at the point P_0 determined by the coordinate x_0, then $|OP_0| = x_0$. (c) $ds^{(ij)}$ is length of an infinitesima part of the crack at the arbitrary point P' (b). $dS^{(ij)}$ is a new infinitesima surface created by release of the energy dW accumulated in the infinitesimal volume dV of an infinitesima prism with the height $|PP''|$ and with the surface area $dS_{ij} = x_{ij}\,d\varphi\,dx_{ij}$ of a basis in the plane $x_i x_j$. Strictly speaking, $dS^{(ij)}$ with the surface area $dS^{(ij)} = ds^{(ij)} x_{ij}\,d\varphi$ represents the crack surface which is related to the infinitesima length $ds^{(ij)}$ (c) at the arbitrary point P' (b,c) of the curve 6 on the crack surface.

for $\varphi = 0$, $\varphi = \pi/2$, respectively. P and P'' both with the coordinate x_{ij} related to the axis x_{ij} are arbitrary points on the axis x_{ij} and the curve 4, respectively.

Stress-strain field acting in the solid continuum, which are analytically determined as functions of a position in the Cartesian system $(Ox_1 x_2 x_3)$ (see Fig. 3.1), induce the energy density $w = w(x_1, x_2, x_3)$. The use of the cylindrical coordinates (x_{ij}, φ, x_k) within the analytical modelling of the crack formation in the plane $x_i x_j$ is assumed to be reasonable, and then we get

$$w = w^{(ij)} = w^{(ij)} \left(x_{ij}, \varphi, x_k \right),$$
$$ij = 12, \ k = 3; \quad ij = 23, \ k = 1; \quad ij = 31, \ k = 2, \tag{8.1}$$

where $x_i x_j \perp x_k$ ($i \neq j \neq k$), and $x_i = x_{ij} \cos \varphi$, $x_j = x_{ij} \sin \varphi$. The superscript ij in Eq. (8.1) thus indicates that the energy density w is derived as a function of the cylindrical coordinates (x_{ij}, φ, x_k).

The energy density w accumulated at an arbitrary point with a position given by the coordinates $(x_1, x_2, x_3) \equiv (x_{ij}, \varphi, x_k)$ results in the energy $W = \int_V w \, dV$ accumulated in the volume V. The energy W tends to be released by the crack formation in the plane $x_i x_j$ ($ij = 12, 23, 31$).

With regard to Fig. 8.1b with the crack formation in the plane $x_i x_j$, the curves *5* and *6* represent the crack shape in the planes $x_i x_k$ and $x_{ij} x_k$ which are perpendicular to $x_i x_j$ of the crack formation, respectively. The curve *7* determines a position of the crack tip in the plane $x_i x_j$. The function $f^{(ij)} = f^{(ij)} \left(x_{ij}, \varphi, x_k \right)$ related to the curve *6* (b) describes the crack shape in the plane $x_{ij} x_k$ with the crack tip at the point P_0 determined by the coordinate x_0, then $|OP_0| = x_0$.

As mentioned above, the energy W tends to be released by the crack formation in the plane $x_i x_j$ ($ij = 12, 23, 31$). With regard to Fig. 8.1c, the same is also valid for the energy $dW = dW \left(w^{(ij)} \right)$ (see Eq. (8.1)) accumulated in the infinitesima volume dV of an infinitesima prism with the height $\left| \overline{PP''} \right|$ (see Fig. 8.1b) and with the surface area $dS_{ij} = x_{ij} \, d\varphi \, dx_{ij}$ of a basis in the plane $x_i x_j$ (see Fig. 8.1c). Consequently, we get

$$dW = x_{ij} \, d\varphi \, dx_{ij} \int_{\overline{PP''}} w^{(ij)} \, dx_k = W_c^{(ij)} \, x_{ij} \, d\varphi \, dx_{ij}, \tag{8.2}$$

and the 'curve' energy density $W_c^{(ij)} = W_c^{(ij)} \left(x_{ij}, \varphi \right)$ as a function of φ, x_{ij} due to $w^{(ij)} = w^{(ij)} \left(x_{ij}, \varphi, x_k \right)$ (see Eq. (8.1)) has the form

$$W_c^{(ij)} = W_c^{(ij)} \left(x_{ij}, \varphi \right) = \int_{\overline{PP''}} w^{(ij)} \, dx_k. \tag{8.3}$$

The crack is formed in the plane $x_i x_j$ provided that the condition

$$\int_{\overline{PP''}} w^{(ij)} \left(x_{ij}, \varphi, x_k \right) dx_k = \int_{\overline{PP''(-)}} w^{(ij)} \left(x_{ij}, \varphi, -x_k \right) dx_k, \quad \varphi \in \langle 0, 2\pi \rangle, \tag{8.4}$$

is valid for $\varphi \in \langle 0, 2\pi \rangle$, where $P''^{(-)}$ is a point of intersection of the line PP'' with a surface of the solid continuum for $x_k < 0$. Strictly speaking, the crack formation plane

$x_i x_j$ is required to represent a plane of symmetry of $W_c^{(ij)} = W_c^{(ij)}(x_{ij}, \varphi)$ regarding $x_k > 0$ and $x_k < 0$.

The energy dW (see Eqs. (8.1)–(8.3)) is in an equilibrium state with the energy

$$dW_{cs}^{(ij)} = \gamma^{(ij)} dS^{(ij)} \tag{8.5}$$

for creation of the new infinitesima surface $dS^{(ij)}$ (see Fig. 8.1c). The crack surface energy per unit surface in the plane $x_i x_j$, $\gamma^{(ij)}$, has the form [12]

$$\gamma^{(ij)} = s_{kk} \left[K_{IC}^{(ij)} \right]^2, \quad s_{kk} = \frac{1}{E_k}, \tag{8.6}$$

where $K_{IC}^{(ij)}$ is fracture toughness along the axis x_{ij}, and s_{kk} is an elastic modulus along the axis x_k which is perpendicular to the plane $x_i x_j$ of the crack formation. E_k is the Young's modulus along the axis x_k.

This new infinitesima surface with the surface area

$$dS^{(ij)} = ds^{(ij)} x_{ij} \, d\varphi \tag{8.7}$$

represents the crack surface which is related to the infinitesima length $ds^{(ij)}$ (see Fig. 8.1c) at the arbitrary point P' (see Fig. 8.1b,c) of the curve 6 on the crack surface.

This infinitesima length representing is length of an infinitesima part of the crack at the arbitrary point P' is derived as [8, p. 324]

$$ds^{(ij)} = dx_{ij} \sqrt{1 + \left[\frac{\partial f^{(ij)}}{\partial x_{ij}} \right]^2}. \tag{8.8}$$

Accordingly, the function $f^{(ij)} = f^{(ij)}(x_{ij}, \varphi)$, strictly speaking $\partial f^{(ij)}/\partial x_{ij}$, describing the crack shape in the plane $x_{ij} x_k$ which is perpendicular to the plane $x_i x_j$ of the crack formation has the form

$$\frac{\partial f^{(ij)}}{\partial x_{ij}} = \pm \frac{1}{s_{kk} \left[K_{IC}^{(ij)} \right]^2} \sqrt{\left[W_c^{(ij)} \right]^2 - \left\{ s_{kk} \left[K_{IC}^{(ij)} \right]^2 \right\}^2}. \tag{8.9}$$

If $W_c^{(ij)} = W_c^{(ij)}(x_{ij}, \varphi)$ is a decreasing or increasing function of the variable x_{ij}, then $f^{(ij)} = f^{(ij)}(x_{ij}, \varphi)$ is assumed to be also a decreasing or increasing function of the variable x_{ij}, and accordingly the sign either "-" or "+" in Eq. (8.9) is considered, respectively.

Consequently, an energy condition for the crack formation has the form

$$W_c^{(ij)} - s_{kk} \left[K_{IC}^{(ij)} \right]^2 \geq 0, \tag{8.10}$$

and results from the condition $\left[W_c^{(ij)} \right]^2 - \left\{ s_{kk} \left[K_{IC}^{(ij)} \right]^2 \right\}^2 = \left\{ W_c^{(ij)} + s_{kk} \left[K_{IC}^{(ij)} \right]^2 \right\}$

$\times \left\{ W_c^{(ij)} - s_{kk} \left[K_{IC}^{(ij)} \right]^2 \right\} \geq 0$. Finally, the term $W_c^{(ij)} - s_{kk} \left[K_{IC}^{(ij)} \right]^2$ is considered only, because $W_c^{(ij)} + s_{kk} \left[K_{IC}^{(ij)} \right]^2 > 0$ due to $w^{(ij)} > 0$ and then $W_c^{(ij)} > 0$ (see Eqs. (8.1), (8.3)).

The same analysis is considered for the crack formation

1. in the plane $x_j x_k$, and then $w^{(ij)} = w^{(ij)}(x_{ij}, \varphi, x_k)$, $W_c^{(ij)} = W_c^{(ij)}(x_{ij}, \varphi)$, $f^{(ij)} = f^{(ij)}(x_{ij}, \varphi)$, s_{kk} are replaced by $w^{(jk)} = w^{(jk)}(x_{jk}, \varphi, x_i)$, $W_c^{(jk)} = W_c^{(jk)}(x_{jk}, \varphi)$, $f^{(jk)} = f^{(jk)}(x_{jk}, \varphi)$, s_{ii}, respectively, where $\varphi = \angle(x_j, x_{jk}) \in \langle 0, 2\pi \rangle$.

2. in the plane $x_j x_k$, and then $w^{(ij)} = w^{(ij)}(x_{ij}, \varphi, x_k)$, $W_c^{(ij)} = W_c^{(ij)}(x_{ij}, \varphi)$, $f^{(ij)} = f^{(ij)}(x_{ij}, \varphi)$, s_{kk} are replaced by $w^{(ki)} = w^{(ki)}(x_{ki}, \varphi, x_j)$, $W_c^{(ki)} = W_c^{(ki)}(x_{ki}, \varphi)$, $f^{(ki)} = f^{(ki)}(x_{ki}, \varphi)$, s_{jj}, respectively, where $\varphi = \angle(x_k, x_{ki}) \in \langle 0, 2\pi \rangle$.

8.2.2 Cell model

The general analysis in Section 8.2.1 with analytical results represented by Eqs. (8.3), (8.9), (8.10) is applicable to a multi-particle-(envelope)-matrix system (see Fig. 2.1a) with spherical particles with the radius R_1 which are periodically distributed in an infinit matrix, without (with) an spherical envelope on their surfaces, where $R_1 < R_2$ are radii of the spherical envelope with the thickness $t = R_2 - R_1$ (see Eq. (2.5)). These multi-particle-matrix and multi-particle-(envelope)-matrix systems correspond to two- and three-component materials define in Items 1–4, Section 2.1, respectively.

In general, if presence of the spherical particles without or with the spherical envelope on their surfaces is a reason of stress-deformation field (e.g. a stress-deformation fiel induced by thermal stresses), these stress-deformation field are expected to be determined within a cell of the multi-particle-(envelope)-matrix system. As analysed in Section 2.1, the cell represents such infinit matrix part which is related to one spherical particle.

If the energy density $w = w^{(ij)}(x_{ij}, \varphi, x_k)$ (see Eq. (8.1)) included in Eq. (8.3) is induced by stress-deformation field which are determined *within the cell*, the curve integral $W_c^{(ij)}$ $(\equiv W_{cq}^{(ij)})$ (see Eq. (8.3)) included in Eqs. (8.12), (8.13) is also required to be determined *within the cell*. Additionally, due to the determination of these stress-deformation field within the cell, the energy density along with the curve integral are also functions of R_1 and R_1, R_2 for the multi-particle-matrix and multi-particle-(envelope)-matrix systems as well as by the particle volume fraction v, i.e. $w^{(ij)} = w^{(ij)}(x_{ij}, \varphi, x_k, R_1, v)$, $W_c^{(ij)} = W_c^{(ij)}(x_{ij}, \varphi, R_1, v)$, and $w^{(ij)} = w^{(ij)}(x_{ij}, \varphi, x_k, R_1, R_2, v)$, $W_c^{(ij)} = W_c^{(ij)}(x_{ij}, \varphi, R_1, R_2, v)$, respectively. Additionally, if $w^{(ij)} = w^{(ij)}(r, \varphi, \nu, R_1, v)$ is derived by the spherical coordinates (r, φ, ν) (see Figs. 3.1) due to the *spherical* particle and the *spherical* envelope, then the relationships

$$r = \sqrt{x_{ij}^2 + x_k^2}, \quad \nu = \arctan\left(\frac{x_{ij}}{x_k}\right) \tag{8.11}$$

are considered for the transformations $w^{(ij)}\left(r,\varphi,\nu,R_1,v\right) \rightarrow w^{(ij)}\left(x_{ij},\varphi,x_k,R_1,v\right)$, $w^{(ij)}\left(r,\varphi,\nu,R_1,R_2,v\right) \rightarrow w^{(ij)}\left(x_{ij},\varphi,x_k,R_1,R_2,v\right)$ from the spherical coordinates (r,φ,ν) to the cylindrical coordinates (x_{ij},φ,x_k).

With regard to an application of Eqs. (8.3), (8.9), (8.10) to the multi-particle-(envelope)-matrix system (see Fig. 2.1a), the subscript $q = p,e,m$ is related to the crack formation in the spherical particle ($q = p$), the spherical envelope ($q = e$) and the cell matrix ($q = m$). Consequently, we get

$$\frac{\partial f_q^{(ij)}}{\partial x_{ij}} = \pm \frac{1}{s_{kkq}\left[K_{ICq}^{(ij)}\right]^2}\sqrt{\left[W_{cq}^{(ij)}\right]^2 - \left\{s_{kkq}\left[K_{ICq}^{(ij)}\right]^2\right\}^2}, \quad s_{kkq} = \frac{1}{E_{kq}}, \quad (8.12)$$

$$W_{cq}^{(ij)} - s_{kkq}\left[K_{ICq}^{(ij)}\right]^2 \geq 0. \quad (8.13)$$

The cracking is sufficien to be investigated within one eighth of the cubic cell (see Fig. 8.2), i.e. for $\varphi, \nu \in \langle 0, \pi/2 \rangle$ (see Fig. 3.1). This is a consequence of symmetry of the multi-particle-matrix and multi-particle-envelope-matrix systems due to the matrix infinit and the periodical distribution of the spherical particles and the spherical envelopes (see Fig. 2.1a).

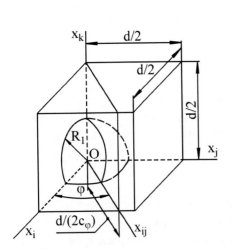 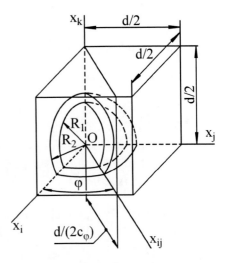

Figure 8.2: One eighth of the cubic cell of with the central spherical particle with the radius R_1, where d is the cubic cell dimension. The cubic cell is related to (a) the multi-particle-matrix system and (b) the multi-particle-envelope-matrix system (see Fig. 2.1a). The general position of the axis x_{ij} of the plane $x_{ij}x_k$ is determined by the angle $\varphi = \angle(x_i, x_{ij}) \in \langle 0, 2\pi \rangle$.

8.2.2.1 Conditions of crack formation

Let the energy density $w_q = w_q(r, \varphi, \nu)$ is induced by the radial; tangential; and shear stresses $\sigma_{rq} = \sigma'_{11q}(r, \varphi, \nu)$; $\sigma_{\varphi q} = \sigma'_{22q}(r, \varphi, \nu)$, $\sigma_{\nu q} = \sigma'_{33q}(r, \varphi, \nu)$; and $\sigma_{r\varphi q} = \sigma'_{12q}(r, \varphi, \nu)$, $\sigma_{r\nu q} = \sigma'_{13q}(r, \varphi, \nu)$, $\sigma_{\varphi\nu q} = \sigma'_{23q}(r, \varphi, \nu)$, respectively, which are determined in the Cartesian system $(Ox_r x_\varphi x_\nu) = (Ox'_1 x'_2 x'_3)$ (see Fig. 3.1) and represent functions of the spherical coordinates (r, φ, ν) in the spherical particle $(q = p)$, the spherical envelope $(q = e)$ and the cell matrix $(q = m)$.

A. Multi-particle-matrix system

In case of the multi-particle-matrix system, we get:

1. If the forces $\vec{F}'_{iip}$ and $\vec{F}'_{iim}$ which are induced by the stresses σ'_{iip} and σ'_{iim} act along the axes $+x'_i$ and $-x'_i$ $(i = 1,2,3)$ (see Fig. 3.1), respectively, then the crack is formed in the spherical particle. If the energy density $w_q = w_q(r, \varphi, \nu, R_1, \nu)$ $(q = p,m)$ is induced by the thermal stresses only, then the crack is formed in the spherical particle (see Fig. 8.2a) provided that $\beta_p > \beta_m$ (see Eqs. (3.100)–(3.102)).

2. If the forces $\vec{F}'_{iip}$ and $\vec{F}'_{iim}$ which are induced by the stresses σ'_{iip} and σ'_{iim} act along the axes $-x'_i$ and $+x'_i$ $(i = 1,2,3)$, respectively, then the crack is formed in the cell matrix. Similarly, if the multi-particle-matrix system is loaded by the thermal stresses only, then the crack is formed in the cell matrix (see Fig. 8.2a) provided that $\beta_p < \beta_m$.

B. Multi-particle-envelope-matrix system

In case of the multi-particle-envelope-matrix system, we get:

1. If the force $\vec{F}'_{iip}$ which is induced by the stress σ'_{iip} acts along the axis $+x'_i$ (see Fig. 3.1), then the crack is formed in the spherical particle. If the multi-particle-envelope-matrix system is loaded by the thermal stresses only, then the crack is formed in the spherical particle on the condition either $\beta_p = \beta_e > \beta_m$ or $\beta_p > \beta_e \gtrless \beta_m$ (see Eqs. (3.100)–(3.102)).

2. If the force $\vec{F}'_{iim}$ which is induced by the stress σ'_{iim} act along the axis $+x'_i$ $(i = 1,2,3)$ (see Fig. 3.1), then the crack is formed in the cell matrix. If the multi-particle-envelope-matrix system is loaded by the thermal stresses only, then the crack is formed in the cell matrix provided that on the condition either $\beta_p < \beta_e = \beta_m$ or $\beta_p \gtrless \beta_e < \beta_m$.

3. In case of the crack formation in the spherical envelope, we get:

   (a) If the forces $\vec{F}'_{iip}$, $\vec{F}'_{iim}$ which are induced by the stresses σ'_{iip}, σ'_{iim} act along the axis $-x'_i$ $(i = 1,2,3)$ (see Fig. 3.1), respectively, then the force $\vec{F}'_{iie}$ which is induced by the stress σ'_{iie} acts along the axis x'_i for $r \in \langle R_1, R_2 \rangle$, $\varphi \in \langle 0, 2\pi \rangle$, $\nu \in \langle 0, \pi \rangle$. The crack is formed in the spherical envelope for $x_{ij} \in \langle R_1, R_2 \rangle$. If the multi-particle-envelope-matrix system is loaded by the thermal stresses only, then the crack is formed in the spherical envelope for $x_{ij} \in \langle R_1, R_2 \rangle$ on the condition either $\beta_p \leq \beta_e > \beta_m$ or $\beta_p < \beta_e \geq \beta_m$, (see Eqs. (3.100)–(3.102)).

(b) Let $r_{0e} = r_{0e}(r, \varphi, \nu, R_1, R_2, v)$ derived by the spherical coordinates (r, φ, ν) (see Fig. 3.1) represents a function in the plane $x_{ij}x_k$. The function $r_{0e} = r_{0e}(r, \varphi, \nu, R_1, R_2, v) \in \langle R_1, R_2 \rangle$ results from the condition $w_e(r, \varphi, \nu, R_1, R_2, v) = 0$, i.e. $[w_e(r, \varphi, \nu, R_1, R_2, v)]_{r=r_{0e}} = 0$. This condition is fulfille for $\left[\sigma'_{ijq}(r, \varphi, \nu)\right]_{r=r_{0e}} = 0$ $(i, j = 1,2,3)$. Consequently we get:

- If $r > r_{0e}$, then the force $\vec{F}'_{iie}$ which is induced by the stress σ'_{iie} acts along the axis either $+x'_i$ or $-x'_i$.
- If $r < r_{0e}$, then the force $\vec{F}'_{iie}$ which is induced by the stress σ'_{iie} acts along the axis either $-x'_i$ or $+x'_i$.

Consequently, we get:

i. Let the forces $\vec{F}'_{iip}$ and $\vec{F}'_{iim}$ which are induced by the stresses σ'_{iip} and σ'_{iim} act along the axes $-x'_i$ and $+x'_i$ $(i = 1,2,3)$, respectively. The force $\vec{F}'_{iie}$ which is induced by the stress σ'_{iie} thus acts along the axes $+x'_i$ and $-x'_i$ for $r \in \langle R_1, r_{0e} \rangle$ and $r \in \langle r_{0e}, R_2 \rangle$, respectively, and for $\varphi \in \langle 0, 2\pi \rangle$, $\nu \in \langle 0, \pi \rangle$. The crack is formed in the spherical envelope for $x_{ij} \in \langle R_1, x_{ij0} \rangle$, where x_{ij0} has the form

$$x_{ij0} = [r_{0e}(r, \varphi, \nu, R_1, R_2, v)]_{\nu=\pi/2} \in \langle R_1, R_2 \rangle. \qquad (8.14)$$

If the multi-particle-envelope-matrix system is loaded by the thermal stresses only, then the crack is formed in the spherical envelope for $x_{ij} \in \langle R_1, x_{ij0} \rangle$ provided that $\beta_p < \beta_e < \beta_m$.

ii. Let the forces $\vec{F}'_{iip}$ and $\vec{F}'_{iim}$ which are induced by the stresses σ'_{iip} and σ'_{iim} act along the axes $+x'_i$ and $-x'_i$ $(i = 1,2,3)$, respectively. The force $\vec{F}'_{iie}$ which is induced by the stress σ'_{iie} thus acts along the axes $-x'_i$ and $+x'_i$ for $r \in \langle R_1, r_{0e} \rangle$ and $r \in \langle r_{0e}, R_2 \rangle$, respectively, and for $\varphi \in \langle 0, 2\pi \rangle$, $\nu \in \langle 0, \pi \rangle$. The crack is formed in the spherical envelope for $x_{ij} \in \langle x_{ij0}, R_2 \rangle$. If the multi-particle-envelope-matrix system is loaded by the thermal stresses only, then the crack is formed in the spherical envelope for $x_{ij} \in \langle x_{ij0}, R_2 \rangle$ provided that $\beta_p > \beta_e > \beta_m$.

C. Conclusions

With regard to Section 8.2.2.1, the crack is formed in the component, or in such part of the component (see Item 3b) of the multi-particle-(envelope)-matrix system, in which the force

$\vec{F'}_{iiq}$ induced by the stress σ'_{iiq} acts along the axis $+x'_i$ ($i = 1,2,3$).

The only exception of this condition, i.e. a crack can be also formed in the component in which the force $\vec{F'}_{iiq}$ induced by the stress σ'_{iiq} acts along the axis $-x'_i$, is analysed in Section 8.2.4. This exception results from a principle that a release of energy of a system considers 'minimal resistance' of the system, i.e. the energy is released through 'minimal resistance' of the system (see Sections 8.2.4.1.B, 8.2.4.2.B).

8.2.2.2 Crack parameters

With regard to Section 8.2.1, the solid continuum with a general shape as shown in Fig. 8.1a is replaced by a cubic cell of the multi-particle-(envelope)-matrix system.

A. Multi-particle-matrix system

A.1 Crack initiation. The crack formation in the spherical particle ($q = p$) or the cell matrix ($q = m$) in the plane $x_i x_j$ of the the multi-particle-matrix system is a consequence of the condition $R_1 = R_{1cq}^{(ij)}$. The critical particle radius $R_{1cq}^{(ij)} = R_{1cq}^{(ij)}(\varphi, v)$ is a reason of an infinitesima crack which is initiated in the spherical particle ($q = p$) or the cell matrix ($q = m$). This infinitesima crack with the length dx_{ij} along the axis x_{ij} is initiated in the position $x_{ij} = x_{ijq}^{max}$ for which the function $W_{cq}^{(ij)} = W_{cq}^{(ij)}(x_{ij}, \varphi, R_1, v)$ of the variable x_{ij} exhibits a maximal value within the interval $x_{ij} \in \langle x_{ijq1}, x_{ijq2} \rangle$.

In case of the spherical particle, we get

$$\langle x_{ijp1}, x_{ijp2} \rangle = \langle 0, R_1 \rangle . \tag{8.15}$$

In case of the cell matrix, we get

$$\langle x_{ijm1}, x_{ijm2} \rangle = \left\langle R_1, \frac{d}{2c_\varphi} \right\rangle = \left\langle R_1, \frac{R_1}{2c_\varphi} \left(\frac{4\pi}{3v} \right)^{1/3} \right\rangle, \quad v \in \left(0, \frac{\pi}{6} \right), \tag{8.16}$$

where the coefficien c_φ is given by Eqs. (2.32), (2.33). The function $d = d(R_1, v)$ is given by Eqs. (2.6), (2.7) for $R_{1ex} = R_1$, $d_{ex} = d$ and $t = 0$.

With regard to Eq. (8.13), the critical particle radius $R_{1cq}^{(ij)} = R_{1cq}^{(ij)}(\varphi, v)$ which is determined with respect to the position $x_{ij} = x_{ijq}^{max}$ represents a root of the condition

$$\left[W_{cq}^{(ij)} \right]_{x_{ij}=x_{ijq}^{max}} - s_{kkq} \left[K_{ICq}^{(ij)} \right]^2 = 0, \quad s_{kkq} = \frac{1}{E_{kq}}. \tag{8.17}$$

A.2 Crack propagation. An analysis of the crack propagation is based on a course of the function $W_{cq}^{(ij)} = W_{cq}^{(ij)}(x_{ij}, \varphi, R_1, v)$ in its definitio interval $x_{ij} \in \langle x_{ijq1}, x_{ijq2} \rangle$. Consequently, the dependence $W_{cq}^{(ij)} - x_{ij}$ results in a course of the function $f_q^{(ij)} = f_q^{(ij)}(x_{ij}, \varphi, R_1, v)$ (see Eq. (8.9)) which is also define for $x_{ij} \in \langle x_{ijq1}, x_{ijq2} \rangle$.

As analysed in Items 1–3 presented below, analyses of the dependence $f_q^{(ij)} - x_{ij}$ and a condition of the crack initiation given by Eq. (8.17) consider

- a decreasing or increasing course of the dependence $W_{cq}^{(ij)} - x_{ij}$ with a maximal value for $x_{ij} = x_{ijq1}^{max} \in \langle x_{ijq1}, x_{ijq2} \rangle$ or $x_{ij} = x_{ijq2}^{max} \in \langle x_{ijq1}, x_{ijq2} \rangle$ and with the critical particle radius $R_{1cq1}^{(ij)}$ or $R_{1cq2}^{(ij)}$, respectively.

- an increasing-decreasing course of the dependence $W_{cq}^{(ij)} - x_{ij}$ with a maximal value for $x_{ij} = x_{ijq3}^{max} \in (x_{ijq1}, x_{ijq2})$ and with the critical particle radius $R_{1cq3}^{(ij)}$.

- a decreasing-increasing course of the dependence $W_{cq}^{(ij)} - x_{ij}$ with a minimal value for $x_{ij} = x_{ijq}^{min} \in (x_{ijq1}, x_{ijq2})$ and with the critical particle radii $R_{1cq1}^{(ij)}$, $R_{1cq2}^{(ij)}$, $R_{1cq4}^{(ij)}$.

The analyses of the dependences $W_{cq}^{(ij)} - x_{ij}$ and $f_q^{(ij)} - x_{ij}$ are as follows.

1. Let $W_{cq}^{(ij)} = W_{cq}^{(ij)}(x_{ij}, \varphi, R_1, v)$ represents a *decreasing* function of $x_{ij} \in \langle x_{ijq1}, x_{ijq2} \rangle$ (see Eqs. (8.15), (8.16)) with a maximal value for $x_{ij} = x_{ijq1}^{max}$, and then we get

$$x_{ijp1}^{max} = 0, \quad x_{ijm1}^{max} = R_1. \tag{8.18}$$

Consequently, $f_{1q}^{(ij)} = f_{1q}^{(ij)}(x_{ij}, \varphi, R_1, v)$ (see Eq. (8.12)) which also represents a *decreasing* function of $x_{ij} \in \langle x_{ijq1}^{max}, x_{0q1} \rangle \subset \langle x_{ijq1}, x_{ijq2} \rangle$ for $R_1 > R_{1cq1}^{(ij)}$ (see Fig. 8.3) is derived as

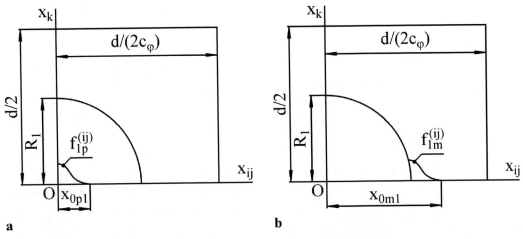

Figure 8.3: A schematic illustration of a shape of cracks in the plane $x_{ij}x_k$ (see Fig. 8.1a,b, 8.2a) which are formed in the plane x_ix_j in (a) the spherical particle ($q=p$) and (b) the cell matrix ($q=m$). The shape in the plane $x_{ij}x_k$ is described by the decreasing function $f_{1q}^{(ij)} = f_{1q}^{(ij)}(x_{ij}, \varphi, R_1, v)$ (see Eqs. (8.19), (8.22)) of the variable $x_{ij} \in \langle x_{ijq1}, x_{0q1} \rangle$ (see Eqs. (8.15), (8.16)) for $R_1 > R_{1cq1}$.

$$f_{1q}^{(ij)} = \frac{1}{s_{kkq}\left[K_{ICq}^{(ij)}\right]^2}\left(C_{1q}^{(ij)} - \int\sqrt{\left[W_{cq}^{(ij)}\right]^2 - \left\{s_{kkq}\left[K_{ICq}^{(ij)}\right]^2\right\}^2}\, dx_{ij}\right),$$
$$x_{ij} \in \langle x_{ijq1}^{max}, x_{0q1}\rangle, \quad R_1 > R_{1cq1}^{(ij)}, \tag{8.19}$$

where the integration constant $C_{1q}^{(ij)}$ is given by Eq. (8.22).

Let $W_{cq}^{(ij)} = W_{cq}^{(ij)}(x_{ij}, \varphi, R_1, v)$ represents an *increasing* function of $x_{ij} \in \langle x_{ijq1}, x_{ijq2}\rangle$ (see Eqs. (8.15), (8.16)) with a maximal value for $x_{ij} = x_{ijq2}^{max}$, and then we get

$$x_{ijp2}^{max} = R_1, \quad x_{ijm2}^{max} = \frac{d}{2c_\varphi}. \tag{8.20}$$

Consequently, $f_{2q}^{(ij)} = f_{2q}^{(ij)}(x_{ij}, \varphi, R_1, v)$ (see Eq. (8.12)) which also represents an *increasing* function of $x_{ij} \in \langle x_{0q2}, x_{ijq2}^{max}\rangle \subset \langle x_{ijq1}, x_{ijq2}\rangle$ for $R_1 > R_{1cq2}^{(ij)}$ (see Fig. 8.4) is derived as

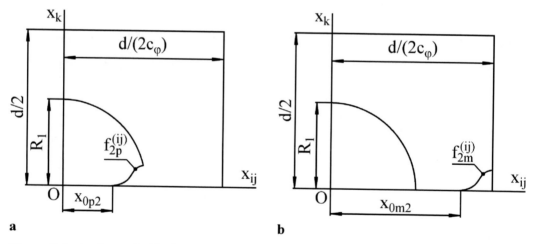

Figure 8.4: A schematic illustration of a shape of cracks in the plane $x_{ij}x_k$ (see Fig. 8.1a,b, 8.2a) which are formed in the plane x_ix_j in (a) the spherical particle ($q=p$) and (b) the cell matrix ($q=m$). The shape in the plane $x_{ij}x_k$ is described by the increasing function $f_{2q}^{(ij)} = f_{2q}^{(ij)}(x_{ij}, \varphi, R_1, v)$ (see Eqs. (8.21), (8.22)) of the variable $x_{ij} \in \langle x_{0q2}, x_{ijq2}\rangle$ (see Eqs. (8.15), (8.16)) for $R_1 > R_{1cq2}$.

$$f_{2q}^{(ij)} = \frac{1}{s_{kkq}\left[K_{ICq}^{(ij)}\right]^2}\left(\int\sqrt{\left[W_{cq}^{(ij)}\right]^2 - \left\{s_{kkq}\left[K_{ICq}^{(ij)}\right]^2\right\}^2}\, dx_{ij} - C_{2q}^{(ij)}\right),$$

$$x_{ij} \in \langle x_{0q2}, x_{ijq2}^{max} \rangle, \quad R_1 > R_{1cq2}^{(ij)}. \tag{8.21}$$

The integration constant $C_{nq}^{(ij)}$ $(n = 1,2)$

$$C_{nq}^{(ij)} = \left[\int \sqrt{\left[W_{cq}^{(ij)} \right]^2 - \left\{ s_{kkq} \left[K_{ICq}^{(ij)} \right]^2 \right\}^2} \, dx_{ij} \right]_{x_{ij}=x_{0qn}} \tag{8.22}$$

is determined by the boundary condition

$$\left[f_{nq}^{(ij)} (x_{ij}, \varphi, R_1, v) \right]_{x_{ij}=x_{0qn}} = 0, \quad R_1 > R_{1cqn}^{(ij)}, \quad n = 1, 2. \tag{8.23}$$

Consequently, the position $x_{0qn} = x_{0qn} (\varphi, R_1, v)$ $(\equiv x_0; n = 1,2)$ (see Fig. 8.1b) of the crack tip for $R_1 > R_{1cqn}^{(ij)}$ represents a root of the condition

$$W_{cq}^{(ij)} - s_{kkq} \left[K_{ICq}^{(ij)} \right]^2 = 0, \quad R_1 > R_{1cqn}^{(ij)}, \tag{8.24}$$

where $x_{0pn} \in \langle 0, R_1 \rangle$ or $x_{0mn} \in \langle R_1, d/ (2c_\varphi) \rangle$ (see Eqs. (2.32), (2.33)) for the crack formation in the plane $x_i x_j$ in the spherical particle $(q = p)$ or the cell matrix $(q = m)$ (see Fig. 8.3), respectively.

2. Let $W_{cq}^{(ij)} = W_{cq}^{(ij)} (x_{ij}, \varphi, R_1, v)$ represents an increasing-decreasing function of $x_{ij} \in \langle x_{ijq1}, x_{ijq2} \rangle$ (see Eqs. (8.15), (8.16)) with a maximal value for $x_{ij} = x_{ijq3}^{max}$, where the function $x_{ijq3}^{max} = x_{ijq3}^{max} (\varphi, R_1, v)$ can be numerically and/or computationally determined. Consequently, we get

$$x_{ijp3}^{max} \in (0, R_1), \quad x_{ijm3}^{max} \in \left(R_1, \frac{d}{2c_\varphi} \right). \tag{8.25}$$

The critical particle radius $R_{1cq3}^{(ij)} = R_{1cq3}^{(ij)} (\varphi, v)$ represents a root of the condition given by Eq. (8.17) for $x_{ij} = x_{ijq3}^{max}$. If $R_1 = R_{1cq3}^{(ij)}$, then two cracks are initiated in the position $x_{ij} = x_{ijq3}^{max}$. These two cracks are given by Eqs. (8.19), (8.22) and (8.21), (8.22) for $x_{ij} \in \langle x_{ijq3}^{max}, x_{0q1} \rangle \subset \langle x_{ijq1}, x_{ijq2} \rangle$ and $x_{ij} \in \langle x_{0q2}, x_{ijq3}^{max} \rangle \subset \langle x_{ijq1}, x_{ijq2} \rangle$ (see Fig. 8.5), respectively, and both for $R_1 \geq R_{1cq3}^{(ij)}$ (see Eqs. (8.19), (8.21)).

3. Let $W_{cq}^{(ij)} = W_{cq}^{(ij)} (x_{ij}, \varphi, R_1, v)$ represents a decreasing-increasing function of $x_{ij} \in \langle x_{ijq1}, x_{ijq2} \rangle$ (see Eqs. (8.15), (8.16)) with a minimal value for $x_{ij} = x_{ijq}^{min}$, where the function $x_{ijq}^{min} = x_{ijq}^{min} (\varphi, R_1, v)$ can be numerically and/or computationally determined. Consequently, we get

$$x_{ijp}^{min} \in (0, R_1), \quad x_{ijm}^{min} \in \left(R_1, \frac{d}{2c_\varphi} \right). \tag{8.26}$$

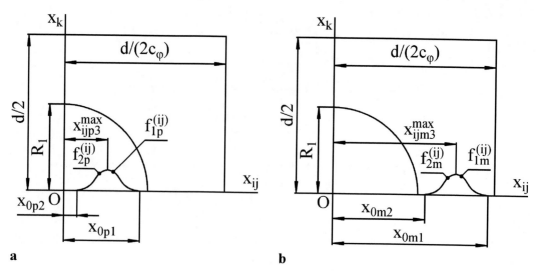

Figure 8.5: A schematic illustration of a shape of cracks in the plane $x_{ij}x_k$ (see Fig. 8.1a,b, 8.2a) which are formed in the plane $x_i x_j$ in (a) the spherical particle ($q = p$) and (b) the cell matrix ($q = m$). The shape in the plane $x_{ij}x_k$ is described by the increasing and decreasing functions $f_{2q}^{(ij)} = f_{2q}^{(ij)}(x_{ij}, \varphi, R_1, v)$ and $f_{1q}^{(ij)} = f_{1q}^{(ij)}(x_{ij}, \varphi, R_1, v)$ (see Eqs. (8.21), (8.22)) of the variable $x_{ij} \in \left\langle x_{0q2}, x_{ijq3}^{max} \right\rangle$ and $x_{ij} \in \left\langle x_{ijq3}^{max}, x_{0q1} \right\rangle$ (see Eqs. (8.15), (8.16)), respectively, and both for $R_1 > R_{1cq3}$.

The critical particle radius $R_{1cq4}^{(ij)} = R_{1cq4}^{(ij)}(\varphi, v)$ represents a root of the condition given by Eq. (8.17) for $x_{ij} = x_{ijq}^{min}(\varphi, R_1, v)$, where $R_{1cq4}^{(ij)} > R_{1cq1}^{(ij)}$ and $R_{1cq4}^{(ij)} > R_{1cq2}^{(ij)}$. Consequently, we get:

(a) If $R_1 \in \left(R_{1cq1}^{(ij)}, R_{1cq4}^{(ij)} \right)$, then the crack propagates from the position $x_{ij} = x_{ijq1}^{max}$ (see Eq. (8.18)) to the position $x_{ij} \in \left(x_{ijq1}^{max}, x_{0q1} \right\rangle \subset \langle x_{ijq1}, x_{ijq2} \rangle$ (see Fig. 8.6) and is described by Eqs. (8.19), (8.22) for $n = 1$.

(b) If $R_1 \in \left(R_{1cq2}^{(ij)}, R_{1cq4}^{(ij)} \right)$, then the crack propagates from the position $x_{ij} = x_{ijq2}^{max}$ (see Eq. (8.20)) to the position $x_{ij} \in \left\langle x_{0q2}, x_{ijq2}^{max} \right) \subset \langle x_{ijq1}, x_{ijq2} \rangle$ (see Fig. 8.6) and is described by Eqs. (8.21), (8.22) for $n = 2$.

(c) If $R_1 \geq R_{1cq4}^{(ij)}$, then these two cracks connected in the position $x_{ij} = x_{ijq}^{min}$ are described by Eqs. (8.19) and (8.21) for $x_{ij} \in \left\langle x_{ijq1}^{max}, x_{ijq}^{min} \right\rangle \subset \langle x_{ijq1}, x_{ijq2} \rangle$ and $x_{ij} \in \left\langle x_{ijq}^{min}, x_{ijq2}^{max} \right\rangle \subset \langle x_{ijq1}, x_{ijq2} \rangle$, respectively, and both for $R_1 \geq R_{1cq4}^{(ij)}$, where $R_{1cq4}^{(ij)} > R_{1cq1}^{(ij)}$ (see Fig. 8.7), $R_{1cq4}^{(ij)} > R_{1cq2}^{(ij)}$. With regard to $\left[f_{1q}^{(ij)} \right]_{x_{ij}=x_{ijq}^{min}} = \left[f_{2q}^{(ij)} \right]_{x_{ij}=x_{ijq}^{min}} \geq 0$ for $R_1 \geq R_{1cq4}^{(ij)}$, the integration constant $C_{nq}^{(ij)}$ ($n = 1,2$) in Eqs. (8.19), (8.21) derived as

$$C_{nq}^{(ij)} = \left[\int \sqrt{\left[W_{cq}^{(ij)}\right]^2 - \left\{s_{kkq}\left[K_{ICq}^{(ij)}\right]^2\right\}^2} \, dx_{ij} \right]_{x_{ij}=x_{ijq}^{min}; R_1=R_{1cq4}^{(ij)}} \quad (8.27)$$

is thus determined by the boundary condition

$$\left[f_{1q}^{(ij)}\right]_{x_{ij}=x_{ijq}^{min}} = \left[f_{2q}^{(ij)}\right]_{x_{ij}=x_{ijq}^{min}} = 0, \quad R_1 = R_{1cq4}^{(ij)}. \quad (8.28)$$

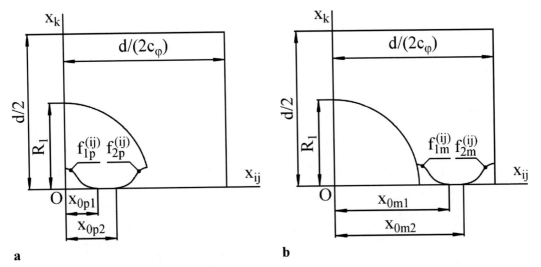

Figure 8.6: A schematic illustration of a shape of cracks in the plane $x_{ij}x_k$ (see Fig. 8.1a,b, 8.2a) which are formed in the plane x_ix_j in (a) the spherical particle ($q = p$) and (b) the cell matrix ($q = m$). The shape in the plane $x_{ij}x_k$ is described by the decreasing and increasing functions $f_{1q}^{(ij)} = f_{1q}^{(ij)}(x_{ij}, \varphi, R_1, v)$ and $f_{2q}^{(ij)} = f_{2q}^{(ij)}(x_{ij}, \varphi, R_1, v)$ (see Eqs. (8.21), (8.22)) of the variable $x_{ij} \in \langle x_{ijq1}, x_{0q1}\rangle$ and $x_{ij} \in \langle x_{0q2}, x_{ijq2}\rangle$ (see Eqs. (8.15), (8.16)) for $R_1 \in \left(R_{1cq1}^{(ij)}, R_{1cq4}^{(ij)}\right)$ and $R_1 \in \left(R_{1cq2}^{(ij)}, R_{1cq4}^{(ij)}\right)$, respectively.

A.3 Limit state. Let $R_{1cq\zeta,min}^{(ij)}$, $R_{1cq\eta,min}^{(jk)}$, $R_{1cq\theta,min}^{(ki)}$ ($\zeta, \eta, \theta = 1,2,3$) (see Items 1, 2, Section 8.2.1) represent minimal values of the functions $R_{1cq\zeta}^{(ij)} = R_{1cq\zeta}^{(ij)}(\varphi, v)$, $R_{1cq\eta}^{(jk)} = R_{1cq\eta}^{(jk)}(\varphi, v)$, $R_{1cq\theta}^{(ki)} = R_{1cq\theta}^{(ki)}(\varphi, v)$ of the variable $\varphi \in \langle 0, 2\pi\rangle$, and then $R_{1cq\zeta,min}^{(ij)} = R_{1cq\zeta,min}^{(ij)}(v)$, $R_{1cq\eta,min}^{(jk)} = R_{1cq\eta,min}^{(jk)}(v)$, $R_{1cq\theta,min}^{(ki)} = R_{1cq\theta,min}^{(ki)}(v)$ defin a *limit state*, i.e. a state of *the crack initiation* in the planes x_ix_j, x_jx_k, x_kx_i, respectively. The minimal values $R_{1cq\zeta,min}^{(ij)}$, $R_{1cq\eta,min}^{(jk)}$, $R_{1cq\theta,min}^{(ki)}$ ($\zeta, \eta, \theta = 1,2,3$) of the functions $R_{1cq\zeta}^{(ij)} = R_{1cq\zeta}^{(ij)}(\varphi, v)$, $R_{1cq\eta}^{(jk)} = R_{1cq\eta}^{(jk)}(\varphi, v)$, $R_{1cq\theta}^{(ki)} = R_{1cq\theta}^{(ki)}(\varphi, v)$ are considered regarding the variable $\varphi \in \langle 0, 2\pi\rangle$.

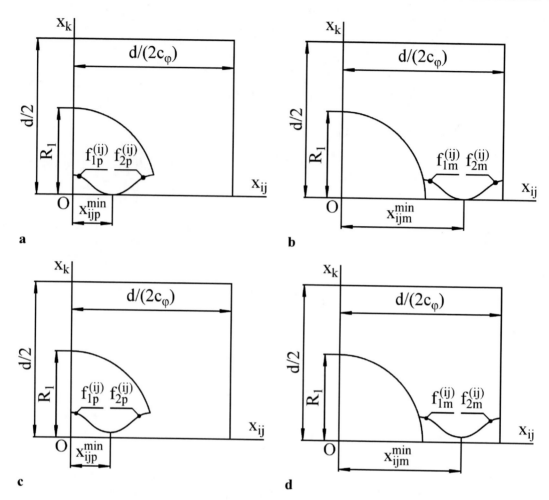

Figure 8.7: A schematic illustration of a shape of cracks in the plane $x_{ij}x_k$ (see Fig. 8.1a,b, 8.2a) which are formed in the plane x_ix_j in (a,c) the spherical particle ($q = p$) and (b,d) the cell matrix ($q = m$). The shape in the plane $x_{ij}x_k$ is described by the decreasing and increasing functions $f_{1q}^{(ij)} = f_{1q}^{(ij)}(x_{ij}, \varphi, R_1, v)$ and $f_{2q}^{(ij)} = f_{2q}^{(ij)}(x_{ij}, \varphi, R_1, v)$ (see Eqs. (8.21), (8.22)) of the variable $x_{ij} \in \langle x_{ijq1}, x_{0q1}\rangle$ and $x_{ij} \in \langle x_{0q2}, x_{ijq2}\rangle$ (see Eqs. (8.15), (8.16)), respectively, and for (a,b) $R_1 = R_{1cq4}^{(ij)}$ and (c,d) $R_1 > R_{1cq4}^{(ij)}$, where $R_{1cq4}^{(ij)} > R_{1cq1}^{(ij)}$, $R_{1cq4}^{(ij)} > R_{1cq2}^{(ij)}$.

Let $R_{1cq\chi, min}^{(ab)}$ ($\chi = 1,2,3$) represents a minimum of the set $\{R_{1cq\zeta, min}^{(ij)}, R_{1cq\eta, min}^{(jk)}, R_{1cq\theta, min}^{(ki)}\}$, then the crack is initiated in the plane $x_a x_b \subset \{x_i x_j, x_j x_k, x_k x_i\}$, and $R_{1cq\chi, min}^{(ab)}$ define a state of *the crack initiation* in the component which is related to the subscript q ($q = p,m$). With regard to the minimal value $R_{1cq\chi, min}^{(ab)}$, the particle volume fraction $v \in (0, \pi/6\rangle$ is considered to be a parameter.

Related phenomena

A.4 Cracking of component. With regard to Item 1 in Section 8.2.2.2.A.2, let $W_{cq}^{(ij)} = W_{cq}^{(ij)}(x_{ij}, \varphi, R_1, v)$ represents a *decreasing* function of $x_{ij} \in \langle x_{ijq1}, x_{ijq2} \rangle$. Let the critical particle radius $R_{1cq2}^{(ij)}$ is determined by the condition given by Eq. (8.17) for $x_{ij} = x_{0q1} = x_{ijq2}$. Let $R_{1cq2, max}^{(ij)} = R_{1cq2, max}^{(ij)}(v)$ represents a maximal value of the function $R_{1cq2}^{(ij)} = R_{1cq2}^{(ij)}(\varphi, v)$ of the variable $\varphi \in \langle 0, 2\pi \rangle$. Consequently, if $R_1 \geq R_{1cq2, max}^{(ij)}$, then a whole surface in the plane $x_i x_j$ of a component related to the subscript q is cracked for $x_{ij} \in \langle x_{ijq1}, x_{ijq2} \rangle$ and $\varphi \in \langle 0, 2\pi \rangle$.

Similarly, with regard to Item 1 in Section 8.2.2.2.A.2, let $W_{cq}^{(ij)} = W_{cq}^{(ij)}(x_{ij}, \varphi, R_1, v)$ represents an *increasing* function of $x_{ij} \in \langle x_{ijq1}, x_{ijq2} \rangle$. Let the critical particle radius $R_1 = R_{1cq1}^{(ij)}$ is determined by the condition given by Eq. (8.17) for $x_{ij} = x_{0q1} = x_{ijq1}$. Let $R_{1cq1, max}^{(ij)} = R_{1cq1, max}^{(ij)}(v)$ represents a maximal value of the function $R_{1cq1}^{(ij)} = R_{1cq1}^{(ij)}(\varphi, v)$ of the variable $\varphi \in \langle 0, 2\pi \rangle$. Consequently, if $R_1 \geq R_{1cq1, max}^{(ij)}$, then a whole surface in the plane $x_i x_j$ of a component related to the subscript q is cracked for $x_{ij} \in \langle x_{ijq1}, x_{ijq2} \rangle$ and $\varphi \in \langle 0, 2\pi \rangle$.

With regard to Item 2 in Section 8.2.2.2.A.2, let $W_{cq}^{(ij)} = W_{cq}^{(ij)}(x_{ij}, \varphi, R_1, v)$ represents an increasing-decreasing function of $x_{ij} \in \langle x_{ijq1}, x_{ijq2} \rangle$. Let the critical particle radii $R_{1cq1}^{(ij)}$ and $R_{1cq2}^{(ij)}$ are determined by the condition given by Eq. (8.17) for $x_{ij} = x_{0q1} = x_{ijq2}$ and $x_{ij} = x_{0q1} = x_{ijq1}$, respectively, where $R_{1cq1}^{(ij)} > R_{1cq3}^{(ij)}$ and $R_{1cq2}^{(ij)} > R_{1cq3}^{(ij)}$ (see Item 2, Section 8.2.2.2.A.2). Let $R_{1cqn, max}^{(ij)} = R_{1cqn, max}^{(ij)}(v)$ $(n = 1,2)$ represents a maximal value of the function $R_{1cqn}^{(ij)} = R_{1cqn}^{(ij)}(\varphi, v)$ of the variable $\varphi \in \langle 0, 2\pi \rangle$. If $R_1 \geq R_{1cq, max}^{(ij)}$, then a whole surface in the plane $x_i x_j$ of a component related to the subscript q is cracked for $x_{ij} \in \langle x_{ijq1}, x_{ijq2} \rangle$ and $\varphi \in \langle 0, 2\pi \rangle$, where $R_{1cq, max}^{(ij)}$ is a maximal value of the set $\left\{ R_{1cq1, max}^{(ij)}, R_{1cq2, max}^{(ij)} \right\}$. With regard to the maximal value $R_{1cq, max}^{(ij)}$, the particle volume fraction $v \in (0, \pi/6\rangle$ is considered to be a parameter.

With regard to Item 3 in Section 8.2.2.2.A.2, let $W_{cq}^{(ij)} = W_{cq}^{(ij)}(x_{ij}, \varphi, R_1, v)$ represents an decreasing-increasing function of $x_{ij} \in \langle x_{ijq1}, x_{ijq2} \rangle$. If $R_1 \geq R_{1cq4, max}^{(ij)}$, then a whole surface in the plane $x_i x_j$ of a component related to the subscript q is cracked for $x_{ij} \in \langle x_{ijq1}, x_{ijq2} \rangle$ and $\varphi \in \langle 0, 2\pi \rangle$, where $R_{1cq4, max}^{(ij)} = R_{1cq4, max}^{(ij)}(v)$ is a maximal value of the function $R_{1cq4}^{(ij)} = R_{1cq4}^{(ij)}(\varphi, v)$ of the variable $\varphi \in \langle 0, 2\pi \rangle$. The function $R_{1cq4}^{(ij)} = R_{1cq4}^{(ij)}(\varphi, v)$ is analysed in Item 3 in Section 8.2.2.2.A.2.

Finally, the maximal value $R_{1cqn, max}^{(ij)} = R_{1cqn, max}^{(ij)}(v)$ $(n = 1,2,4)$ of the function $R_{1cqn}^{(ij)} = R_{1cqn}^{(ij)}(\varphi, v)$ is considered regarding the variable $\varphi \in \langle 0, 2\pi \rangle$.

B. Multi-particle-envelope-matrix system

The analyses in Sections 8.2.2.2.A.1–8.2.2.2.A.4 concerning the crack initiation, the crack propagation, the limit state and the cracking of a component are also valid for the multi-particle-envelope-matrix system with the radii $R_1 < R_2$ and the thickness $t = R_2 - R_1$ (see Eq. (2.5)) of the spherical envelope.

In case of the cell matrix of the multi-particle-envelope-matrix system and with regard to Eqs. (8.25), (8.26), we get

$$\langle x_{ijm1}, x_{ijm2} \rangle = \left\langle R_2, \frac{d}{2c_\varphi} \right\rangle$$

$$= \left\langle R_2, \frac{R_1}{2c_\varphi} \left(\frac{4\pi}{3v} \right)^{1/3} \right\rangle, \quad v \in \left(0, \frac{\pi}{6} \left(1 - \frac{t}{R_2} \right)^3 \right), \tag{8.29}$$

$$x_{ijm3}^{max} \in \left(R_2, \frac{d}{2c_\varphi} \right), \quad x_{ijm}^{min} \in \left(R_2, \frac{d}{2c_\varphi} \right). \tag{8.30}$$

Additionally, the analyses in Sections 8.2.2.2.A.3, 8.2.2.2.A.4 concerning the critical radii $R_{1cms,min}^{(ij)} = R_{1cms,min}^{(ij)}(v)$, $R_{1cmt,min}^{(jk)} = R_{1cmt,min}^{(jk)}(v)$, $R_{1cmu,min}^{(ki)} = R_{1cmu,min}^{(ki)}(v)$, $R_{1cmz,min}^{(ab)} = R_{1cmz,min}^{(ab)}(v)$, $R_{1cm1,max}^{(ij)} = R_{1cm1,max}^{(ij)}(v)$, $R_{1cm2,max}^{(ij)} = R_{1cm2,max}^{(ij)}(v)$, $R_{1cm,max}^{(ij)} = R_{1cm,max}^{(ij)}(v)$, $R_{1cm4,max}^{(ij)} = R_{1cm4,max}^{(ij)}(v)$ $(\zeta, \eta, \theta, \chi = 1,2,3; ab = ij, jk, ki; i, j, k = 1,2,3, i \neq j \neq k)$ and the functions $R_{1cms}^{(ij)} = R_{1cms}^{(ij)}(\varphi, v)$, $R_{1cmt}^{(jk)} = R_{1cmt}^{(jk)}(\varphi, v)$, $R_{1cmu}^{(ki)} = R_{1cmu}^{(ki)}(\varphi, v)$, $R_{1cmn}^{(ij)} = R_{1cmn}^{(ij)}(\varphi, v)$ $(n = 1, \ldots, 4)$ of the variable $\varphi \in \langle 0, 2\pi \rangle$ are considered for $R_{2cms,min}^{(ij)} = R_{2cms,min}^{(ij)}(t, v)$, $R_{2cmt,min}^{(jk)} = R_{2cmt,min}^{(jk)}(t, v)$, $R_{2cmu,min}^{(ki)} = R_{2cmu,min}^{(ki)}(t, v)$, $R_{2cmz,min}^{(ab)} = R_{2cmz,min}^{(ab)}(t, v)$, $R_{2cm1,max}^{(ij)} = R_{2cm1,max}^{(ij)}(t, v)$, $R_{2cm2,max}^{(ij)} = R_{2cm2,max}^{(ij)}(t, v)$, $R_{2cm,max}^{(ij)} = R_{2cm,max}^{(ij)}(t, v)$, $R_{2cm4,max}^{(ij)} = R_{2cm4,max}^{(ij)}(t, v)$ and $R_{2cms}^{(ij)} = R_{2cms}^{(ij)}(\varphi, t, v)$, $R_{2cmt}^{(jk)} = R_{2cmt}^{(jk)}(\varphi, t, v)$, $R_{2cmu}^{(ki)} = R_{2cmu}^{(ki)}(\varphi, t, v)$, $R_{2cmn}^{(ij)} = R_{2cmn}^{(ij)}(\varphi, t, v)$, respectively, due to the 'transformation' $R_1 \to R_2$ regarding different intervals of $x_{ij} \in \langle x_{ijm1}, x_{ijm2} \rangle$ for the multi-particle-matrix and multi-particle-envelope-matrix systems (see Eqs. (8.16), (8.29)). Additionally, these dependences on the envelope thickness $t = R_2 - R_1 \in (0, (d/2) - R_1 \rangle$ result from $W_{cq}^{(ij)} = W_{cq}^{(ij)}(x_{ij}, \varphi, R_1, R_2, v) = W_{cq}^{(ij)}(x_{ij}, \varphi, R_1, t, v) = W_{cq}^{(ij)}(x_{ij}, \varphi, R_2, t, v)$ $(q = p, e, m)$.

Accordingly, in case of the spherical particle of the multi-particle-envelope-matrix system, we get $R_{1cp\ldots}^{(ij)}(v) \to R_{1cp\ldots}^{(ij)}(t, v)$ and $R_{1cp\ldots}^{(ij)}(\varphi, v) \to R_{1cp\ldots}^{(ij)}(\varphi, t, v)$.

With regard to the spherical envelope $(q = e)$, we get:

1. Let the conditions analysed in Item 3a in Section 8.2.2.1.B are valid. Consequently, the analyses in Items 1–3 in Section 8.2.2.2.A.2 are also considered for the spherical envelope $(q = e)$. Accordingly, the dependence $W_{ce}^{(ij)} - x_{ij}$ for $x_{ij} \in \langle x_{ije1}, x_{ije2} \rangle$ exhibits one of the decreasing, increasing, increasing-decreasing or decreasing-increasing courses analysed in Items 1–3 in Section 8.2.2.2.A.2, where

$$\langle x_{ije1}, x_{ije2} \rangle = \langle R_1, R_2 \rangle. \tag{8.31}$$

Related phenomena

(a) Let $W_{ce}^{(ij)} = W_{ce}^{(ij)}(x_{ij}, \varphi, R_1, t, v)$ represents a *decreasing* function of $x_{ij} \in \langle x_{ije1}, x_{ije2} \rangle$ (see Eq. (8.31)) with a maximal value for $x_{ij} = x_{ije1}^{max}$, and then

$$x_{ije1}^{max} = R_1. \tag{8.32}$$

Consequently, with regard to the analyses in Sections 8.2.2.2.A.3, 8.2.2.2.A.4, we get $R_{1cq1}^{(ij)}(v) \to R_{1ce}^{(ij)}(t, v)$ (see Eq. (8.19)); $R_{1cq\zeta, min}^{(ij)}(v) \to R_{1ce, min}^{(ij)}(t, v)$, $R_{1cq\eta, min}^{(jk)}(v) \to R_{1ce, min}^{(jk)}(t, v)$, $R_{1cq\theta, min}^{(ki)}(v) \to R_{1ce, min}^{(ki)}(t, v)$, $R_{1cq\chi, min}^{(ab)}(v) \to R_{1ce, min}^{(ab)}(t, v)$ ($\zeta, \eta, \theta, \chi = 1,2,3$; $ab = ij, jk, ki$; $i, j, k = 1, 2,3$, $i \neq j \neq k$) (see Section 8.2.2.2.A.3); $R_{1cq2, max}^{(ij)} \to R_{1ce2, max}^{(ij)}$ (see Section 8.2.2.2.A.4).

Let $W_{ce}^{(ij)} = W_{ce}^{(ij)}(x_{ij}, \varphi, R_2, t, v)$ represents an *increasing* function of $x_{ij} \in \langle x_{ije1}, x_{ije2} \rangle$ (see Eq. (8.31)) with a maximal value for $x_{ij} = x_{ije2}^{max}$, and then

$$x_{ije2}^{max} = R_2. \tag{8.33}$$

Consequently, with regard to the analyses in Sections 8.2.2.2.A.3, 8.2.2.2.A.4, we get $R_{1cq2}^{(ij)}(v) \to R_{2ce}^{(ij)}(t, v)$ (see Eq. (8.21)); $R_{1cq\zeta, min}^{(ij)}(v) \to R_{2ce, min}^{(ij)}(t, v)$, $R_{1cq\eta, min}^{(jk)}(v) \to R_{2ce, min}^{(jk)}(t, v)$, $R_{1cq\theta, min}^{(ki)}(v) \to R_{2ce, min}^{(ki)}(t, v)$, $R_{1cq\chi, min}^{(ab)}(v) \to R_{2ce, min}^{(ab)}(t, v)$ (see Section 8.2.2.2.A.3); $R_{1cq2, max}^{(ij)}(v) \to R_{2ce1, max}^{(ij)}(t, v)$ (see Section 8.2.2.2.A.4).

(b) Let $W_{ce}^{(ij)} = W_{ce}^{(ij)}(x_{ij}, \varphi, R_1, t, v)$ represents an increasing-decreasing function of $x_{ij} \in \langle x_{ije1}, x_{ije2} \rangle$ (see Eq. (8.31)) with a maximal value for $x_{ij} = x_{ije3}^{max}$, where the function $x_{ije3}^{max} = x_{ije3}^{max}(\varphi, R_1, t, v)$ can be numerically and/or computationally determined, and

$$x_{ije3}^{max} \in (R_1, R_2). \tag{8.34}$$

Consequently, with regard to Item 2, Section 8.2.2.2.A.2, we get $R_{1ce1}^{(ij)}(\varphi, v) \to R_{1ce}^{(ij)}(\varphi, t, v)$, $R_{1ce2}^{(ij)}(\varphi, v) \to R_{2ce}^{(ij)}(\varphi, t, v)$, $R_{1ce3}^{(ij)}(\varphi, v) \to R_{1ce3}^{(ij)}(\varphi, t, v)$, $R_{1ce1, max}^{(ij)}(v) \to R_{1ce, max}^{(ij)}(t, v)$, $R_{1ce2, max}^{(ij)}(v) \to R_{2ce, max}^{(ij)}(t, v)$, $R_{1ce, max}^{(ij)}(v) \to R_{1ce, max}^{(ij)}(t, v)$.

(c) Let $W_{ce}^{(ij)} = W_{ce}^{(ij)}(x_{ij}, \varphi, R_1, t, v)$ represents a decreasing-increasing function of $x_{ij} \in \langle x_{ije1}, x_{ije2} \rangle$ (see Eq. (8.31)) with a minimal value for $x_{ij} = x_{ije}^{min}$, where the function $x_{ije}^{min} = x_{ije}^{min}(\varphi, R_1, t, v)$ can be numerically and/or computationally determined, and

$$x_{ije}^{min} \in (R_1, R_2). \tag{8.35}$$

Consequently, with regard to Item 2, Section 8.2.2.2.A.2, we get $R_{1cq4}^{(ij)}(\varphi, v) \to R_{1cq4}^{(ij)}(\varphi, t, v)$, $R_{1cq4, max}^{(ij)}(v) \to R_{1cq4, max}^{(ij)}(t, v)$.

2. Let the conditions analysed in Item 3(b)i in Section 8.2.2.1.B are valid. Consequently, $W_{ce}^{(ij)} = W_{ce}^{(ij)}(x_{ij}, \varphi, R_1, R_2, v)$ representing a decreasing function of $x_{ij} \in \langle x_{ije1}, x_{ije2} \rangle$ is define for

$$\langle x_{ije1}, x_{ije2} \rangle = \langle R_1, x_{ij0} \rangle \subset \langle R_1, R_2 \rangle, \tag{8.36}$$

where x_{ij0} is given by Eq. (8.14).

Consequently, the analysis in Item 1 in Section 8.2.2.2.A.2 (see Eqs. (8.19), (8.22)–(8.24)) is considered for $R_{1cqn}^{(ij)} \to R_{1cen}^{(ij)}$ ($n = 1,2$) and x_{ije1}^{max} (see Eq. (8.17)) given by Eq. (8.32).

3. Let the conditions analysed in Item 3(b)ii in Section 8.2.2.1.B are valid. Consequently, the function $W_{ce}^{(ij)} = W_{ce}^{(ij)}(x_{ij}, \varphi, R_1, R_2, v)$ of $x_{ij} \in \langle x_{ije1}, x_{ije2} \rangle$ exhibiting one of the decreasing, increasing, increasing-decreasing or decreasing-increasing courses analysed in Items 1–3 in Section 8.2.2.2.A.2 is define for

$$\langle x_{ije1}, x_{ije2} \rangle = \langle r_{0e}, R_2 \rangle \subset \langle R_1, R_2 \rangle. \tag{8.37}$$

Consequently, the analysis in Item 1 in Section 8.2.2.2.A.2 (see Eqs. (8.21)–(8.24)) is considered for $R_{1cqn}^{(ij)} \to R_{2cen}^{(ij)}$ ($n = 1,2$) and x_{ije2}^{max} (see Eq. (8.17)) given by Eq. (8.33).

8.2.2.3 Modificatio of stress-deformation field

The stress-deformation field

- which induce the energy density $w^{(ij)} = w^{(ij)}(x_{ij}, \varphi, x_k, R_1, v)$,

- which act in the cell for $R_1 \leq R_{1cq}^{(ij)}$ ($\equiv R_{1cqn}^{(ij)}; n = 1,2$),

- and which are a reason of the crack initiation for $R_1 = R_{1cq}^{(ij)}$ ($\equiv R_{1cqn}^{(ij)}; n = 1,2$),

are modifie (changed) due to the crack propagation in a component of the multi-particle-matrix system for $R_1 > R_{1cq}^{(ij)}$. Consequently, a modificatio of the stress-deformationfield during the crack propagation results in a modificatio of $w^{(ij)} = w^{(ij)}(x_{ij}, \varphi, x_k, R_1, v)$ and $W_c^{(ij)} = W_c^{(ij)}(x_{ij}, \varphi, R_1, v)$ for $R_1 > R_{1cq}^{(ij)}$.

If the modificatio of $W_c^{(ij)} = W_c^{(ij)}(x_{ij}, \varphi, R_1, v)$ for $R_1 > R_{1cq}^{(ij)}$ can not be considered, i.e. if $W_c^{(ij)} = W_c^{(ij)}(x_{ij}, \varphi, R_1, v)$ can not be analytically and/or computationally determined during the crack propagation for $R_1 > R_{1cq}^{(ij)}$, then a determination of the crack parameters $f_q^{(ij)} = f_q^{(ij)}(x_{ij}, \varphi, R_1, v)$ and $x_{0q} = x_{0q}(\varphi, R_1, v)$ is valid for a two-component material with ceramic components. In general, a ceramic component exhibits a high-speed crack propagation during which the modificatio (change) of $W_c^{(ij)} = W_c^{(ij)}(x_{ij}, \varphi, R_1, v)$ can be assumed to be neglected.

8.2.3 Crack formation I

As mentioned at the beginning of Section 8.2.2.2, the solid continuum with a general shape shown in Fig. 8.1a is replaced by the cubic cell of the multi-particle-(envelope)-matrix system.

Section 8.2.3 deals with a determination of $W_{cq}^{(ij)}$ (see Eq. (8.3)) regarding the crack formation in the spherical particle ($q = p$), the spherical envelope ($q = e$) and the cell matrix ($q = m$). The curve integral $W_{cq}^{(ij)}$ of the energy density $w = w\,(x_1, x_2, x_3, R_1, v)$ and $w = w\,(x_1, x_2, x_3, R_1, R_2, v)$ for the multi-particle-matrix and multi-particle-envelope-matrix systems is determined in the cubic cell on the condition $w\,(x_1, x_2, x_3, R_1, v) \neq 0$ and $w\,(x_1, x_2, x_3, R_1, R_2, v) \neq 0$ both for $x_i \in \langle 0, d/2 \rangle$ ($i = 1,2,3$) (see Sections 8.2.3.1.A, 8.2.3.2.A), respectively.

With regard to this condition, the crack is formed only in one of the planes $x_i x_j$, $x_j x_k$, $x_k x_i$. Additionally, the analysis in Section 8.2.2.2.A.3 is considered regarding the crack formation only in one of the planes $x_i x_j$, $x_j x_k$, $x_k x_i$.

Considering the spherical coordinates (r, φ, ν) (see Fig. 3.1), if the energy density $w = w\,(r, \varphi, \nu, R_1, v)$ and $w = w\,(r, \varphi, \nu, R_1, R_2, v)$ is induced by thermal stresses only, then the determination of $W_{cq1}^{(ij)} = W_{cq1}^{(ij)}\,(x_{ij}, \varphi, R_1, v)$ and $W_{cq2}^{(ij)} = W_{cq2}^{(ij)}\,(x_{ij}, \varphi, R_1, R_2, v)$ ($q_1, q_2 = p,e,m$) in Sections 8.2.3.1 and 8.2.3.2 is valid

1. for the multi-particle-matrix system on the condition $\beta_p \neq \beta_m$ (see Eqs. (3.100)–(3.102)) for $\varphi \in \langle 0, 2\pi \rangle$ and $\nu \in \langle 0, \pi \rangle$ (see Fig. 3.1),

2. for the multi-particle-(envelope)-matrix system on the conditions $\beta_p \neq \beta_e = \beta_m$, $\beta_p \neq \beta_e \neq \beta_m$, $\beta_p = \beta_e \neq \beta_m$ for $\varphi \in \langle 0, 2\pi \rangle$ and $\nu \in \langle 0, \pi \rangle$.

The only exception of the condition $w\,(x_1, x_2, x_3, R_1, R_2, v) \neq 0$ for $x_i \in \langle 0, d/2 \rangle$ ($i = 1,2,3$) is analysed in Item 3b, Section 8.2.2.1.B regarding the condition $w_e\,(r, \varphi, \nu, R_1, R_2, v) = 0$ for the energy density $w_e = w_e\,(r, \varphi, \nu, R_1, R_2, v)$ of the spherical envelope of the multi-particle-envelope-matrix system. If the multi-particle-envelope-matrix system is loaded by the thermal stresses only, the condition $w_e\,(r, \varphi, \nu, R_1, R_2, v) = 0$, derived e.g. by the spherical coordinates (r, φ, ν) (see Fig. 3.1), is valid for either $\beta_p < \beta_e < \beta_m$ or $\beta_p > \beta_e > \beta_m$, both for $\varphi \in \langle 0, 2\pi \rangle$ and $\nu \in \langle 0, \pi \rangle$ (see Items 3(b)i, 3(b)ii, Section 8.2.2.1.B).

8.2.3.1 Determination of $W_{cp}^{(ij)}$, $W_{cm}^{(ij)}$ in multi-particle-matrix system

With regard to Fig. 8.2a, the curve integral $W_{cq}^{(ij)} = W_{cq}^{(ij)}\,(x_{ij}, \varphi, R_1, v)$ ($q = p,m$) is determined within the plane $x_{ij} x_k$ for $\varphi \in \langle 0, 2\pi \rangle$.

A. Spherical particle

If the condition in Item 1, Section 8.2.2.1.A is valid, then the curve integral $W_{cp}^{(ij)} = W_{cp}^{(ij)}\,(x_{ij}, \varphi, R_1, v)$ for the crack formation in the spherical particle for $x_{ij} \in \langle 0, R_1 \rangle$

is determined along the abscissa $\overline{P_1P_2P_3}$ in Fig. 8.8. If the multi-particle-matrix system is loaded by the thermal stresses only, then this condition is valid for $\beta_p > \beta_m$ (see Eqs. (3.100)–(3.102)).

With regard to Eq. (8.3), we get

$$W_{cp}^{(ij)} = \int_{\overline{P_1P_2}} w_p^{(ij)}\, dx_k + \int_{\overline{P_2P_3}} w_m^{(ij)}\, dx_k = \int_0^{\sqrt{R_1^2-x_{ij}^2}} w_p^{(ij)}\, dx_k + \int_{\sqrt{R_1^2-x_{ij}^2}}^{d/2} w_m^{(ij)}\, dx_k,$$

$$x_{ij} \in \langle 0, R_1 \rangle. \tag{8.38}$$

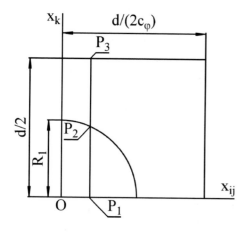

Figure 8.8: The abscissa $\overline{P_1P_2P_3}$ along which the curve integral $W_{cp}^{(ij)} = W_{cp}^{(ij)}(x_{ij}, \varphi, R_1, v)$ (see Eqs. (8.3), (8.38)) for $x_{ij} \in \langle 0, R_1 \rangle$ is determined due to the crack formation in the spherical particle on the condition in Item 1, Section 8.2.2.1.A. If the multi-particle-matrix system is loaded by the thermal stresses only, then this condition is valid for $\beta_p > \beta_m$ (see Eqs. (3.100)–(3.102)).

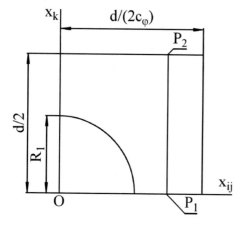

Figure 8.9: The abscissa $\overline{P_1P_2}$ along which the curve integral $W_{cm}^{(ij)} = W_{cm}^{(ij)}(x_{ij}, \varphi, R_1, v)$ (see Eqs. (8.3), (8.39)) for $x_{ij} \in \langle R_1, d/(2c_\varphi) \rangle$ (see Eqs. (2.32), (2.33)) is determined due to the crack formation in the cell matrix on the condition in Item 2, Section 8.2.2.1.A. If the multi-particle-matrix system is loaded by the thermal stresses only, then this condition is valid for $\beta_p < \beta_m$ (see Eqs. (3.100)–(3.102)).

B. Cell matrix

If the condition in Item 2, Section 8.2.2.1.A is valid, then the curve integral $W_{cm}^{(ij)} = W_{cm}^{(ij)}(x_{ij}, \varphi, R_1, v)$ for the crack formation in the cell matrix for $x_{ij} \in \langle R_1, d/(2c_\varphi) \rangle$

(see Eqs. (2.32), (2.33)) is determined along the abscissa $\overline{P_1 P_2}$ in Fig. 8.9. If the multi-particle-matrix system is loaded by the thermal stresses only, then this condition is valid for $\beta_p < \beta_m$ (see Eqs. (3.100)–(3.102)).

With regard to Eq. (8.3), we get

$$W_{cm}^{(ij)} = \int_{\overline{P_1 P_2}} w_m^{(ij)} \, dx_k = \int_0^{d/2} w_m^{(ij)} \, dx_k, \quad x_{ij} \in \left\langle R_1, \frac{d}{2c_\varphi} \right\rangle. \tag{8.39}$$

8.2.3.2 Multi-particle-envelope-matrix system

With regard to Fig. 8.2b, the curve integral $W_{cq}^{(ij)} = W_{cq}^{(ij)}(x_{ij}, \varphi, R_1, R_2, v) = W_{cq}^{(ij)}(x_{ij}, \varphi, R_1, t, v) = W_{cq}^{(ij)}(x_{ij}, \varphi, R_2, t, v)$ ($q = p,e,m$) is determined within the plane $x_{ij} x_k$ for $\varphi \in \langle 0, 2\pi \rangle$.

A. Spherical particle

If the conditions in Items 1, 3a, Section 8.2.2.1.B are valid, then the curve integral $W_{cp}^{(ij)} = W_{cp}^{(ij)}(x_{ij}, \varphi, R_1, R_2, v) = W_{cp}^{(ij)}(x_{ij}, \varphi, R_1, t, v) = W_{cp}^{(ij)}(x_{ij}, \varphi, R_2, t, v)$ for the crack formation in the spherical particle for $x_{ij} \in \langle 0, R_1 \rangle$ is determined along the abscissa $\overline{P_1 P_2 P_3 P_4}$ in Fig. 8.10. If the multi-particle-envelope-matrix system is loaded by the thermal stresses only, then these conditions are valid for either $\beta_p = \beta_e > \beta_m$ or $\beta_p > \beta_e \gtreqless \beta_m$ (see Eqs. (3.100)–(3.102)).

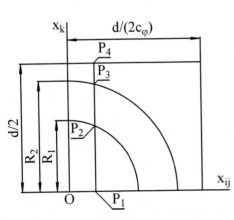

Figure 8.10: The abscissa $\overline{P_1 P_2 P_3 P_4}$ along which the curve integral $W_{cp}^{(ij)} = W_{cp}^{(ij)}(x_{ij}, \varphi, R_1, R_2, v) = W_{cp}^{(ij)}(x_{ij}, \varphi, R_1, t, v) = W_{cp}^{(ij)}(x_{ij}, \varphi, R_2, t, v)$ (see Eqs. (8.3), (8.40)) for $x_{ij} \in \langle 0, R_1 \rangle$ is determined due to the crack formation in the spherical particle on the conditions in Items 1, 3a, Section 8.2.2.1.B. If the multi-particle-envelope-matrix system is loaded by the thermal stresses only, then these conditions are valid for either $\beta_p = \beta_e > \beta_m$ or $\beta_p > \beta_e \gtreqless \beta_m$ (see Eqs. (3.100)–(3.102)).

With regard to Eq. (8.3), we get

$$
\begin{aligned}
W_{cp}^{(ij)} &= \underbrace{\int}_{\overline{P_1P_2}} w_p^{(ij)}\, dx_k + \underbrace{\int}_{\overline{P_2P_3}} w_e^{(ij)}\, dx_k + \underbrace{\int}_{\overline{P_3P_4}} w_m^{(ij)}\, dx_k \\
&= \int_0^{\sqrt{R_1^2 - x_{ij}^2}} w_p^{(ij)}\, dx_k + \int_{\sqrt{R_1^2 - x_{ij}^2}}^{\sqrt{R_2^2 - x_{ij}^2}} w_e^{(ij)}\, dx_k + \int_{\sqrt{R_2^2 - x_{ij}^2}}^{d/2} w_m^{(ij)}\, dx_k,
\end{aligned}
$$

$$
x_{ij} \in \langle 0, R_1 \rangle, \tag{8.40}
$$

If the conditions in Items 1, 3(b)ii, Section 8.2.2.1.B are valid, then the curve integral $W_{cp}^{(ij)} = W_{cp}^{(ij)}(x_{ij}, \varphi, R_1, R_2, v) = W_{cp}^{(ij)}(x_{ij}, \varphi, R_1, t, v) = W_{cp}^{(ij)}(x_{ij}, \varphi, R_2, t, v)$ for the crack formation in the spherical particle for $x_{ij} \in \langle 0, R_1 \rangle$ is determined along the abscissa $\overline{P_5P_6P_7}$ in Fig. 8.11. If the multi-particle-envelope-matrix system is loaded by the thermal stresses only, then these conditions are valid for $\beta_p > \beta_e > \beta_m$ (see Eqs. (3.100)–(3.102)).

With regard to Eq. (8.3), we get

$$
W_{cp}^{(ij)} = \underbrace{\int}_{\overline{P_5P_6}} w_p^{(ij)}\, dx_k + \underbrace{\int}_{\overline{P_6P_7}} w_e^{(ij)}\, dx_k = \int_0^{\sqrt{R_1^2 - x_{ij}^2}} w_p^{(ij)}\, dx_k + \int_{\sqrt{R_1^2 - x_{ij}^2}}^{x_{k0e}} w_e^{(ij)}\, dx_k,
$$

$$
x_{ij} \in \langle 0, R_1 \rangle. \tag{8.41}
$$

With regard to Eq. (8.11), the function $r_{0e} = r_{0e}(r, \varphi, v, R_1, R_2, v)$ described in Item 3b, Section 8.2.2.1.A is transformed to the function $x_{k0e} = x_{k0e}(x_{ij}, \varphi, R_1, R_2, v) = x_{k0e}(x_{ij}, \varphi, R_1, t, v) = x_{k0e}(x_{ij}, \varphi, R_2, t, v)$ of the variable $x_{ij} \in \langle 0, x_{ij0} \rangle$, where x_{ij0} given by Eq. (8.14) also results from the condition $x_{k0e}(x_{ij}, \varphi, R_1, R_2, v) = 0$.

B. Spherical envelope

If the condition in Item 3a, Section 8.2.2.1.B is valid, then the curve integral $W_{ce}^{(ij)} = W_{ce}^{(ij)}(x_{ij}, \varphi, R_1, R_2, v) = W_{ce}^{(ij)}(x_{ij}, \varphi, R_1, t, v) = W_{ce}^{(ij)}(x_{ij}, \varphi, R_2, t, v)$ for the crack formation in the spherical envelope for $x_{ij} \in \langle R_1, R_2 \rangle$ is determined along the abscissa $\overline{P_1P_2P_3}$ in Fig. 8.12. If the multi-particle-envelope-matrix system is loaded by the thermal stresses only, then this condition is valid for either $\beta_p = \beta_e > \beta_m$ or $\beta_p > \beta_e \gtrless \beta_m$ (see Eqs. (3.100)–(3.102)).

With regard to Eq. (8.3), we get

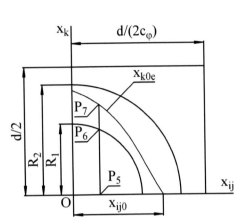

Figure 8.11: The abscissa $\overline{P_5P_6P_7}$ along which the curve integral $W_{cp}^{(ij)} = W_{cp}^{(ij)}(x_{ij},\varphi,R_1,R_2,v) = W_{cp}^{(ij)}(x_{ij},\varphi,R_1,t,v) = W_{cp}^{(ij)}(x_{ij},\varphi,R_2,t,v)$ (see Eqs. (8.3), (8.41)) for $x_{ij} \in \langle 0, R_1 \rangle$ is determined due to the crack formation in the spherical particle on the conditions in Items 1, 3(b)ii, Section 8.2.2.1.B. If the multi-particle-envelope-matrix system is loaded by the thermal stresses only, then these conditions are valid for $\beta_p > \beta_e > \beta_m$ (see Eqs. (3.100)–(3.102)). The function $x_{k0e} = x_{k0e}(x_{ij},\varphi,R_1,R_2,v)$ of the variable $x_{ij} \in \langle 0, x_{ij0} \rangle$ (see Eq. (8.14)) is described in Section 8.2.3.2.A.

$$W_{ce}^{(ij)} = \int_{\overline{P_1P_2}} w_e^{(ij)}\,dx_k + \int_{\overline{P_2P_3}} w_m^{(ij)}\,dx_k = \int_0^{\sqrt{R_2^2-x_{ij}^2}} w_e^{(ij)}\,dx_k + \int_{\sqrt{R_2^2-x_{ij}^2}}^{d/2} w_m^{(ij)}\,dx_k,$$

$$x_{ij} \in \langle R_1, R_2 \rangle. \tag{8.42}$$

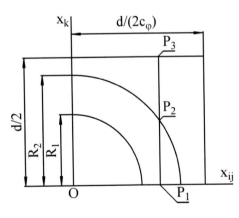

Figure 8.12: The abscissa $\overline{P_1P_2P_3}$ along which the curve integral $W_{ce}^{(ij)} = W_{ce}^{(ij)}(x_{ij},\varphi,R_1,R_2,v) = W_{ce}^{(ij)}(x_{ij},\varphi,R_1,t,v) = W_{ce}^{(ij)}(x_{ij},\varphi,R_2,t,v)$ (see Eqs. (8.3), (8.42)) for $x_{ij} \in \langle R_1, R_2 \rangle$ is determined due to the crack formation in the spherical envelope on the condition in Item 3a, Section 8.2.2.1.B. If the multi-particle-envelope-matrix system is loaded by the thermal stresses only, then this condition is valid for either $\beta_p = \beta_e > \beta_m$ or $\beta_p > \beta_e \gtreqless \beta_m$ (see Eqs. (3.100)–(3.102)).

If the condition in Item 3(b)i, Section 8.2.2.1.B is valid, then the curve integral $W_{ce}^{(ij)} = W_{ce}^{(ij)}(x_{ij},\varphi,R_1,R_2,v) = W_{ce}^{(ij)}(x_{ij},\varphi,R_1,t,v) = W_{ce}^{(ij)}(x_{ij},\varphi,R_2,t,v)$ for the crack formation in the spherical envelope for $x_{ij} \in \langle R_1, x_{ij0} \rangle$ (see Eq. (8.14)) is determined along the abscissa $\overline{P_3P_4}$ in Fig. 8.13. If the multi-particle-envelope-matrix system is loaded by the thermal stresses only, then this condition is valid for $\beta_p < \beta_e < \beta_m$ (see Eqs. (3.100)–(3.102)).

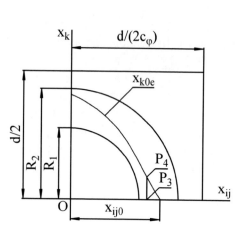

Figure 8.13: The abscissa $\overline{P_3 P_4}$ along which the curve integral $W_{ce}^{(ij)} = W_{ce}^{(ij)}(x_{ij}, \varphi, R_1, R_2, v) = W_{ce}^{(ij)}(x_{ij}, \varphi, R_1, t, v) = W_{ce}^{(ij)}(x_{ij}, \varphi, R_2, t, v)$ (see Eqs. (8.3), (8.43)) for $x_{ij} \in \langle R_1, x_{ij0} \rangle$ (see Eq. (8.14)) is determined due to the crack formation in the spherical envelope on the condition in Item 3(b)i, Section 8.2.2.1.B. If the multi-particle-envelope-matrix system is loaded by the thermal stresses only, then this condition is valid for $\beta_p < \beta_e < \beta_m$ (see Eqs. (3.100)–(3.102)). the function $x_{k0e} = x_{k0e}(x_{ij}, \varphi, R_1, R_2, v)$ of the variable $x_{ij} \in \langle 0, x_{ij0} \rangle$ (see Eq. (8.14)) is described in Section 8.2.3.2.A.

With regard to Eq. (8.3), we get

$$W_{ce}^{(ij)} = \int_{\overline{P_3 P_4}} w_e^{(ij)} \, dx_k = \int_0^{x_{k0e}} w_e^{(ij)} \, dx_k, \quad x_{ij} \in \langle R_1, x_{ij0} \rangle. \qquad (8.43)$$

If the condition in Item 3(b)ii, Section 8.2.2.1.B is valid, then the curve integral $W_{ce}^{(ij)} = W_{ce}^{(ij)}(x_{ij}, \varphi, R_1, R_2, v) = W_{ce}^{(ij)}(x_{ij}, \varphi, R_1, t, v) = W_{ce}^{(ij)}(x_{ij}, \varphi, R_2, t, v)$ for the crack formation in the spherical envelope for $x_{ij} \in \langle x_{ij0}, R_2 \rangle$ (see Eq. (8.14)) is determined along the abscissa $\overline{P_5 P_6}$ in Fig. 8.14. If the multi-particle-envelope-matrix system is loaded by the thermal stresses only, then this condition is valid for $\beta_p > \beta_e > \beta_m$ (see Eqs. (3.100)–(3.102)).

With regard to Eq. (8.3), we get

$$W_{ce}^{(ij)} = \int_{\overline{P_5 P_6}} w_e^{(ij)} \, dx_k = \int_0^{\sqrt{R_2^2 - x_{ij}^2}} w_e^{(ij)} \, dx_k, \quad x_{ij} \in \langle x_{ij0}, R_2 \rangle. \qquad (8.44)$$

With regard to Figs. 8.13, 8.14, the function $x_{k0e} = x_{k0e}(x_{ij}, \varphi, R_1, R_2, v)$ of the variable $x_{ij} \in \langle 0, x_{ij0} \rangle$ (see Eq. (8.14)) is described in Section 8.2.3.2.A.

C. Cell matrix

If the condition in Item 2, Section 8.2.2.1.B is valid, then the curve integral $W_{cm}^{(ij)} = W_{cm}^{(ij)}(x_{ij}, \varphi, R_1, v)$ for the crack formation in the cell matrix for $x_{ij} \in \langle R_2, d/(2c_\varphi) \rangle$ (see

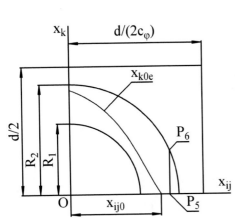

Figure 8.14: The abscissa $\overline{P_5P_6}$ along which the curve integral $W_{ce}^{(ij)} = W_{ce}^{(ij)}(x_{ij}, \varphi, R_1, R_2, v) = W_{ce}^{(ij)}(x_{ij}, \varphi, R_1, t, v) = W_{ce}^{(ij)}(x_{ij}, \varphi, R_2, t, v)$ (see Eqs. (8.3), (8.44)) for $x_{ij} \in \langle x_{ij0}, R_2 \rangle$ (see Eq. (8.14)) is determined due to the crack formation in the spherical envelope on the condition in Item 3(b)ii, Section 8.2.2.1.B. If the multi-particle-envelope-matrix system is loaded by the thermal stresses only, then this condition is valid for $\beta_p > \beta_e > \beta_m$ (see Eqs. (3.100)–(3.102)). the function $x_{k0e} = x_{k0e}(x_{ij}, \varphi, R_1, R_2, v)$ of the variable $x_{ij} \in \langle 0, x_{ij0} \rangle$ (see Eq. (8.14)) is described in Section 8.2.3.2.A.

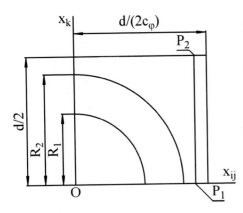

Figure 8.15: The abscissa $\overline{P_1P_2}$ along which the curve integral $W_{cm}^{(ij)} = W_{cm}^{(ij)}(x_{ij}, \varphi, R_1, v)$ for $x_{ij} \in \langle R_2, d/(2c_\varphi) \rangle$ (see Eqs. (2.32), (2.33)) is determined due to the crack formation in the cell matrix on the condition in Item 2, Section 8.2.2.1.B. If the multi-particle-envelope-matrix system is loaded by the thermal stresses only, then the crack is formed in the cell matrix provided that on the condition either $\beta_p < \beta_e = \beta_m$ or $\beta_p \gtreqless \beta_e < \beta_m$ (see Eqs. (3.100)–(3.102)).

Eqs. (2.32), (2.33)) is determined along the abscissa $\overline{P_1P_2}$ in Fig. 8.15. If the multi-particle-envelope-matrix system is loaded by the thermal stresses only, then the crack is formed in the cell matrix provided that on the condition either $\beta_p < \beta_e = \beta_m$ or $\beta_p \gtreqless \beta_e < \beta_m$ (see Eqs. (3.100)–(3.102)).

With regard to Eq. (8.3), we get

$$W_{cm}^{(ij)} = \int_{\overline{P_1P_2}} w_m^{(ij)} dx_k = \int_0^{d/2} w_m^{(ij)} dx_k, \quad x_{ij} \in \left\langle R_2, \frac{d}{2c_\varphi} \right\rangle. \tag{8.45}$$

8.2.4 Crack formation II

8.2.4.1 Determination of $W_{cp}^{(ij)}$, $W_{cm}^{(ij)}$ in multi-particle-matrix system

Section 8.2.4.1 deals with a determination of $W_{cq1}^{(ij)} = W_{cq1}^{(ij)}(x_{ij}, \varphi, R_1, v)$ (see Eq. (8.3)) for the crack formation in the spherical particle ($q_1 = p$) and the cell matrix ($q_1 = m$) on the condition $w_{q2}^{(ij)} = 0$ for the energy density $w_{q2} = w_{q2}^{(ij)}(x_{ij}, \varphi, x_k, R_1, v)$ (see Eq. (8.1)) of the spherical particle ($q_2 = p$) and the cell matrix ($q_2 = m$).

Consequently, the conditions $w_p^{(ij)}(x_{ij}, \varphi, x_k, R_1, v) = w_p^{(ij)}(r, \varphi, \nu, R_1, v) = 0$ and $w_m^{(ij)}(x_{ij}, \varphi, x_k, R_1, v) = w_m^{(ij)}(r, \varphi, \nu, R_1, v) = 0$, which are valid for $\sigma'_{n_1 n_2 p}(x_{ij}, \varphi, x_k, R_1, v) = \sigma'_{n_1 n_2 p}(r, \varphi, \nu, R_1, v) = 0$ and $\sigma'_{n_1 n_2 m}(x_{ij}, \varphi, x_k, R_1, v) = \sigma'_{n_1 n_2 m}(r, \varphi, \nu, R_1, v) = 0$ ($n_1, n_2 = 1,2,3$), result in the functions $x_{k0p} = x_{k0p}(x_{ij}, \varphi, R_1, v)$ and $x_{k0m} = x_{k0m}(x_{ij}, \varphi, R_1, v)$ of the variables $x_{ij} \in \langle 0, x_{ijp} \rangle$ and $x_{ij} \in \langle x_{ijp}, x_{ijm} \rangle$, (see Fig. 8.16), respectively, and both for $\varphi \in \langle \varphi_{1pm}, \varphi_{2pm} \rangle$. The interval $\langle \varphi_{1pm}, \varphi_{2pm} \rangle$ represents an interval of the validity of the condition $w_q(x_{ij}, \varphi, x_k, R_1, v) = w_q(r, \varphi, \nu, R_1, v) = 0$ regarding the variable φ.

Additionally, with regard to Fig. 8.16, we get $[x_{k0p}(x_{ij}, \varphi, R_1, v)]_{x_{ij}=x_{ijp}} = [x_{k0m}(x_{ij}, \varphi, R_1, v)]_{x_{ij}=x_{ijp}}$; $x_{ijp} \in \langle 0, R_1 \rangle$; and either $x_{ijm} \in \langle 0, R_1 \rangle$ (see Fig. 8.16a) or $x_{ijm} \in (R_1, d/(2c_\varphi))$ (see Fig. 8.16b) or $x_{ijm} = d/(2c_\varphi)$ (see Fig. 8.16c; Eqs. (2.32), (2.33)). Finally, with regard to Fig. 8.16a,b, x_{ijm} is determined by the condition $[x_{k0m}(x_{ij}, \varphi, R_1, v)]_{x_k=x_{ijm}} = d/2$.

The stresses $\sigma'_{1nq} = \sigma'_{1nq}(r, \varphi, \nu, R_1, v)$, $\sigma'_{2nq} = \sigma'_{2nq}(r, \varphi, \nu, R_1, v)$ and $\sigma'_{3nq} = \sigma'_{3nq}(r, \varphi, \nu, R_1, v)$ ($n = 1,2,3$) which induce the energy density $w_q = w_q(r, \varphi, \nu, R_1, v)$ act along the axes x'_1, x'_2 and x'_3 (see Fig. 3.1), respectively.

With regard to $[\sigma'_{n_1 n_2 q}(x_{ij}, \varphi, x_k, R_1, v)]_{x_k=x_{k0q}} = 0$ ($q = p,m$; $n_1, n_2 = 1,2,3$), we get

1. If the force $\vec{F}'_{nnp}$ induced by the stress σ'_{nnp} for positions within the surface $O12O$, i.e. for the coordinates $(x_{ij}, x_k) \in O12O$ (see Fig. 8.16a,b,c) acts along the axis $+x'_n$ (see Fig. 3.1), and the force $\vec{F}'_{nnm}$ induced by σ'_{nnm} for $(x_{ij}, x_k) \in 145621$ (see Fig. 8.16a,b) or for $(x_{ij}, x_k) \in 14521$ (see Fig. 8.16c) acts along the axis $-x'_n$, then the force $\vec{F}'_{nnp}$ induced by the stress σ'_{nnp} for $(x_{ij}, x_k) \in O23O$ (see Fig. 8.16a,b,c) acts along the axis $-x'_n$ and the force $\vec{F}'_{nnm}$ induced by the stress σ'_{nnm} for $(x_{ij}, x_k) \in 26732$ (see Fig. 8.16a,b) or for $(x_{ij}, x_k) \in 256732$ (see Fig. 8.16c) acts along the axis $+x'_n$. If the multi-particle-matrix system is loaded by the thermal stresses only, this condition concerning $\vec{F}'_{nnq}$ is valid provided that $\beta_p \geq \beta_m$ (see Eqs. (3.100)–(3.102)) for $(x_{ij}, x_k) \in O14562O$ (see Fig. 8.16a,b) or $(x_{ij}, x_k) \in O14520$ (see Fig. 8.16c). Consequently, we get $\beta_p \leq \beta_m$ for $(x_{ij}, x_k) \in O26730$ (see Fig. 8.16a,b) or $(x_{ij}, x_k) \in O256730$ (see Fig. 8.16c).

2. If the force $\vec{F}'_{nnp}$ induced by the stress σ'_{nnp} for $(x_{ij}, x_k) \in O12O$ (see Fig. 8.16a,b,c) acts along the axis $-x'_n$ (see Fig. 3.1), and the force $\vec{F}'_{nnm}$ induced by the stress σ'_{nnm} for $(x_{ij}, x_k) \in 145621$ (see Fig. 8.16a,b) or for $(x_{ij}, x_k) \in 14521$

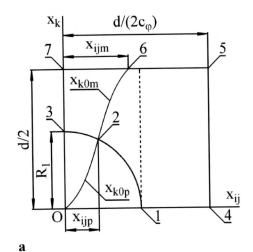
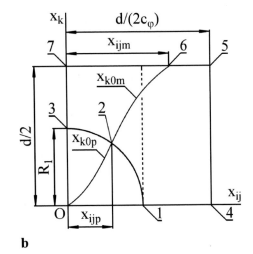
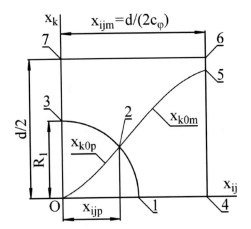

Figure 8.16: The functions $x_{k0p} = x_{k0p}(x_{ij}, \varphi, R_1, v)$ and $x_{k0m} = x_{k0m}(x_{ij}, \varphi, R_1, v)$ of the variables $x_{ij} \in \langle 0, x_{ijp} \rangle$ and $x_{ij} \in \langle x_{ijp}, x_{ijm} \rangle$, respectively, where $[x_{k0p}(x_{ij}, \varphi, R_1, v)]_{x_{ij}=x_{ijp}} = [x_{k0m}(x_{ij}, \varphi, R_1, v)]_{x_{ij}=x_{ijp}}$; $x_{ijp} \in \langle 0, R_1 \rangle$; and either (a) $x_{ijm} \in \langle 0, R_1 \rangle$ or (b) $x_{ijm} \in (R_1, d/(2c_\varphi))$ (see Eqs. (2.32), (2.33)) or (c) $x_{ijm} = d/(2c_\varphi)$ (see Eqs. (2.32), (2.33)), where the condition (c) is related to $[x_{k0m}(x_{ij}, \varphi, R_1, v)]_{x_{ij}=x_{ijm}} \in (0, d/2)$. The function $x_{k0q} = x_{k0q}(x_{ij}, \varphi, R_1, v)$ ($q = p, m$) results from the condition $w_q(x_{ij}, \varphi, x_k, R_1, v) = w_q(r, \varphi, \nu, R_1, v) = 0$ which is valid for $\sigma'_{n_1 n_2 q}(x_{ij}, \varphi, x_k, R_1, v) = \sigma'_{n_1 n_2 q}(r, \varphi, \nu, R_1, v) = 0$ ($n_1, n_2 = 1, 2, 3$). The stresses $\sigma'_{1nq} = \sigma'_{1nq}(r, \varphi, \nu, R_1, v)$, $\sigma'_{2nq} = \sigma'_{2nq}(r, \varphi, \nu, R_1, v)$ and $\sigma'_{3nq} = \sigma'_{3nq}(r, \varphi, \nu, R_1, v)$ ($n = 1, 2, 3$) which induce the energy density $w_q = w_q(r, \varphi, \nu, R_1, v)$ act along the axes x'_1, x'_2 and x'_3 (see Fig. 3.1), respectively.

(see Fig. 8.16c) acts along the axis $+x'_n$, then the force $\vec{F}'_{nnp}$ induced by the stress σ'_{nnp} for $(x_{ij}, x_k) \in O23O$ (see Fig. 8.16a,b,c) acts along the axis $+x'_n$ and the force

$\vec{F}'_{nnm}$ induced by the stress σ'_{nnm} for $(x_{ij}, x_k) \in 26732$ (see Fig. 8.16a,b) or for $(x_{ij}, x_k) \in 256732$ (see Fig. 8.16c) acts along the axis $-x'_n$. If the multi-particle-matrix system is loaded by the thermal stresses only, this condition concerning $\vec{F}'_{nnq}$ is valid provided that $\beta_p \leq \beta_m$ (see Eqs. (3.100)–(3.102)) for $(x_{ij}, x_k) \in 0145620$ (see Fig. 8.16a,b) or $(x_{ij}, x_k) \in 014520$ (see Fig. 8.16c). Consequently, we get $\beta_p \geq \beta_m$ for $(x_{ij}, x_k) \in 026730$ (see Fig. 8.16a,b) or $(x_{ij}, x_k) \in 0256730$ (see Fig. 8.16c).

The surface 0120 in Fig. 8.16a,b,c is define for $x_{ij} \in \langle 0, x_{ijp} \rangle$, $x_k \in \langle 0, x_{k0p} \rangle$; and for $x_{ij} \in \langle x_{ijp}, R_1 \rangle$, $x_k \in \left\langle 0, \sqrt{R_1^2 - x_{ij}^2} \right\rangle$.

The surface 0230 in Fig. 8.16a,b,c is define for $x_{ij} \in \langle 0, x_{ijp} \rangle$, $x_k \in \left\langle x_{k0p}, \sqrt{R_1^2 - x_{ij}^2} \right\rangle$.

The surface 145621 in Fig. 8.16a is define for $x_{ij} \in \langle x_{ijp}, x_{ijm} \rangle$, $x_k \in \left\langle \sqrt{R_1^2 - x_{ij}^2}, x_{k0m} \right\rangle$; for $x_{ij} \in \langle x_{ijm}, R_1 \rangle$, $x_k \in \left\langle \sqrt{R_1^2 - x_{ij}^2}, d/2 \right\rangle$; and for $x_{ij} \in \langle R_1, d/(2c_\varphi) \rangle$, $x_k \in \langle 0, d/2 \rangle$ (see Eqs. (2.32), (2.33)).

The surface 145621 in Fig. 8.16b is define for $x_{ij} \in \langle x_{ijp}, R_1 \rangle$, $x_k \in \left\langle \sqrt{R_1^2 - x_{ij}^2}, x_{k0m} \right\rangle$; for $x_{ij} \in \langle R_1, x_{ijm} \rangle$, $x_k \in \langle 0, x_{k0m} \rangle$; and for $x_{ij} \in \langle x_{ijm}, d/(2c_\varphi) \rangle$, $x_k \in \langle 0, d/2 \rangle$ (see Eqs. (2.32), (2.33)).

The surface 14521 in Fig. 8.16c is define for $x_{ij} \in \langle x_{ijp}, R_1 \rangle$, $x_k \in \left\langle \sqrt{R_1^2 - x_{ij}^2}, x_{k0m} \right\rangle$; and for $x_{ij} \in \langle R_1, d/(2c_\varphi) \rangle$, $x_k \in \langle 0, x_{k0m} \rangle$ (see Eqs. (2.32), (2.33)).

The surface 26732 in Fig. 8.16a,b is define for $x_{ij} \in \langle 0, x_{ijp} \rangle$, $x_k \in \left\langle \sqrt{R_1^2 - x_{ij}^2}, d/2 \right\rangle$; and for $x_{ij} \in \langle x_{ijp}, x_{ijm} \rangle$, $x_k \in \langle x_{k0m}, d/2 \rangle$.

The surface 256732 in Fig. 8.16c is define for $x_{ij} \in \langle 0, x_{ijp} \rangle$, $x_k \in \left\langle \sqrt{R_1^2 - x_{ij}^2}, d/2 \right\rangle$; and for $x_{ij} \in \langle x_{ijp}, d/(2c_\varphi) \rangle$, $x_k \in \langle x_{k0m}, d/2 \rangle$ (see Eqs. (2.32), (2.33)).

If the multi-particle-matrix system is loaded by the thermal stresses only, then the conditions $w_p(x_{ij}, \varphi, x_k, R_1, v) = w_p(r, \varphi, \nu, R_1, v) = 0$ and $w_m(x_{ij}, \varphi, x_k, R_1, v) = w_m(r, \varphi, \nu, R_1, v) = 0$ are transformed to the condition $\beta_p = \beta_m$ (see Eqs. (3.100)–(3.102)). Additionally, if the thermal stresses originate as a consequence of a difference in thermal expansion coefficient of the spherical particle and the cell matrix, i.e. the phase-transformation induced strain $\varepsilon_{11tq} = 0$ (see Eqs. (3.100), (3.104)), then $x_{k0p} = x_{k0p}(x_{ij}, \varphi, R_1, v)$ for $x_{ij} \in \langle 0, x_{ijp} \rangle$ and $x_{k0m} = x_{k0m}(x_{ij}, \varphi, R_1, v)$ for $x_{ij} \in \langle x_{ijp}, x_{ijm} \rangle$ are derived as

$$x_{k0q} = x_{ij} \cot \nu_{pm}, \quad q = p, m, \tag{8.46}$$

where the angle $\nu_{pm} = \nu_{pm}(\varphi, R_1, v)$ is given by Eq. (3.106) for the multi-particle-matrix system with anisotropic components only, or with anisotropic and isotropic components, respectively.

The analyses in Items 1, 2 are also valid for the crack formation in the plane $x_j x_k$. In this case, $x_{k0q} = x_{k0q} (x_{ij}, \varphi, R_1, v)$ and x_{ijq} $(q = p, m)$ are replaced by $x_{i0q} = x_{i0q} (x_{jk}, \varphi, R_1, v)$ and x_{jkq}, respectively, where $\varphi = \angle (x_j, x_{jk})$, $x_{jk} \subset x_j x_k$. Similarly, $x_{k0q} = x_{k0q} (x_{ij}, \varphi, R_1, v)$ and x_{ijq} $(q = p, m)$ are replaced by $x_{j0q} = x_{j0q} (x_{ki}, \varphi, R_1, v)$ and x_{kiq} in case of the crack formation in the plane $x_k x_i$, respectively, where $\varphi = \angle (x_k, x_{ki})$, $x_{ki} \subset x_k x_i$.

A. Method 1

Let $W_{O14562O}$ or W_{O1452O} is energy accumulated in the cell volume which is related to the surface $O14562O$ (see Fig. 8.17a,b) or $O1452O$ (see Fig. 8.17c), respectively, and both are related to the angle $\varphi \in \langle \varphi_{1pm}, \varphi_{2pm} \rangle$.

Consequently, let W_{O26730} or $W_{O256730}$ is energy accumulated in the cell volume which is related to the surface $O26730$ (see Fig. 8.17a,b) or $O256730$ (see Fig. 8.17c), and both are also related to the angle $\varphi \in \langle \varphi_{1pm}, \varphi_{2pm} \rangle$.

This method of determination of $W_{cp}^{(ij)} = W_{cp}^{(ij)} (x_{ij}, \varphi, R_1, v)$ and $W_{cm}^{(ij)} = W_{cm}^{(ij)} (x_{ij}, \varphi, R_1, v)$ is based on an assumption that energy $W_{O14562O}$ or W_{O1452O} is released by the crack formation in the plane $x_i x_j$. Accordingly, W_{O26730} or $W_{O256730}$ is assumed to be released

(a) either in the plane $x_j x_k$ provided that $R_{1cq, max}^{(jk)} < R_{1cq, max}^{(ki)}$,

(b) or in the plane $x_k x_i$ provided that $R_{1cq, max}^{(jk)} > R_{1cq, max}^{(ki)}$,

(c) or in one of the planes $x_j x_k$, $x_k x_i$ provided that $R_{1cq, max}^{(jk)} = R_{1cq, max}^{(ki)}$ $(q = p, m)$.

The critical particle radii $R_{1cq, max}^{(jk)} = R_{1cq, max}^{(jk)} (v)$ and $R_{1cq, max}^{(ki)} = R_{1cq, max}^{(ki)} (v)$ represent maximal values of the functions $R_{1cq}^{(jk)} = R_{1cq}^{(jk)} (\varphi, v)$ and $R_{1cq}^{(ki)} = R_{1cq}^{(ki)} (\varphi, v)$ of the variable $\varphi = \angle (x_j, x_{jk})$ and $\varphi = \angle (x_k, x_{ki})$, respectively, where $R_{1cq\,max}^{(jk)} \equiv R_{1cqn\,max}^{(jk)}$, $R_{1cq\,max}^{(ki)} \equiv R_{1cqn\,max}^{(ki)}$ $(n = 1, 2, 4)$ (see Section 8.2.2.2.A.4).

Spherical particle. If the condition in Item 1, Section 8.2.4.1 is valid, then a crack is formed in the spherical particle, i.e. for $x_{ij} \in \langle 0, R_1 \rangle$. Additionally, the conditions

- $x_{ijm} \in \langle 0, R_1 \rangle$ (see Fig. 8.16a),

- $x_{ijm} \in (R_1, d/ (2c_\varphi))$ (see Fig. 8.16b; Eqs. (2.32), (2.33)),

- $x_{ijm} = d/ (2c_\varphi)$ (see Fig. 8.16c)

are required to be considered for the determination of the curve integral $W_{cp}^{(ij)} = W_{cp}^{(ij)} (x_{ij}, \varphi, R_1, v)$ for $x_{ij} \in \langle 0, R_1 \rangle$.

Consequently, if $x_{ijm} \in \langle 0, R_1 \rangle$ (see Fig. 8.16a), then $W_{cp}^{(ij)} = W_{cp}^{(ij)}(x_{ij}, \varphi, R_1, v)$ is determined along the abscissae $\overline{P_1 P_2}$, $\overline{P_3 P_4 P_5}$ and $\overline{P_6 P_7 P_8}$ for $x_{ij} \in \langle 0, x_{ijp} \rangle$, $x_{ij} \in \langle x_{ijp}, x_{ijm} \rangle$ and $x_{ij} \in \langle x_{ijm}, R_1 \rangle$ in Fig. 8.17, respectively.

With regard to Eq. (8.3), we get

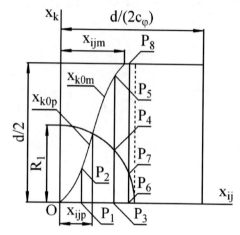

Figure 8.17: The abscissae $\overline{P_1 P_2}$, $\overline{P_3 P_4 P_5}$ and $\overline{P_6 P_7 P_8}$ for $x_{ij} \in \langle 0, x_{ijp} \rangle$, $x_{ij} \in \langle x_{ijp}, x_{ijm} \rangle$ and $x_{ij} \in \langle x_{ijm}, R_1 \rangle$, respectively, along which the curve integral $W_{cp}^{(ij)} = W_{cp}^{(ij)}(x_{ij}, \varphi, R_1, v)$ (see Eqs. (8.3), (8.47)–(8.49)) for $x_{ijm} \in \langle 0, R_1 \rangle$ (see Fig. 8.16a) is determined due to the crack formation in the spherical particle on the condition in Item 1, Section 8.2.4.1.

$$W_{cp}^{(ij)} = \int_{\overline{P_1 P_2}} w_p^{(ij)} \, dx_k = \int_0^{x_{k0p}} w_p^{(ij)} \, dx_k, \quad x_{ij} \in \langle 0, x_{ijp} \rangle, \quad x_{ijm} \in \langle 0, R_1 \rangle, \quad (8.47)$$

$$W_{cp}^{(ij)} = \int_{\overline{P_3 P_4}} w_p^{(ij)} \, dx_k + \int_{\overline{P_4 P_5}} w_m^{(ij)} \, dx_k = \int_0^{\sqrt{R_1^2 - x_{ij}^2}} w_p^{(ij)} \, dx_k + \int_{\sqrt{R_1^2 - x_{ij}^2}}^{x_{k0m}} w_m^{(ij)} \, dx_k,$$
$$x_{ij} \in \langle x_{ijp}, x_{ijm} \rangle, \quad x_{ijm} \in \langle 0, R_1 \rangle, \quad (8.48)$$

$$W_{cp}^{(ij)} = \int_{\overline{P_6 P_7}} w_p^{(ij)} \, dx_k + \int_{\overline{P_7 P_8}} w_m^{(ij)} \, dx_k = \int_0^{\sqrt{R_1^2 - x_{ij}^2}} w_p^{(ij)} \, dx_k + \int_{\sqrt{R_1^2 - x_{ij}^2}}^{d/2} w_m^{(ij)} \, dx_k,$$
$$x_{ij} \in \langle x_{ijm}, R_1 \rangle, \quad x_{ijm} \in \langle 0, R_1 \rangle. \quad (8.49)$$

Finally, if $x_{ijm} \in (R_1, d/(2c_\varphi))$ (see Fig. 8.16b,c; Eqs. (2.32), (2.33)), then $W_{cp}^{(ij)} = W_{cp}^{(ij)}(x_{ij}, \varphi, R_1, v)$ is determined along the abscissae $\overline{P_1 P_2}$ and $\overline{P_3 P_4 P_5}$ for $x_{ij} \in \langle 0, x_{ijp} \rangle$ and $x_{ij} \in \langle x_{ijp}, R_1 \rangle$ in Fig. 8.18, respectively.

With regard to Eq. (8.3), we get

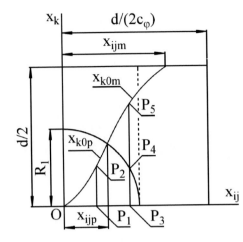

Figure 8.18: The abscissae $\overline{P_1P_2}$ and $\overline{P_3P_4P_5}$ for $x_{ij} \in \langle 0, x_{ijp} \rangle$ and $x_{ij} \in \langle x_{ijp}, R_1 \rangle$, respectively, along which the curve integral $W_{cp}^{(ij)} = W_{cp}^{(ij)}(x_{ij}, \varphi, R_1, v)$ (see Eqs. (8.3), (8.50), (8.51)) is determined due to the crack formation in the spherical particle on the condition in Item 1, Section 8.2.4.1. This determination is also considered for $x_{ijm} \in (R_1, d/(2c_\varphi))$ (see Fig. 8.16b,c; Eqs. (2.32), (2.33)), although Fig. 8.18 is related to the condition $x_{ijm} \in (R_1, d/(2c_\varphi))$ only (see Fig. 8.16b).

$$W_{cp}^{(ij)} = \int_{\overline{P_1P_2}} w_p^{(ij)} \, dx_k = \int_0^{x_{k0p}} w_p^{(ij)} \, dx_k,$$

$$x_{ij} \in \langle 0, x_{ijp} \rangle, \quad x_{ijm} \in \left(R_1, \frac{d}{2c_\varphi} \right), \tag{8.50}$$

$$W_{cp}^{(ij)} = \int_{\overline{P_3P_4}} w_p^{(ij)} \, dx_k + \int_{\overline{P_4P_5}} w_m^{(ij)} \, dx_k = \int_0^{\sqrt{R_1^2 - x_{ij}^2}} w_p^{(ij)} \, dx_k + \int_{\sqrt{R_1^2 - x_{ij}^2}}^{x_{k0m}} w_m^{(ij)} \, dx_k,$$

$$x_{ij} \in \langle x_{ijp}, R_1 \rangle, \quad x_{ijm} \in \left(R_1, \frac{d}{2c_\varphi} \right). \tag{8.51}$$

Cell matrix. Similarly, if the condition in Item 2, Section 8.2.4.1 is valid, then a crack is formed in the cell matrix, i.e. for $x_{ij} \in \langle R_1, d/(2c_\varphi) \rangle$ (see Eqs. (2.32), (2.33)). Consequently, the conditions

- $x_{ijm} \in \langle 0, R_1 \rangle$ (see Fig. 8.16a),
- $x_{ijm} \in (R_1, d/(2c_\varphi))$ (see Fig. 8.16b; Eqs. (2.32), (2.33)),
- $x_{ijm} = d/(2c_\varphi)$ (see Fig. 8.16c)

are also required to be considered for the determination of the curve integral $W_{cm}^{(ij)} = W_{cm}^{(ij)}(x_{ij}, \varphi, R_1, v)$ for $x_{ij} \in \langle R_1, d/(2c_\varphi) \rangle$.

If $x_{ijm} \in \langle 0, R_1 \rangle$ (see Fig. 8.16a), then $W_{cm}^{(ij)} = W_{cm}^{(ij)}(x_{ij}, \varphi, R_1, v)$ is determined along the abscissa $\overline{P_1P_2}$ for $x_{ij} \in \langle R_1, d/(2c_\varphi) \rangle$ in Fig. 8.19.

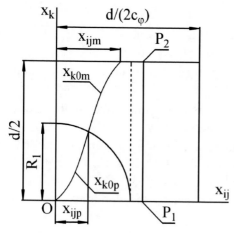

Figure 8.19: The abscissa $\overline{P_1P_2}$ for $x_{ij} \in \langle R_1, d/(2c_\varphi)\rangle$ (see Eqs. (2.32), (2.33)) along which the curve integral $W_{cm}^{(ij)} = W_{cm}^{(ij)}(x_{ij}, \varphi, R_1, v)$ (see Eqs. (8.3), (8.52)) for $x_{ijm} \in \langle 0, R_1\rangle$ (see Fig. 8.16a) is determined due to the crack formation in the cell matrix on the condition in Item 2, Section 8.2.4.1.

With regard to Eq. (8.3), we get

$$W_{cm}^{(ij)} = \int_{\overline{P_1P_2}} w_m^{(ij)}\,dx_k = \int_0^{d/2} w_m^{(ij)}\,dx_k, \quad x_{ij} \in \left\langle R_1, \frac{d}{2c_\varphi}\right\rangle, \quad x_{ijm} \in \langle 0, R_1\rangle. \quad (8.52)$$

Similarly, if $x_{ijm} \in (R_1, d/(2c_\varphi))$ (see Fig. 8.16b; Eqs. (2.32), (2.33)), then $W_{cm}^{(ij)} = W_{cm}^{(ij)}(x_{ij}, \varphi, R_1, v)$ is determined along the abscissae $\overline{P_1P_2}$ and $\overline{P_3P_4}$ for $x_{ij} \in \langle R_1, x_{ijm}\rangle$ and $x_{ij} \in \langle x_{ijm}, d/(2c_\varphi)\rangle$ in Fig. 8.20, respectively.

With regard to Eq. (8.3), we get

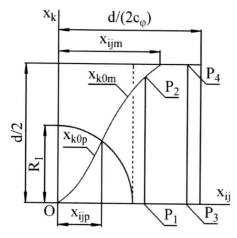

Figure 8.20: The abscissae $\overline{P_1P_2}$ and $\overline{P_3P_4}$ for $x_{ij} \in \langle R_1, x_{ijm}\rangle$ and $x_{ij} \in \langle x_{ijm}, d/(2c_\varphi)\rangle$ (see Eqs. (2.32), (2.33)), respectively, along which the curve integral $W_{cm}^{(ij)} = W_{cm}^{(ij)}(x_{ij}, \varphi, R_1, v)$ (see Eqs. (8.3), (8.53), (8.54)) for $x_{ijm} \in (R_1, d/(2c_\varphi))$ (see Fig. 8.16b) is determined due to the crack formation in the cell matrix on the condition in Item 2, Section 8.2.4.1.

$$W_{cm}^{(ij)} = \int_{\overline{P_1P_2}} w_m^{(ij)}\,dx_k = \int_0^{x_{k0m}} w_m^{(ij)}\,dx_k,$$

$$x_{ij} \in \langle R_1, x_{ijm}\rangle, \quad x_{ijm} \in \left(R_1, \frac{d}{2c_\varphi}\right), \tag{8.53}$$

$$W_{cm}^{(ij)} = \int_{\overline{P_3 P_4}} w_m^{(ij)} \, dx_k = \int_0^{d/2} w_m^{(ij)} \, dx_k,$$

$$x_{ij} \in \left\langle x_{ijm}, \frac{d}{2c_\varphi}\right\rangle, \quad x_{ijm} \in \left(R_1, \frac{d}{2c_\varphi}\right). \tag{8.54}$$

Finally, if $x_{ijm} = d/(2c_\varphi)$ (see Fig. 8.16c; Eqs. (2.32), (2.33)), then $W_{cm}^{(ij)} = W_{cm}^{(ij)}(x_{ij}, \varphi, R_1, v)$ is determined along the abscissa $\overline{P_1 P_2}$ for $x_{ij} \in \langle R_1, d/(2c_\varphi)\rangle$ in Fig. 8.21, respectively.

With regard to Eq. (8.3), we get

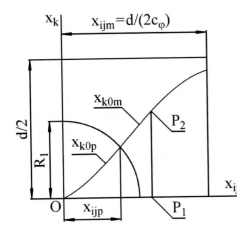

Figure 8.21: The abscissa $\overline{P_1 P_2}$ for $x_{ij} \in \langle R_1, d/(2c_\varphi)\rangle$ (see Eqs. (2.32), (2.33)) along which the curve integral $W_{cm}^{(ij)} = W_{cm}^{(ij)}(x_{ij}, \varphi, R_1, v)$ (see Eqs. (8.3), (8.55)) for $x_{ijm} = d/(2c_\varphi)$ (see Fig. 8.16c) is determined due to the crack formation in the cell matrix on the condition in Item 2, Section 8.2.4.1.

$$W_{cm}^{(ij)} = \int_{\overline{P_1 P_2}} w_m^{(ij)} \, dx_k = \int_0^{x_{k0m}} w_m^{(ij)} \, dx_k, \quad x_{ij} \in \left\langle R_1, \frac{d}{2c_\varphi}\right\rangle, \quad x_{ijm} = \frac{d}{2c_\varphi}. \tag{8.55}$$

B. Method 2

This method of determination of the curve integrals $W_{cp}^{(ij)} = W_{cp}^{(ij)}(x_{ij}, \varphi, R_1, v)$ and $W_{cm}^{(ij)} = W_{cm}^{(ij)}(x_{ij}, \varphi, R_1, v)$ is based on an assumption that energy W_{O2673O} (see Fig. 8.17b) or $W_{O25673O}$ (see Fig. 8.17c) is released by the crack formation in the plane $x_i x_j$ instead of the crack formation in the planes $x_j x_k$, $x_k x_i$ as assumed in Items (a)–(c), Section 8.2.4.1.A.

Finally, the method 2 results from a principle that a release of energy of a system considers 'minimal resistance' of the system, i.e. the energy is released through 'minimal resistance' of the system.

Analyses of the determination of $W_{cp}^{(ij)} = W_{cp}^{(ij)}(x_{ij}, \varphi, R_1, v)$ and $W_{cm}^{(ij)} = W_{cm}^{(ij)}(x_{ij}, \varphi, R_1, v)$ by the method 2 for the crack formation in the spherical particle and the cell matrix are as follows, respectively.

Spherical particle. Let the conditions in Item 2, Section 8.2.4.1 are valid.

Let the condition $x_{ijm} \in (R_1, d/(2c_\varphi))$ (see Eqs. (2.32), (2.33)) is valid, i.e. $x_{ijm} \in (R_1, d/(2c_\varphi))$ (see Fig. 8.17b) or $x_{ijm} = d/(2c_\varphi)$ (see Fig. 8.17c).

Let the critical particle radii $R_{1cp\,max}^{(jk)}$ and $R_{1cp\,max}^{(ki)}$ analysed in Items (a)–(c), Section 8.2.4.1.A are determined by the curve integrals $W_{cp}^{(jk)} = W_{cp}^{(jk)}(x_{jk}, \varphi, R_1, v)$ for $\varphi = \angle(x_j, x_{jk})$, $x_{jk} \subset x_j x_k$ and $W_{cp}^{(ki)} = W_{cp}^{(ki)}(x_{ki}, \varphi, R_1, v)$ for $\varphi = \angle(x_k, x_{ki})$, $x_{ki} \subset x_k x_i$, respectively, where $W_{cp}^{(jk)}$, $W_{cp}^{(ki)}$ are determined by the **method 1**, and a condition given by Eq. (8.17) for $q = p$ is considered for the determination of $R_{1cp\,max}^{(jk)}$, $R_{1cp\,max}^{(ki)}$.

Similarly, let the critical particle radius $R_{1cm\,max}^{(ij)}$ is determined by the curve integral $W_{cm}^{(ij)} = W_{cm}^{(ij)}(x_{ij}, \varphi, R_1, v)$ for $\varphi = \angle(x_i, x_{ij})$, $x_{ij} \subset x_i x_j$, where $W_{cm}^{(ij)}$ is also determined by the **method 1**, and a condition given by Eq. (8.17) for $q = m$ is considered for the determination of $R_{1cm\,max}^{(ij)}$.

Finally, let the critical particle radius $R_{1cp\,max}^{(ij)}$ is determined by the curve integral $W_{cp}^{(ij)} = W_{cp}^{(ij)}(x_{ij}, \varphi, R_1, v)$ for $\varphi = \angle(x_i, x_{ij})$, $x_{ij} \subset x_i x_j$, where $W_{cp}^{(ij)}$ is determined by the **method 2**, and a condition given by Eq. (8.17) for $q = p$ is considered for the determination of $R_{1cp\,max}^{(ij)}$.

With regard to Fig. 8.22, the curve integral $W_{cp}^{(ij)} = W_{cp}^{(ij)}(x_{ij}, \varphi, R_1, v)$ determined by the **method 2** for $x_{ij} \in \langle 0, x_{ijp} \rangle$ and $x_{ij} \in \langle x_{ijp}, R_1 \rangle$ is derived as

$$
W_{cp}^{(ij)} = \int_{\overline{P_2 P_6}} w_p^{(ij)}\, dx_k + \int_{\overline{P_6 P_7}} w_m^{(ij)}\, dx_k - \int_{\overline{P_1 P_2}} w_p^{(ij)}\, dx_k
$$

$$
= \int_{x_{k0p}}^{\sqrt{R_1^2 - x_{ij}^2}} w_p^{(ij)}\, dx_k + \int_{\sqrt{R_1^2 - x_{ij}^2}}^{d/2} w_m^{(ij)}\, dx_k - \int_{0}^{x_{k0p}} w_p^{(ij)}\, dx_k \geq 0,
$$

$$
x_{ij} \in \langle 0, x_{ijp} \rangle, \quad x_{ijm} \in \left(R_1, \frac{d}{2c_\varphi} \right),
\tag{8.56}
$$

$$
W_{cp}^{(ij)} = \int_{\overline{P_5 P_8}} w_m^{(ij)}\, dx_k - \left[\int_{\overline{P_3 P_4}} w_p^{(ij)}\, dx_k + \int_{\overline{P_4 P_5}} w_m^{(ij)}\, dx_k \right]
$$

$$= \int_{x_{k0m}}^{d/2} w_m^{(ij)} \, dx_k - \left[\int_0^{\sqrt{R_1^2 - x_{ij}^2}} w_p^{(ij)} \, dx_k + \int_{\sqrt{R_1^2 - x_{ij}^2}}^{x_{k0m}} w_m^{(ij)} \, dx_k \right] \geq 0,$$

$$x_{ij} \in \langle x_{ijp}, R_1 \rangle, \quad x_{ijm} \in \left(R_1, \frac{d}{2c_\varphi} \right). \qquad (8.57)$$

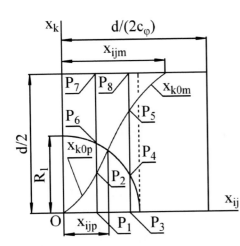

Figure 8.22: The abscissae $\overline{P_1 P_2 P_6 P_7}$ and $\overline{P_3 P_4 P_5 P_8}$ for $x_{ij} \in \langle 0, x_{ijp} \rangle$ and $x_{ij} \in \langle x_{ijp}, R_1 \rangle$, respectively, along which the curve integral $W_{cp}^{(ij)} = W_{cp}^{(ij)}(x_{ij}, \varphi, R_1, v)$ (see Eqs. (8.3), (8.56), (8.57)) for $x_{ijm} \in (R_1, d/(2c_\varphi))$ is determined by the method 2 due to the crack formation in the spherical particle in spite of the fact that the conditions in Item 2, Section 8.2.4.1 are valid. This determination is also considered for $x_{ijm} = d/(2c_\varphi)$ (see Eqs. (2.32), (2.33); Fig. 8.16c), although Fig. 8.22 is related to the condition $x_{ijm} \in (R_1, d/(2c_\varphi))$ (see Fig. 8.16b).

Additionally, $W_{cp}^{(ij)} = W_{cp}^{(ij)}(x_{ij}, \varphi, R_1, v)$ is define for such interval $x_{ij} \in \langle x_{ijp1}, x_{ijp0} \rangle \subset \langle 0, R_1 \rangle$ or $x_{ij} \in \langle x_{ijp0}, x_{ijp2} \rangle \subset \langle 0, R_1 \rangle$ for which the condition $W_{cp}^{(ij)}(x_{ij}, \varphi, R_1, v) \geq 0$ is valid, where x_{ijp0} is determined by the condition $W_{cp}^{(ij)}(x_{ij}, \varphi, R_1, v) = 0$.

If the conditions

$$R_{1cp\,max}^{(ij)} < R_{1cp\,max}^{(jk)}, \quad R_{1cp\,max}^{(ij)} < R_{1cp\,max}^{(ki)}, \quad R_{1cp\,max}^{(ij)} < R_{1cm\,max}^{(ij)} \qquad (8.58)$$

are simultaneously valid for the critical particle radii $R_{1cp\,max}^{(jk)}$, $R_{1cp\,max}^{(ki)}$, $R_{1cm\,max}^{(ij)}$ and $R_{1cp\,max}^{(ij)}$ determined by the methods 1 and 2, respectively, then the crack is formed in the plane $x_i x_j$ in the spherical particle

- in spite of the fact that the conditions in Item 2, Section 8.2.4.1 are valid,

- and in spite of the fact that the crack would be expected to be formed in the cell matrix regarding these conditions in Item 2, Section 8.2.4.1.

In this case, the 'minimal resistance' is represented by the critical particle radius $R_{1cp\,max}^{(ij)}$ as a minimal value of the set $\left\{ R_{1cp\,max}^{(ij)}, R_{1cp\,max}^{(jk)}, R_{1cp\,max}^{(ki)}, R_{1cm\,max}^{(ij)} \right\}$.

Cell matrix. Let the conditions in Item 1, Section 8.2.4.1 are valid. Let the condition $x_{ijm} = d/(2c_\varphi)$ (see Eqs. (2.32), (2.33)) is valid (see Fig. 8.17c).

Let the critical particle radii $R^{(jk)}_{1cm\,max}$ and $R^{(ki)}_{1cm\,max}$ analysed in Items (a)–(c), Section 8.2.4.1.A are determined by the curve integrals $W^{(jk)}_{cm} = W^{(jk)}_{cm}(x_{jk}, \varphi, R_1, v)$ for $\varphi = \angle(x_j, x_{jk})$, $x_{jk} \subset x_j x_k$ and $W^{(ki)}_{cm} = W^{(ki)}_{cm}(x_{ki}, \varphi, R_1, v)$ for $\varphi = \angle(x_k, x_{ki})$, $x_{ki} \subset x_k x_i$, respectively, where $W^{(jk)}_{cm}$, $W^{(ki)}_{cm}$ are determined by the **method 1**, and a condition given by Eq. (8.17) for $q = m$ is considered for the determination of $R^{(jk)}_{1cm\,max}$, $R^{(ki)}_{1cm\,max}$.

Similarly, let the critical particle radius $R^{(ij)}_{1cp\,max}$ is determined by the curve integral $W^{(ij)}_{cp} = W^{(ij)}_{cp}(x_{ij}, \varphi, R_1, v)$ for $\varphi = \angle(x_i, x_{ij})$, $x_{ij} \subset x_i x_j$, where $W^{(ij)}_{cp}$ is also determined by the **method 1**, and a condition given by Eq. (8.17) for $q = p$ is considered for the determination of $R^{(ij)}_{1cp\,max}$.

Finally, let the critical particle radius $R^{(ij)}_{1cm\,max}$ is determined by the curve integral $W^{(ij)}_{cm} = W^{(ij)}_{cm}(x_{ij}, \varphi, R_1, v)$ for $\varphi = \angle(x_i, x_{ij})$, $x_{ij} \subset x_i x_j$, where $W^{(ij)}_{cm}$ is determined by the **method 2**, and a condition given by Eq. (8.17) for $q = m$ is considered for the determination of $R^{(ij)}_{1cm\,max}$.

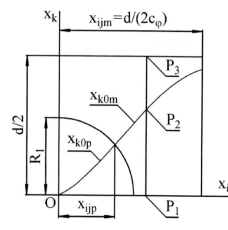

Figure 8.23: The abscissa $\overline{P_1 P_2 P_3}$ for $x_{ij} \in \langle R_1, d/(2c_\varphi)\rangle$ (see Eqs. (2.32), (2.33)) along which the curve integral $W^{(ij)}_{cm} = W^{(ij)}_{cm}(x_{ij}, \varphi, R_1, v)$ (see Eqs. (8.3), (8.59)) for $x_{ijm} = d/(2c_\varphi)$ (see Fig. 8.16c) is determined by the method 2 due to the crack formation in the cell matrix in spite of the fact that the conditions in Item 1, Section 8.2.4.1 are valid.

With regard to Fig. 8.23, the curve integral $W^{(ij)}_{cm} = W^{(ij)}_{cm}(x_{ij}, \varphi, R_1, v)$ determined by the **method 2** for $x_{ij} \in \langle R_1, d/(2c_\varphi)\rangle$ (see Eqs. (2.32), (2.33)) is derived as

$$W^{(ij)}_{cm} = \int_{\overline{P_2 P_3}} w^{(ij)}_m \, dx_k - \int_{\overline{P_1 P_2}} w^{(ij)}_m \, dx_k = \int_{x_{k0m}}^{d/2} w^{(ij)}_m \, dx_k - \int_0^{x_{k0m}} w^{(ij)}_m \, dx_k,$$

$$x_{ij} \in \left\langle R_1, \frac{d}{2c_\varphi} \right\rangle, \quad x_{ijm} = \frac{d}{2c_\varphi}. \tag{8.59}$$

Additionally, $W^{(ij)}_{cm} = W^{(ij)}_{cm}(x_{ij}, \varphi, R_1, v)$ is define for such interval $x_{ij} \in \langle x_{ijm1}, x_{ijm0}\rangle \subset \langle R_1, d/(2c_\varphi)\rangle$ or $x_{ij} \in \langle x_{ijm0}, x_{ijm2}\rangle \subset \langle R_1, d/(2c_\varphi)\rangle$ for which the condition $W_{cm}^{(ij)}\left(x_{ij}, \varphi, R_1, v\right) \geq 0$ is valid, where x_{ijm0} is determined by the condition $W_{cm}^{(ij)}\left(x_{ij}, \varphi, R_1, v\right) = 0$.

If the conditions

$$R_{1cm\,max}^{(ij)} < R_{1cm\,max}^{(jk)}, \quad R_{1cm\,max}^{(ij)} < R_{1cm\,max}^{(ki)}, \quad R_{1cm\,max}^{(ij)} < R_{1cp\,max}^{(ij)} \qquad (8.60)$$

are simultaneously valid for the critical particle radii $R_{1cm\,max}^{(jk)}$, $R_{1cm\,max}^{(ki)}$, $R_{1cp\,max}^{(ij)}$ and $R_{1cm\,max}^{(ij)}$ determined by the methods 1 and 2, respectively, then the crack is formed in the plane $x_i x_j$ in the cell matrix

- in spite of the fact that the conditions in Item 1, Section 8.2.4.1 are valid,

- and in spite of the fact that the crack would be expected to be formed in the spherical particle regarding these conditions in Item 1, Section 8.2.4.1.

In this case, the 'minimal resistance' is represented by the critical particle radius $R_{1cm\,max}^{(ij)}$ as a minimal value of the set $\left\{ R_{1cm\,max}^{(ij)}, R_{1cm\,max}^{(jk)}, R_{1cm\,max}^{(ki)}, R_{1cp\,max}^{(ij)} \right\}$.

8.2.4.2 Determination of $W_{cp}^{(ij)}$, $W_{ce}^{(ij)}$, $W_{cm}^{(ij)}$ in multi-particle-envelope-matrix system

The determination of $W_{cp}^{(ij)} = W_{cp}^{(ij)}\left(x_{ij}, \varphi, R_1, t, v\right)$, $W_{ce}^{(ij)} = W_{ce}^{(ij)}\left(x_{ij}, \varphi, R_1, t, v\right) = W_{ce}^{(ij)}\left(x_{ij}, \varphi, R_2, t, v\right)$ and $W_{cm}^{(ij)} = W_{cm}^{(ij)}\left(x_{ij}, \varphi, R_2, t, v\right)$ (see Eq. (8.3)) for the crack formation in the spherical particle, the spherical envelope and the cell matrix considers the the condition $w_q^{(ij)} = 0$ $(q=p,e,m)$ for the energy density $w_p = w_p^{(ij)}\left(x_{ij}, \varphi, x_k, R_1, t, v\right)$, $w_e = w_e^{(ij)}\left(x_{ij}, \varphi, x_k, R_1, t, v\right) = w_e^{(ij)}\left(x_{ij}, \varphi, x_k, R_2, t, v\right)$ and $w_m = w_m^{(ij)}\left(x_{ij}, \varphi, x_k, R_2, t, v\right)$ (see Eq. (8.1)), respectively.

Consequently, the conditions $w_p^{(ij)}\left(x_{ij}, \varphi, x_k, R_1, t, v\right) = w_p^{(ij)}\left(r, \varphi, \nu, R_1, t, v\right) = 0$, $w_e^{(ij)}\left(x_{ij}, \varphi, x_k, R_1, t, v\right) = w_e^{(ij)}\left(r, \varphi, \nu, R_1, t, v\right) = 0$ and $w_m^{(ij)}\left(x_{ij}, \varphi, x_k, R_1, t, v\right) = w_m^{(ij)}\left(r, \varphi, \nu, R_1, t, v\right) = 0$, which are valid for $\sigma'_{n_1 n_2 p}\left(x_{ij}, \varphi, x_k, R_1, t, v\right) = \sigma'_{n_1 n_2 p}\left(r, \varphi, \nu, R_1, t, v\right) = 0$, $\sigma'_{n_1 n_2 e}\left(x_{ij}, \varphi, x_k, R_1, t, v\right) = \sigma'_{n_1 n_2 e}\left(r, \varphi, \nu, R_1, t, v\right) = 0$ and $\sigma'_{n_1 n_2 m}\left(x_{ij}, \varphi, x_k, R_1, v\right) = \sigma'_{n_1 n_2 m}\left(r, \varphi, \nu, R_1, v\right) = 0$ $(n_1, n_2 = 1,2,3)$, result in the functions $x_{k0p} = x_{k0p}\left(x_{ij}, \varphi, R_1, t, v\right)$, $x_{k0e} = x_{k0e}\left(x_{ij}, \varphi, R_1, t, v\right)$ and $x_{k0m} = x_{k0m}\left(x_{ij}, \varphi, R_1, t, v\right)$ of the variables $x_{ij} \in \langle 0, x_{ijp} \rangle$, $x_{ij} \in \langle x_{ijp}, x_{ije} \rangle$ and $x_{ij} \in \langle x_{ije}, x_{ijm} \rangle$, (see Figs. 8.24, 8.25), respectively, where these functions are considered for $\varphi \in \langle \varphi_{1pe}, \varphi_{2pe} \rangle = \langle \varphi_{1em}, \varphi_{2em} \rangle$. The interval $\langle \varphi_{1pe}, \varphi_{2pe} \rangle = \langle \varphi_{1em}, \varphi_{2em} \rangle$ represents an interval of the validity of the condition $w_q\left(x_{ij}, \varphi, x_k, R_1, t, v\right) = 0$ regarding the variable φ.

Figures 8.24 and 8.25 are related to the conditions $x_{ije} \in \langle 0, R_1 \rangle$ and $x_{ije} \in (R_1, R_2)$, respectively.

Additionally, with regard to Figs. 8.24, 8.25, we get $\left[x_{k0p}\left(x_{ij}, \varphi, R_1, t, v\right)\right]_{x_{ij}=x_{ijp}} = \left[x_{k0e}\left(x_{ij}, \varphi, R_1, t, v\right)\right]_{x_{ij}=x_{ijp}}$ and $\left[x_{k0e}\left(x_{ij}, \varphi, R_1, t, v\right)\right]_{x_{ij}=x_{ije}} =$

$[x_{k0m}(x_{ij}, \varphi, R_1, t, v)]_{x_{ij}=x_{ije}}$; $x_{ijp} \in \langle 0, R_1 \rangle$; either $x_{ije} \in \langle 0, R_1 \rangle$ (see Fig. 8.24) or $x_{ije} \in (R_1, R_2)$ (see Fig. 8.25); and either $x_{ijm} \in \langle 0, R_1 \rangle$ (see Fig. 8.24a) or $x_{ijm} \in (R_1, R_2 \rangle$ (see Figs. 8.24b, 8.25a) or $x_{ijm} \in (R_2, d/(2c_\varphi))$ (see Figs. 8.24c, 8.25b) or $x_{ijm} = d/(2c_\varphi)$ (see Figs. 8.24d, 8.25c; Eqs. (2.32), (2.33)). Finally, with regard to Figs. 8.24d, 8.25c, x_{ijm} is determined by the condition $[x_{k0m}(x_{ij}, \varphi, R_1, t, v)]_{x_k=x_{ijm}} = d/2$.

The stresses $\sigma'_{1nq} = \sigma'_{1nq}(r, \varphi, \nu, R_1, t, v)$, $\sigma'_{2nq} = \sigma'_{2nq}(r, \varphi, \nu, R_1, t, v)$ and $\sigma'_{3nq} = \sigma'_{3nq}(r, \varphi, \nu, R_1, t, v)$ ($n = 1,2,3$) which induce the energy density $w_q = w_q(r, \varphi, \nu, R_1, t, v)$ act along the axes x'_1, x'_2 and x'_3 (see Fig. 3.1), respectively.

With regard to $[\sigma'_{n_1 n_2 q}(x_{ij}, \varphi, x_k, R_1, t, v) = 0]_{x_k=x_{k0q}}$ ($q = p,e,m$; $n_1, n_2 = 1,2,3$), the curve integrals $W_{cp}^{(ij)}$, $W_{ce}^{(ij)}$, $W_{cm}^{(ij)}$ are determined on the following conditions:

1. If the force $\vec{F'}_{nnp}$ induced by the stress σ'_{nnp} for positions within the surface $O12O$, i.e. for the coordinates $(x_{ij}, x_k) \in O12O$ (see Figs. 8.24, 8.25) acts along the axis $+x'_n$ (see Fig. 3.1), and the force $\vec{F'}_{nne}$ induced by the stress σ'_{nne} for $(x_{ij}, x_k) \in 14521$ (see Figs. 8.24, 8.25) acts along the axis $-x'_n$, and the force $\vec{F'}_{nnm}$ induced by the stress σ'_{nnm} for $(x_{ij}, x_k) \in 478954$ (see Figs. 8.24a,b,c, 8.25a,b) or for $(x_{ij}, x_k) \in 47854$ (see Figs. 8.24d, 8.25c) acts along the axis $+x'_n$, then the force $\vec{F'}_{nnp}$ induced by the stress σ'_{nnp} for $(x_{ij}, x_k) \in O23O$ (see Fig. 8.16a,b,c) (see Figs. 8.24, 8.25) acts along the axis $-x'_n$, and the force $\vec{F'}_{nne}$ induced by the stress σ'_{nne} for $(x_{ij}, x_k) \in 25632$ (see Figs. 8.24, 8.25) acts along the axis $+x'_n$, and the force $\vec{F'}_{nnm}$ induced by the stress σ'_{nnm} for $(x_{ij}, x_k) \in 59\,10\,65$ (see Figs. 8.24a,b,c, 8.25a,b) or for $(x_{ij}, x_k) \in 589\,10\,65$ (see Figs. 8.24d, 8.25c) acts along the axis $-x'_n$.

   If the multi-particle-matrix system is loaded by the thermal stresses only, this condition concerning $\vec{F'}_{nnq}$ is valid provided that $\beta_p \geq \beta_e \leq \beta_m$ (see Eqs. (3.100)–(3.102)) for $(x_{ij}, x_k) \in O14789520$ (see Figs. 8.24a,b,c, 8.25a,b) or $(x_{ij}, x_k) \in O14785 2O$ (see Figs. 8.24d, 8.25c). Consequently, we get $\beta_p \leq \beta_e \geq \beta_m$ for $(x_{ij}, x_k) \in O2591063O$ (see Figs. 8.24a,b,c, 8.25a,b) or $(x_{ij}, x_k) \in O2891063O$ (see Figs. 8.24d, 8.25c).

2. If the force $\vec{F'}_{nnp}$ induced by the stress σ'_{nnp} for positions within the surface $O12O$, i.e. for the coordinates $(x_{ij}, x_k) \in O12O$ (see Figs. 8.24, 8.25) acts along the axis $-x'_n$, and the force $\vec{F'}_{nne}$ induced by the stress σ'_{nne} for $(x_{ij}, x_k) \in 14521$ (see Figs. 8.24, 8.25) acts along the axis $+x'_n$, and the force $\vec{F'}_{nnm}$ induced by the stress σ'_{nnm} for $(x_{ij}, x_k) \in 478954$ (see Figs. 8.24a,b,c, 8.25a,b) or for $(x_{ij}, x_k) \in 47854$ (see Figs. 8.24d, 8.25c) acts along the axis $-x'_n$, then the force $\vec{F'}_{nnp}$ induced by the stress σ'_{nnp} for $(x_{ij}, x_k) \in O23O$ (see Fig. 8.16a,b,c) (see Figs. 8.24, 8.25) acts along the axis $+x'_n$, and the force $\vec{F'}_{nne}$ induced by the stress σ'_{nne} for $(x_{ij}, x_k) \in 25632$ (see Figs. 8.24, 8.25) acts along the axis $-x'_n$, and the force $\vec{F'}_{nnm}$ induced by the stress σ'_{nnm} for $(x_{ij}, x_k) \in 59\,10\,65$ (see Figs. 8.24a,b,c, 8.25a,b) or for $(x_{ij}, x_k) \in 589\,10\,65$ (see Figs. 8.24d, 8.25c) acts along the axis $+x'_n$.

Related phenomena   127

If the multi-particle-matrix system is loaded by the thermal stresses only, this condition concerning $\vec{F}'_{nnq}$ is valid provided that $\beta_p \leq \beta_e \geq \beta_m$ (see Eqs. (3.100)–(3.102)) for $(x_{ij}, x_k) \in O1478952O$ (see Figs. 8.24a,b,c, 8.25a,b) or $(x_{ij}, x_k) \in O147852O$ (see Figs. 8.24d, 8.25c). Consequently, we get $\beta_p \geq \beta_e \leq \beta_m$ for $(x_{ij}, x_k) \in O2591063O$ (see Figs. 8.24a,b,c, 8.25a,b) or $(x_{ij}, x_k) \in O2891063O$ (see Figs. 8.24d, 8.25c).

The surface $O12O$ in Figs. 8.24, 8.25 is define for $x_{ij} \in \langle 0, x_{ijp} \rangle$, $x_k \in \langle 0, x_{k0p} \rangle$; and for $x_{ij} \in \langle x_{ijp}, R_1 \rangle$, $x_k \in \left\langle 0, \sqrt{R_1^2 - x_{ij}^2} \right\rangle$.

The surface $O23O$ in Figs. 8.24, 8.25 is define for $x_{ij} \in \langle 0, x_{ijp} \rangle$, $x_k \in \left\langle x_{k0p}, \sqrt{R_1^2 - x_{ij}^2} \right\rangle$.

The surface 14521 in Figs. 8.24, 8.25 is define for $x_{ij} \in \langle x_{ijp}, x_{ije} \rangle$, $x_k \in \left\langle \sqrt{R_1^2 - x_{ij}^2}, x_{k0e} \right\rangle$; for $x_{ij} \in \langle x_{ije}, R_1 \rangle$, $x_k \in \left\langle \sqrt{R_1^2 - x_{ij}^2}, \sqrt{R_2^2 - x_{ij}^2} \right\rangle$; and for $x_{ij} \in \langle R_1, R_2 \rangle$, $x_k \in \left\langle 0, \sqrt{R_2^2 - x_{ij}^2} \right\rangle$.

The surface 25632 in Figs. 8.24, 8.25 is define for $x_{ij} \in \langle 0, x_{ijp} \rangle$, $x_k \in \left\langle \sqrt{R_1^2 - x_{ij}^2}, \sqrt{R_2^2 - x_{ij}^2} \right\rangle$; and for $x_{ij} \in \langle x_{ijp}, x_{ije} \rangle$, $x_k \in \left\langle x_{k0e}, \sqrt{R_2^2 - x_{ij}^2} \right\rangle$.

The surface 478954 in Figs. 8.24a,b,c, 8.25a,b is define for $x_{ij} \in \langle x_{ije}, x_{ijm} \rangle$, $x_k \in \left\langle \sqrt{R_2^2 - x_{ij}^2}, x_{k0m} \right\rangle$; for $x_{ij} \in \langle x_{ijm}, R_2 \rangle$, $x_k \in \left\langle \sqrt{R_2^2 - x_{ij}^2}, d/2 \right\rangle$; and for $x_{ij} \in \langle R_2, d/(2c_\varphi) \rangle$, $x_k \in \langle 0, d/2 \rangle$ (see Eqs. (2.32), (2.33)).

The surface 47854 in Figs. 8.24d, 8.25c is define for $x_{ij} \in \langle x_{ije}, R_2 \rangle$, $x_k \in \left\langle \sqrt{R_2^2 - x_{ij}^2}, x_{k0m} \right\rangle$; and for $x_{ij} \in \langle R_2, d/(2c_\varphi) \rangle$, $x_k \in \langle 0, x_{k0m} \rangle$ (see Eqs. (2.32), (2.33)).

The surface 591065 in Figs. 8.24a,b,c, 8.25a,b is define for $x_{ij} \in \langle 0, x_{ije} \rangle$, $x_k \in \left\langle \sqrt{R_2^2 - x_{ij}^2}, d/2 \right\rangle$; and for $x_{ij} \in \langle x_{ije}, x_{ijm} \rangle$, $x_k \in \langle x_{k0m}, d/2 \rangle$.

The surface 5891065 in Figs. 8.24d, 8.25c is define for $x_{ij} \in \langle 0, x_{ije} \rangle$, $x_k \in \left\langle \sqrt{R_2^2 - x_{ij}^2}, d/2 \right\rangle$; and for $x_{ij} \in \langle x_{ije}, d/(2c_\varphi) \rangle$, $x_k \in \langle x_{k0m}, d/2 \rangle$ (see Eqs. (2.32), (2.33)).

If the multi-particle-envelope-matrix system is loaded by the thermal stresses only, then the conditions $w_p(x_{ij}, \varphi, x_k, R_1, v) = w_p(r, \varphi, \nu, R_1, v) = 0$ and $w_m(x_{ij}, \varphi, x_k, R_1, v) = w_m(r, \varphi, \nu, R_1, v) = 0$ are transformed to the condition $\beta_p = \beta_e = \beta_m$ (see Eqs. (3.100)–(3.102)). Additionally, if the thermal stresses originate as a consequence of a difference in thermal expansion coefficient of the spherical particle, the spherical envelope and the cell matrix, i.e. the phase-transformation induced strain $\varepsilon_{11tq} = 0$ (see Eqs. (3.100), (3.104)), then $x_{k0p} = x_{k0p}(x_{ij}, \varphi, R_1, t, v)$ for $x_{ij} \in \langle 0, x_{ijp} \rangle$, $x_{k0e} = x_{k0e}(x_{ij}, \varphi, R_1, t, v)$ for $x_{ij} \in \langle x_{ijp}, x_{ije} \rangle$ and $x_{k0m} = x_{k0m}(x_{ij}, \varphi, R_1, t, v)$ for $x_{ij} \in \langle x_{ije}, x_{ijm} \rangle$ are derived as

$$x_{k0q} = x_{ij} \cot \nu_{pe}, \quad q = p, e, m, \tag{8.61}$$

where the angle $\nu_{pe} = \nu_{pe}(\varphi, R_1, v) = \nu_{em}(\varphi, R_1, v)$ is given by Eq. (3.106) for the multi-particle-envelope-matrix system with anisotropic components only, or with anisotropic and

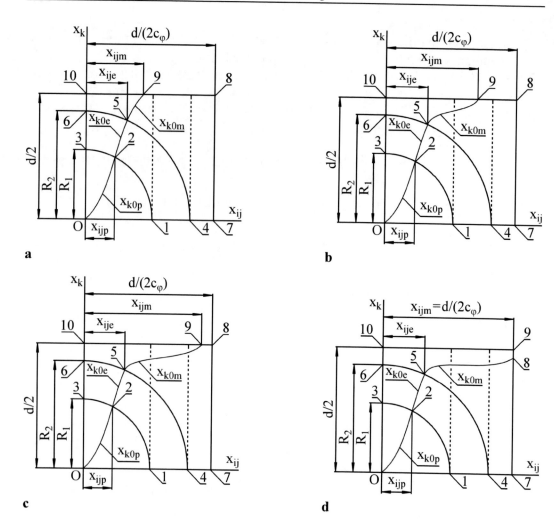

Figure 8.24: The functions $x_{k0p} = x_{k0p}(x_{ij}, \varphi, R_1, t, v)$, $x_{k0e} = x_{k0e}(x_{ij}, \varphi, R_1, t, v)$ and $x_{k0m} = x_{k0m}(x_{ij}, \varphi, R_1, t, v)$ of the variables $x_{ij} \in \langle 0, x_{ijp} \rangle$, $x_{ij} \in \langle x_{ijp}, x_{ije} \rangle$ and $x_{ij} \in \langle x_{ije}, x_{ijm} \rangle$, respectively, where $[x_{k0p}(x_{ij}, \varphi, R_1, t, v)]_{x_{ij}=x_{ijp}} = [x_{k0e}(x_{ij}, \varphi, R_1, t, v)]_{x_{ij}=x_{ijp}}$, $[x_{k0e}(x_{ij}, \varphi, R_1, t, v)]_{x_{ij}=x_{ije}} = [x_{k0m}(x_{ij}, \varphi, R_1, t, v)]_{x_{ij}=x_{ije}}$; $x_{ijq} \in \langle 0, R_1 \rangle$ $(q = p, e)$; and either (a) $x_{ijm} \in \langle 0, R_1 \rangle$ or (b) $x_{ijm} \in (R_1, R_2)$ or (c) $x_{ijm} \in (R_2, d/(2c_\varphi))$ (see Eqs. (2.32), (2.33)) or (d) $x_{ijm} = d/(2c_\varphi)$. The function $x_{k0q} = x_{k0q}(x_{ij}, \varphi, R_1, t, v)$ $(q = p, e, m)$ results from the condition $w_q(x_{ij}, \varphi, x_k, R_1, t, v) = w_q(r, \varphi, \nu, R_1, t, v) = 0$ which is valid for $\sigma'_{n_1 n_2 q}(x_{ij}, \varphi, x_k, R_1, t, v) = \sigma'_{n_1 n_2 q}(r, \varphi, \nu, R_1, t, v) = 0$ $(n_1, n_2 = 1, 2, 3)$. The stresses $\sigma'_{1nq} = \sigma'_{1nq}(r, \varphi, \nu, R_1, t, v)$, $\sigma'_{2nq} = \sigma'_{2nq}(r, \varphi, \nu, R_1, t, v)$ and $\sigma'_{3nq} = \sigma'_{3nq}(r, \varphi, \nu, R_1, t, v)$ $(n = 1, 2, 3)$ which induce the energy density $w_q = w_q(r, \varphi, \nu, R_1, t, v)$ act along the axes x'_1, x'_2 and x'_3 (see Fig. 3.1), respectively.

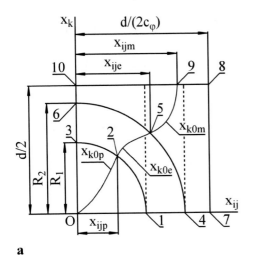

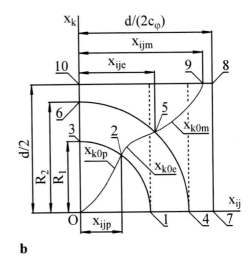

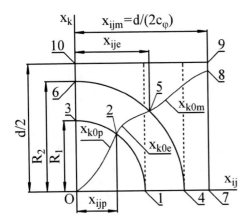

Figure 8.25: The functions $x_{k0p} = x_{k0p}(x_{ij}, \varphi, R_1, t, v)$, $x_{k0e} = x_{k0e}(x_{ij}, \varphi, R_1, t, v)$ and $x_{k0m} = x_{k0m}(x_{ij}, \varphi, R_1, t, v)$ of the variables $x_{ij} \in \langle 0, x_{ijp} \rangle$, $x_{ij} \in \langle x_{ijp}, x_{ije} \rangle$ and $x_{ij} \in \langle x_{ije}, x_{ijm} \rangle$, respectively, where $x_{ijp} \in \langle 0, R_1 \rangle$; $x_{ije} \in (R_1, R_2)$; and either (a) $x_{ijm} \in (R_1, R_2)$ or (b) $x_{ijm} \in (R_2, d/(2c_\varphi))$ (see Eqs. (2.32), (2.33)) or (c) $x_{ijm} = d/(2c_\varphi)$.

isotropic components, respectively.

The analyses in Items 1, 2 are also valid for the crack formation in the plane $x_j x_k$. In this case, $x_{k0q} = x_{k0q}(x_{ij}, \varphi, R_1, v)$ and x_{ijq} ($q = p,e,m$) are replaced by $x_{i0q} = x_{i0q}(x_{jk}, \varphi, R_1, t, v)$ and x_{jkq}, respectively, where $\varphi = \angle(x_j, x_{jk})$, $x_{jk} \subset x_j x_k$. Similarly, $x_{k0q} = x_{k0q}(x_{ij}, \varphi, R_1, t, v)$ and x_{ijq} ($q = p,m$) are replaced by $x_{j0q} = x_{j0q}(x_{ki}, \varphi, R_1, t, v)$ and x_{kiq} in case of the crack formation in the plane $x_k x_i$, respectively, where $\varphi = \angle(x_k, x_{ki})$, $x_{ki} \subset x_k x_i$.

A. Method 1

Let $W_{O14789520}$ or $W_{O1478520}$ is energy accumulated in the cell volume which is related to the surface $O14789520$ (see Fig. 8.25a,b,c, Fig. 8.26a,b) or $O1478520$ (see Fig. 8.25d, Fig. 8.26c), respectively, and both are related to the angle $\varphi \in \langle \varphi_{1pm}, \varphi_{2pm} \rangle$.

Consequently, let $W_{O259\,10\,630}$ or $W_{O2589\,10\,630}$ is energy accumulated in the cell volume which is related to the surface $O259\,10\,630$ (see Fig. 8.25a,b,c, Fig. 8.26a,b) or

$O2589\,10\,630$ (see Fig. 8.25d, Fig. 8.26c), and both are also related to the angle $\varphi \in \langle \varphi_{1pm}, \varphi_{2pm} \rangle$.

As presented in Section 8.2.4, this method of determination of $W_{cp}^{(ij)} = W_{cp}^{(ij)}(x_{ij}, \varphi, R_1, t, v)$, $W_{ce}^{(ij)} = W_{ce}^{(ij)}(x_{ij}, \varphi, R_1, t, v) = W_{ce}^{(ij)}(x_{ij}, \varphi, R_2, t, v)$ and $W_{cm}^{(ij)} = W_{cm}^{(ij)}(x_{ij}, \varphi, R_2, t, v)$ is based on an assumption that energy $W_{O14789520}$ or $W_{O1478520}$ is released by the crack formation in the plane $x_i x_j$. Accordingly, $W_{O259\,10\,630}$ or $W_{O2589\,10\,630}$ is assumed to be released

(a) either in the plane $x_j x_k$ provided that $R_{\eta cq_\eta, max}^{(jk)} < R_{\eta cq_\eta, max}^{(ki)}$,

(b) or in the plane $x_k x_i$ provided that $R_{\eta cq_\eta, max}^{(jk)} > R_{\eta cq_\eta, max}^{(ki)}$,

(c) or in one of the planes $x_j x_k$, $x_k x_i$ provided that $R_{\eta cq_\eta, max}^{(jk)} = R_{\eta cq_\eta, max}^{(ki)}$, ($\eta = 1,2$; $q_1 = p,e$; $q_2 = e,m$).

The critical particle radii $R_{\eta cq_\eta, max}^{(jk)} = R_{\eta cq_\eta, max}^{(jk)}(t, v)$ and $R_{\eta cq_\eta, max}^{(ki)} = R_{\eta cq_\eta, max}^{(ki)}(t, v)$ represent maximal values of the functions $R_{\eta cq_\eta}^{(jk)} = R_{\eta cq_\eta}^{(jk)}(\varphi, t, v)$ and $R_{\eta cq_\eta}^{(ki)} = R_{\eta cq_\eta}^{(ki)}(\varphi, t, v)$ of the variable $\varphi = \angle(x_j, x_{jk})$ and $\varphi = \angle(x_k, x_{ki})$, respectively, where $R_{\eta cq_\eta\,max}^{(jk)} \equiv R_{\eta cq_\eta n\,max}^{(jk)}$, $R_{\eta cq_\eta\,max}^{(ki)} \equiv R_{\eta cq_\eta n\,max}^{(ki)}$ ($\eta = 1,2$; $n = 1,2,4$; $q_1 = p,e$; $q_2 = e,m$) (see Section 8.2.2.2.A.4).

Spherical particle. If the condition in Item 1, Section 8.2.4.1 is valid, then a crack is formed in the spherical particle, i.e. for $x_{ij} \in \langle 0, R_1 \rangle$. Additionally, the intervals

- $x_{ije} \in \langle 0, R_1 \rangle$ (see Fig. 8.24) for $x_{ijm} \in \langle 0, R_1 \rangle$ (see Fig. 8.24a) or $x_{ijm} \in (R_1, R_2)$ (see Fig. 8.24b) or $x_{ijm} \in (R_2, d/(2c_\varphi))$ (see Eqs. (2.32), (2.33); Fig. 8.24c) or $x_{ijm} = d/(2c_\varphi)$ (see Fig. 8.24d),

- $x_{ije} \in (R_1, R_2)$ (see Fig. 8.25) for $x_{ijm} \in (R_1, R_2)$ (see Fig. 8.25a) or $x_{ijm} \in (R_2, d/(2c_\varphi))$ (see Fig. 8.25b) or $x_{ijm} = d/(2c_\varphi)$ (see Fig. 8.25c)

are required to be considered for the determination of the curve integral $W_{cp}^{(ij)} = W_{cp}^{(ij)}(x_{ij}, \varphi, R_1, t, v)$ for $x_{ij} \in \langle 0, R_1 \rangle$.

Consequently, if $x_{ije} \in \langle 0, R_1 \rangle$ and $x_{ijm} \in \langle 0, R_1 \rangle$ (see Fig. 8.24a), then $W_{cp}^{(ij)} = W_{cp}^{(ij)}(x_{ij}, \varphi, R_1, t, v)$ is determined along the abscissae $\overline{P_1 P_2}$, $\overline{P_3 P_4 P_5}$, $\overline{P_6 P_7 P_8 P_9}$ and $\overline{P_{10} P_{11} P_{12} P_{13}}$ for $x_{ij} \in \langle 0, x_{ijp} \rangle$, $x_{ij} \in \langle x_{ijp}, x_{ije} \rangle$, $x_{ij} \in \langle x_{ije}, x_{ijm} \rangle$ and $x_{ij} \in \langle x_{ijm}, R_1 \rangle$ in Fig. 8.26, respectively.

With regard to Eq. (8.3), we get

$$W_{cp}^{(ij)} = \int_{\overline{P_1 P_2}} w_p^{(ij)} \, dx_k = \int_0^{x_{k0p}} w_p^{(ij)} \, dx_k,$$

$$x_{ij} \in \langle 0, x_{ijp} \rangle, \quad x_{ije} \in \langle 0, R_1 \rangle, \quad x_{ijm} \in \langle 0, R_1 \rangle, \tag{8.62}$$

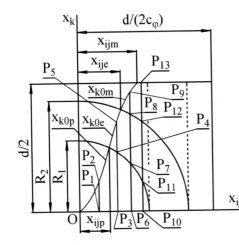

Figure 8.26: The abscissae $\overline{P_1P_2}$, $\overline{P_3P_4P_5}$, $\overline{P_6P_7P_8P_9}$ and $\overline{P_{10}P_{11}P_{12}P_{13}}$ for $x_{ij} \in \langle 0, x_{ijp} \rangle$, $x_{ij} \in \langle x_{ijp}, x_{ije} \rangle$, $x_{ij} \in \langle x_{ije}, x_{ijm} \rangle$ and $x_{ij} \in \langle x_{ijm}, R_1 \rangle$, respectively, along which the curve integral $W_{cp}^{(ij)} = W_{cp}^{(ij)}(x_{ij}, \varphi, R_1, t, v)$ (see Eqs. (8.3), (8.62)–(8.65)) for $x_{ije} \in \langle 0, R_1 \rangle$ and $x_{ijm} \in \langle 0, R_1 \rangle$ (see Fig. 8.24a) is determined due to the crack formation in the spherical particle on the condition in Item 1, Section 8.2.4.2.

$$W_{cp}^{(ij)} = \int_{\overline{P_3P_4}} w_p^{(ij)} dx_k + \int_{\overline{P_4P_5}} w_e^{(ij)} dx_k = \int_0^{\sqrt{R_1^2-x_{ij}^2}} w_p^{(ij)} dx_k + \int_{\sqrt{R_1^2-x_{ij}^2}}^{x_{k0e}} w_e^{(ij)} dx_k,$$

$$x_{ij} \in \langle x_{ijp}, x_{ije} \rangle, \quad x_{ije} \in \langle 0, R_1 \rangle, \quad x_{ijm} \in \langle 0, R_1 \rangle, \tag{8.63}$$

$$W_{cp}^{(ij)} = \int_{\overline{P_6P_7}} w_p^{(ij)} dx_k + \int_{\overline{P_7P_8}} w_e^{(ij)} dx_k + \int_{\overline{P_8P_9}} w_m^{(ij)} dx_k$$
$$= \int_0^{\sqrt{R_1^2-x_{ij}^2}} w_p^{(ij)} dx_k + \int_{\sqrt{R_1^2-x_{ij}^2}}^{\sqrt{R_2^2-x_{ij}^2}} w_e^{(ij)} dx_k + \int_{\sqrt{R_2^2-x_{ij}^2}}^{x_{k0m}} w_m^{(ij)} dx_k,$$

$$x_{ij} \in \langle x_{ije}, x_{ijm} \rangle, \quad x_{ije} \in \langle 0, R_1 \rangle, \quad x_{ijm} \in \langle 0, R_1 \rangle, \tag{8.64}$$

$$W_{cp}^{(ij)} = \int_{\overline{P_{10}P_{11}}} w_p^{(ij)} dx_k + \int_{\overline{P_{11}P_{12}}} w_e^{(ij)} dx_k + \int_{\overline{P_{12}P_{13}}} w_m^{(ij)} dx_k$$
$$= \int_0^{\sqrt{R_1^2-x_{ij}^2}} w_p^{(ij)} dx_k + \int_{\sqrt{R_1^2-x_{ij}^2}}^{\sqrt{R_2^2-x_{ij}^2}} w_e^{(ij)} dx_k + \int_{\sqrt{R_2^2-x_{ij}^2}}^{d/2} w_m^{(ij)} dx_k,$$

$$x_{ij} \in \langle x_{ijm}, R_2 \rangle, \quad x_{ije} \in \langle 0, R_1 \rangle, \quad x_{ijm} \in \langle 0, R_1 \rangle. \tag{8.65}$$

If $x_{ije} \in \langle 0, R_1 \rangle$ and $x_{ijm} \in (R_1, d/(2c_\varphi))$ (see Eqs. (2.32), (2.33); Fig. 8.24b,c,d), then $W_{cp}^{(ij)} = W_{cp}^{(ij)}(x_{ij}, \varphi, R_1, t, v)$ for $x_{ij} \in \langle 0, x_{ijp} \rangle$ and $x_{ij} \in \langle x_{ijp}, x_{ije} \rangle$ is given

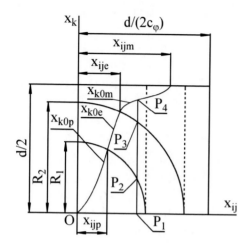

Figure 8.27: The abscissae $\overline{P_1P_2P_3P_4}$ for $x_{ij} \in \langle x_{ije}, R_2 \rangle$ along which the curve integral $W_{cp}^{(ij)} = W_{cp}^{(ij)}(x_{ij}, \varphi, R_1, t, v)$ (see Eqs. (8.3), (8.66)) for $x_{ije} \in \langle 0, R_1 \rangle$ is determined due to the crack formation in the spherical particle on the condition in Item 1, Section 8.2.4.2. This determination is also considered for $x_{ijm} \in (R_1, d/(2c_\varphi))$ (see Eqs. (2.32), (2.33); Fig. 8.24b,c,d), although Fig. 8.27 is related to the condition $x_{ijm} \in (R_1, R_2)$ (see Fig. 8.24b).

by Eqs. (8.62) and (8.63), respectively. Consequently, $W_{cp}^{(ij)} = W_{cp}^{(ij)}(x_{ij}, \varphi, R_1, t, v)$ is determined along the abscissae $\overline{P_1P_2P_3P_4}$ for $x_{ij} \in \langle x_{ije}, R_2 \rangle$ in Fig. 8.27.

With regard to Eq. (8.3), we get

$$W_{cp}^{(ij)} = \int_{\overline{P_1P_2}} w_p^{(ij)}\, dx_k + \int_{\overline{P_2P_3}} w_e^{(ij)}\, dx_k + \int_{\overline{P_3P_4}} w_m^{(ij)}\, dx_k$$

$$= \int_0^{\sqrt{R_1^2-x_{ij}^2}} w_p^{(ij)}\, dx_k + \int_{\sqrt{R_1^2-x_{ij}^2}}^{\sqrt{R_2^2-x_{ij}^2}} w_e^{(ij)}\, dx_k + \int_{\sqrt{R_2^2-x_{ij}^2}}^{x_{k0m}} w_m^{(ij)}\, dx_k,$$

$$x_{ij} \in \langle x_{ije}, R_2 \rangle, \quad x_{ije} \in \langle 0, R_1 \rangle, \quad x_{ijm} \in (R_1, R_2). \tag{8.66}$$

If $x_{ije} \in (R_1, R_2)$ and $x_{ijm} \in (R_1, d/(2c_\varphi))$ (see Eqs. (2.32), (2.33); Fig. 8.25), then $W_{cp}^{(ij)} = W_{cp}^{(ij)}(x_{ij}, \varphi, R_1, t, v)$ for $x_{ij} \in \langle 0, x_{ijp} \rangle$ and $x_{ij} \in \langle x_{ijp}, R_1 \rangle$ is given by Eqs. (8.62) and (8.63), respectively.

Spherical envelope. If the condition in Item 1, Section 8.2.4.1 is valid, then a crack is formed in the spherical envelope, i.e. for $x_{ij} \in \langle R_1, R_2 \rangle$. Additionally, the intervals

- $x_{ije} \in \langle 0, R_1 \rangle$ (see Fig. 8.24) for $x_{ijm} \in \langle 0, R_1 \rangle$ (see Fig. 8.24a) or $x_{ijm} \in (R_1, R_2)$ (see Fig. 8.24b) or $x_{ijm} \in (R_2, d/(2c_\varphi))$ (see Eqs. (2.32), (2.33); Fig. 8.24c) or $x_{ijm} = d/(2c_\varphi)$ (see Fig. 8.24d),

- $x_{ije} \in (R_1, R_2)$ (see Fig. 8.25) for $x_{ijm} \in (R_1, R_2)$ (see Fig. 8.25a) or $x_{ijm} \in (R_2, d/(2c_\varphi))$ (see Fig. 8.25b) or $x_{ijm} = d/(2c_\varphi)$ (see Fig. 8.25c)

are also required to be considered for the determination of the curve integral $W_{ce}^{(ij)} = W_{ce}^{(ij)}(x_{ij}, \varphi, R_1, t, v)$ for $x_{ij} \in \langle R_1, R_2 \rangle$.

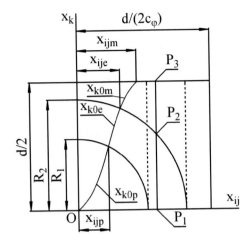

Figure 8.28: The abscissa $\overline{P_1P_2P_3}$ for $x_{ij} \in \langle R_1, R_2 \rangle$ along which the curve integral $W_{ce}^{(ij)} = W_{ce}^{(ij)}(x_{ij}, \varphi, R_1, t, v)$ (see Eqs. (8.3), (8.67)) for $x_{ije} \in \langle 0, R_1 \rangle$ and $x_{ijm} \in \langle 0, R_1 \rangle$ is determined due to the crack formation in the spherical particle on the condition in Item 1, Section 8.2.4.2.

Consequently, if $x_{ije} \in \langle 0, R_1 \rangle$ and $x_{ijm} \in \langle 0, R_1 \rangle$ (see Fig. 8.24a), then $W_{ce}^{(ij)} = W_{ce}^{(ij)}(x_{ij}, \varphi, R_1, t, v)$ is determined along the abscissa $\overline{P_1P_2P_3}$ for $x_{ij} \in \langle R_1, R_2 \rangle$ in Fig. 8.28.

With regard to Eq. (8.3), we get

$$W_{ce}^{(ij)} = \int_{\overline{P_1P_2}} w_e^{(ij)}\, dx_k + \int_{\overline{P_2P_3}} w_m^{(ij)}\, dx_k = \int_0^{\sqrt{R_2^2-x_{ij}^2}} w_e^{(ij)}\, dx_k + \int_{\sqrt{R_2^2-x_{ij}^2}}^{d/2} w_m^{(ij)}\, dx_k,$$

$$x_{ij} \in \langle R_1, R_2 \rangle, \quad x_{ije} \in \langle 0, R_1 \rangle, \quad x_{ijm} \in \langle 0, R_1 \rangle. \tag{8.67}$$

If $x_{ije} \in \langle 0, R_1 \rangle$ and $x_{ijm} \in (R_1, R_2)$ (see Fig. 8.24b), then $W_{ce}^{(ij)} = W_{ce}^{(ij)}(x_{ij}, \varphi, R_1, t, v)$ for $x_{ij} \in \langle x_{ijm}, R_2 \rangle$ is given by Eq. (8.67). Consequently, $W_{ce}^{(ij)} = W_{ce}^{(ij)}(x_{ij}, \varphi, R_1, t, v)$ is determined along the abscissae $\overline{P_1P_2P_3}$ for $x_{ij} \in \langle R_1, x_{ijm} \rangle$ in Fig. 8.29.

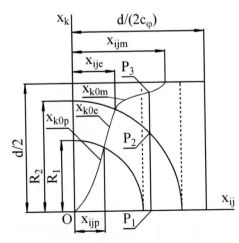

Figure 8.29: The abscissa $\overline{P_1P_2P_3}$ for $x_{ij} \in \langle x_{ijm}, R_2 \rangle$ along which the curve integral $W_{ce}^{(ij)} = W_{ce}^{(ij)}(x_{ij}, \varphi, R_1, t, v)$ (see Eqs. (8.3), (8.68)) for $x_{ije} \in \langle 0, R_1 \rangle$ and $x_{ijm} \in (R_1, R_2)$ is determined due to the crack formation in the spherical particle on the condition in Item 1, Section 8.2.4.2.

With regard to Eq. (8.3), we get

$$W_{ce}^{(ij)} = \int_{\overline{P_1 P_2}} w_e^{(ij)} dx_k + \int_{\overline{P_2 P_3}} w_m^{(ij)} dx_k = \int_0^{\sqrt{R_2^2 - x_{ij}^2}} w_e^{(ij)} dx_k + \int_{\sqrt{R_2^2 - x_{ij}^2}}^{x_{k0m}} w_m^{(ij)} dx_k,$$

$$x_{ij} \in \langle x_{ijm}, R_2 \rangle, \quad x_{ije} \in \langle 0, R_1 \rangle, \quad x_{ijm} \in \langle R_1, R_2 \rangle. \qquad (8.68)$$

If $x_{ije} \in \langle 0, R_1 \rangle$ and $x_{ijm} \in \langle R_2, d/(2c_\varphi) \rangle$ (see Eqs. (2.32), (2.33); Fig. 8.24c,d), then $W_{ce}^{(ij)} = W_{ce}^{(ij)}(x_{ij}, \varphi, R_1, t, v)$ for $x_{ij} \in \langle R_1, R_2 \rangle$ is given by Eq. (8.66).

If $x_{ije} \in \langle R_1, R_2 \rangle$ and $x_{ijm} \in \langle R_1, R_2 \rangle$ (see Fig. 8.25a), then $W_{ce}^{(ij)} = W_{ce}^{(ij)}(x_{ij}, \varphi, R_1, t, v)$ for $x_{ij} \in \langle x_{ije}, x_{ijm} \rangle$ and $x_{ij} \in \langle x_{ijm}, R_2 \rangle$ is given by Eqs. (8.66) and (8.67), respectively. Consequently, $W_{ce}^{(ij)} = W_{ce}^{(ij)}(x_{ij}, \varphi, R_1, t, v)$ is determined along the abscissae $\overline{P_1 P_2}$ for $x_{ij} \in \langle R_1, x_{ije} \rangle$ in Fig. 8.30.

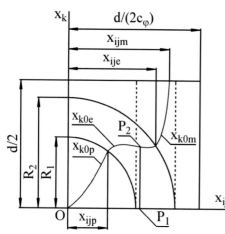

Figure 8.30: The abscissa $\overline{P_1 P_2}$ for $x_{ij} \in \langle R_1, x_{ije} \rangle$ along which the curve integral $W_{ce}^{(ij)} = W_{ce}^{(ij)}(x_{ij}, \varphi, R_1, t, v)$ (see Eqs. (8.3), (8.69)) for $x_{ije} \in \langle R_1, R_2 \rangle$ and $x_{ijm} \in \langle R_1, R_2 \rangle$ is determined due to the crack formation in the spherical particle on the condition in Item 1, Section 8.2.4.2.

With regard to Eq. (8.3), we get

$$W_{ce}^{(ij)} = \int_{\overline{P_1 P_2}} w_e^{(ij)} dx_k = \int_0^{x_{k0e}} w_e^{(ij)} dx_k,$$

$$x_{ij} \in \langle R_1, x_{ije} \rangle, \quad x_{ije} \in \langle R_1, R_2 \rangle, \quad x_{ijm} \in \langle R_1, R_2 \rangle. \qquad (8.69)$$

If $x_{ije} \in \langle R_1, R_2 \rangle$ and $x_{ijm} \in \langle R_2, d/(2c_\varphi) \rangle$ (see Eqs. (2.32), (2.33); Fig. 8.25b,c), then $W_{ce}^{(ij)} = W_{ce}^{(ij)}(x_{ij}, \varphi, R_1, t, v)$ for $x_{ij} \in \langle R_1, x_{ije} \rangle$ and $x_{ij} \in \langle x_{ije}, R_2 \rangle$ is given by Eqs. (8.69) and (8.68), respectively.

Cell matrix. If the condition in Item 1, Section 8.2.4.1 is valid, then a crack is formed in the cell matrix, i.e. for $x_{ij} \in \langle R_2, d/(2c_\varphi) \rangle$ (see Eqs. (2.32), (2.33)). Additionally, disregarding $x_{ije} \in \langle 0, R_1 \rangle$ (see Fig. 8.24) and $x_{ije} \in \langle R_1, R_2 \rangle$ (see Fig. 8.25), the intervals $x_{ijm} \in \langle 0, R_1 \rangle$ (see Fig. 8.24a), $x_{ijm} \in (R_1, R_2)$ (see Figs. 8.24b, 8.25a), $x_{ijm} \in (R_2, d/(2c_\varphi))$ (see Eqs. (2.32), (2.33); Fig. 8.24c, 8.25b), $x_{ijm} = d/(2c_\varphi)$ (see Fig. 8.24d, 8.25c) are required to be considered for the determination of the curve integral $W_{cm}^{(ij)} = W_{cm}^{(ij)}(x_{ij}, \varphi, R_1, t, v)$ for $x_{ij} \in \langle R_2, d/(2c_\varphi) \rangle$.

Consequently, if $x_{ijm} \in \langle 0, R_1 \rangle$ (see Fig. 8.24a) or $x_{ijm} \in (R_1, R_2)$ (see Figs. 8.24b, 8.25a), then $W_{cm}^{(ij)} = W_{cm}^{(ij)}(x_{ij}, \varphi, R_1, t, v)$ is determined along the abscissa $\overline{P_1 P_2}$ for $x_{ij} \in \langle R_2, d/(2c_\varphi) \rangle$ in Fig. 8.31.

With regard to Eq. (8.3), we get

$$W_{cm}^{(ij)} = \int_{\overline{P_1 P_2}} w_m^{(ij)} dx_k = \int_0^{d/2} w_m^{(ij)} dx_k, \quad x_{ij} \in \langle R_2, d/(2c_\varphi) \rangle;$$

$$x_{ije} \in \langle 0, R_1 \rangle \text{ or } x_{ije} \in (R_1, R_2); \quad x_{ijm} \in \langle 0, R_1 \rangle \text{ or } x_{ijm} \in (R_1, R_2). \quad (8.70)$$

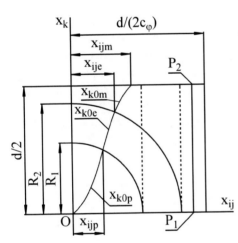

Figure 8.31: The abscissa $\overline{P_1 P_2}$ for $x_{ij} \in \langle R_2, d/(2c_\varphi) \rangle$ (see Eqs. (2.32), (2.33)) along which the curve integral $W_{cm}^{(ij)} = W_{cm}^{(ij)}(x_{ij}, \varphi, R_1, t, v)$ (see Eqs. (8.3), (8.70)) is determined due to the crack formation in the spherical particle on the condition in Item 1, Section 8.2.4.2. This determination is also considered for $x_{ije} \in \langle 0, R_1 \rangle$ (see Fig. 8.24) and $x_{ije} \in (R_1, R_2)$ (see Fig. 8.25a) as well as for $x_{ijm} \in \langle 0, R_1 \rangle$ (see Fig. 8.24a) or $x_{ijm} \in (R_1, R_2)$ (see Figs. 8.24b, 8.25a), although Fig. 8.31 is related to the conditions $x_{ije} \in \langle 0, R_1 \rangle$ and $x_{ijm} \in \langle 0, R_1 \rangle$ only (see Fig. 8.24a).

If $x_{ijm} \in (R_2, d/(2c_\varphi))$ (see Eqs. (2.32), (2.33); Fig. 8.24c, 8.25b), then $W_{cm}^{(ij)} = W_{cm}^{(ij)}(x_{ij}, \varphi, R_1, t, v)$ for $x_{ij} \in \langle x_{ijm}, d/(2c_\varphi) \rangle$ is given by Eq. (8.70). Consequently, $W_{cm}^{(ij)} = W_{cm}^{(ij)}(x_{ij}, \varphi, R_1, t, v)$ is determined along the abscissa $\overline{P_1 P_2}$ for $x_{ij} \in \langle R_2, x_{ijm} \rangle$ in Fig. 8.32.

With regard to Eq. (8.3), we get

$$W_{cm}^{(ij)} = \int_{\overline{P_1 P_2}} w_m^{(ij)} dx_k = \int_0^{x_{k0m}} w_m^{(ij)} dx_k, \quad x_{ij} \in \langle R_2, x_{ijm} \rangle;$$

$$x_{ije} \in \langle 0, R_1 \rangle \text{ or } x_{ije} \in (R_1, R_2); \quad x_{ijm} \in \left(R_2, \frac{d}{2c_\varphi} \right). \quad (8.71)$$

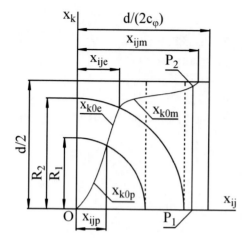

Figure 8.32: The abscissa $\overline{P_1 P_2}$ for $x_{ij} \in \langle R_2, x_{ijm} \rangle$ (see Eqs. (2.32), (2.33)) along which the curve integral $W_{cm}^{(ij)} = W_{cm}^{(ij)}(x_{ij}, \varphi, R_1, t, v)$ (see Eqs. (8.3), (8.71)) is determined due to the crack formation in the spherical particle on the condition in Item 1, Section 8.2.4.2. This determination is also considered for $x_{ije} \in (R_1, R_2)$ (see Fig. 8.25b), although Fig. 8.32 is related to the condition $x_{ije} \in \langle 0, R_1 \rangle$ only (see Fig. 8.24c).

If $x_{ijm} = d/(2c_\varphi)$ (see Eqs. (2.32), (2.33); Fig. 8.24d, 8.25c), then $W_{cm}^{(ij)} = W_{cm}^{(ij)}(x_{ij}, \varphi, R_1, t, v)$ for $x_{ij} \in \langle R_2, d/(2c_\varphi) \rangle$ is given by Eq. (8.71).

B. Method 2

This method of determination of the curve integrals $W_{cp}^{(ij)} = W_{cp}^{(ij)}(x_{ij}, \varphi, R_1, t, v)$, $W_{ce}^{(ij)} = W_{ce}^{(ij)}(x_{ij}, \varphi, R_1, t, v)$ and $W_{cm}^{(ij)} = W_{cm}^{(ij)}(x_{ij}, \varphi, R_1, t, v)$ is based on an assumption that energy $W_{O2591063O}$ (see Fig. 8.25b,c, 8.26a,b) or $W_{O2589 10 63O}$ (see Fig. 8.25d, 8.26c) is released by the crack formation in the plane $x_i x_j$ instead of the crack formation in the planes $x_j x_k$, $x_k x_i$ as assumed in Items (a)–(c), Section 8.2.4.2.A.

As analysed in Section 8.2.4.1.B, the method 2 results from a principle that a release of energy of a system considers 'minimal resistance' of the system, i.e. the energy is released through 'minimal resistance' of the system.

Analyses of the determination of $W_{cp}^{(ij)} = W_{cp}^{(ij)}(x_{ij}, \varphi, R_1, t, v)$, $W_{ce}^{(ij)} = W_{ce}^{(ij)}(x_{ij}, \varphi, R_1, t, v)$ and $W_{cm}^{(ij)} = W_{cm}^{(ij)}(x_{ij}, \varphi, R_1, t, v)$ by the method 2 for the crack formation in the spherical particle, the spherical envelope and the cell matrix are as follows, respectively.

Spherical particle. Let the conditions in Item 2, Section 8.2.4.2 are valid.

Let the condition $x_{ije} \in \langle 0, R_1 \rangle$ or $x_{ije} \in (R_1, R_2)$ is valid which is related to $x_{ijm} \in (R_1, R_2\rangle$ (see Fig. 8.24b, 8.25a) or $x_{ijm} \in (R_2, d/(2c_\varphi))$ (see Eqs. (2.32), (2.33); Fig. 8.24c, 8.25b) or $x_{ijm} = d/(2c_\varphi)$ (see Fig. 8.24d, 8.25c).

Let the critical radii $R_{\eta cp, max}^{(jk)} = R_{\eta cp, max}^{(jk)}(t, v)$, $R_{\eta cp, max}^{(ki)} = R_{\eta cp, max}^{(ki)}(t, v)$ ($\eta = 1, 2$) analysed in Items (a)–(c), Section 8.2.4.2.A are determined by the curve integrals $W_{cp}^{(jk)} = W_{cp}^{(jk)}(x_{jk}, \varphi, R_1, t, v)$ for $\varphi = \angle(x_j, x_{jk})$, $x_{jk} \subset x_j x_k$ and $W_{cp}^{(ki)} = W_{cp}^{(ki)}(x_{ki}, \varphi, R_1, t, v)$ for $\varphi = \angle(x_k, x_{ki})$, $x_{ki} \subset x_k x_i$, respectively, where $W_{cp}^{(jk)}$, $W_{cp}^{(ki)}$ are determined by the **method 1**, and $R_2 = R_1 + t$ (see Eq. (2.5)). Additionally, a condition given by Eq. (8.17) for $q = p$ is considered for the determination of $R_{\eta cp, max}^{(jk)}$, $R_{\eta cp, max}^{(ki)}$.

Similarly, let the critical radius $R^{(ij)}_{\eta ce, max} = R^{(ij)}_{\eta ce, max}(t, v)$ is determined by the curve integral $W^{(ij)}_{ce} = W^{(ij)}_{ce}(x_{ij}, \varphi, R_1, t, v)$ for $\varphi = \angle(x_i, x_{ij})$, $x_{ij} \subset x_i x_j$, where $W^{(ij)}_{ce}$ is also determined by the **method 1**, and a condition given by Eq. (8.17) for $q = e$ is considered for the determination of $R^{(ij)}_{\eta cm, max}$.

Finally, let the critical radius $R^{(ij)}_{\eta cp, max} = R^{(ij)}_{\eta cp, max}(t, v)$ is determined by the curve integral $W^{(ij)}_{cp} = W^{(ij)}_{cp}(x_{ij}, \varphi, R_1, t, v)$ for $\varphi = \angle(x_i, x_{ij})$, $x_{ij} \subset x_i x_j$, where $W^{(ij)}_{cp}$ is determined by the **method 2**, and a condition given by Eq. (8.17) for $q = p$ is considered for the determination of $R^{(ij)}_{\eta cp, max}$.

If $x_{ije} \in \langle 0, R_1 \rangle$ and $x_{ijm} \in (R_2, d/(2c_\varphi)\rangle$ (see Eqs. (2.32), (2.33)), then $W^{(ij)}_{cp} = W^{(ij)}_{cp}(x_{ij}, \varphi, R_1, t, v)$ is determined by the **method 2** along the abscissae $\overline{P_1 P_2 P_{10} P_{11} P_{12}}$, $\overline{P_3 P_4 P_5 P_{13} P_{14}}$ and $\overline{P_6 P_7 P_8 P_9 P_{15}}$ for $x_{ij} \in \langle 0, x_{ijp} \rangle$, $x_{ij} \in \langle x_{ijp}, x_{ije} \rangle$ and $x_{ij} \in \langle x_{ije}, R_1 \rangle$ in Fig. 8.33, respectively.

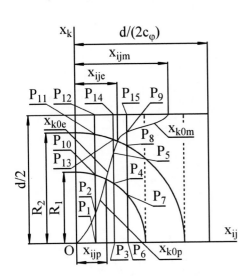

Figure 8.33: The abscissae $\overline{P_1 P_2 P_{10} P_{11} P_{12}}$, $\overline{P_3 P_4 P_5 P_{13} P_{14}}$ and $\overline{P_6 P_7 P_8 P_9 P_{15}}$ for $x_{ij} \in \langle 0, x_{ijp} \rangle$, $x_{ij} \in \langle x_{ijp}, x_{ije} \rangle$ and $x_{ij} \in \langle x_{ije}, R_1 \rangle$, respectively, along which the curve integral $W^{(ij)}_{cp} = W^{(ij)}_{cp}(x_{ij}, \varphi, R_1, t, v)$ (see Eqs. (8.3), (8.72)–(8.74)) for $x_{ije} \in \langle 0, R_1 \rangle$ and $x_{ijm} \in (R_1, R_2\rangle$ is determined by the method 2 due to the crack formation in the spherical particle in spite of the fact that the conditions in Item 2, Section 8.2.4.2 are valid. This determination is also considered for $x_{ijm} \in (R_2, d/(2c_\varphi))$ (see Eqs. (2.32), (2.33); Fig. 8.24c) or $x_{ijm} = d/(2c_\varphi)$ (see Fig. 8.24d), although Fig. 8.33 is related to the condition $x_{ije} \in \langle 0, R_1 \rangle$ (see Fig. 8.24b).

With regard to Eq. (8.3), we get

$$W^{(ij)}_{cp} = \int_{\overline{P_2 P_{10}}} w^{(ij)}_p \, dx_k + \int_{\overline{P_{10} P_{11}}} w^{(ij)}_e \, dx_k + \int_{\overline{P_{11} P_{12}}} w^{(ij)}_m \, dx_k - \int_{\overline{P_1 P_2}} w^{(ij)}_p \, dx_k$$

$$= \int_{x_{k0p}}^{\sqrt{R_1^2 - x_{ij}^2}} w^{(ij)}_p \, dx_k + \int_{\sqrt{R_1^2 - x_{ij}^2}}^{\sqrt{R_2^2 - x_{ij}^2}} w^{(ij)}_e \, dx_k + \int_{\sqrt{R_2^2 - x_{ij}^2}}^{d/2} w^{(ij)}_m \, dx_k$$

$$- \int_0^{x_{k0p}} w^{(ij)}_p \, dx_k \geq 0,$$

$$x_{ij} \in \langle 0, x_{ijp} \rangle, \quad x_{ije} \in \langle 0, R_1 \rangle, \quad x_{ijm} \in \left(R_1, \frac{d}{2c_\varphi} \right), \qquad (8.72)$$

$$W_{cp}^{(ij)} = \int\limits_{\overline{P_5 P_{13}}} w_e^{(ij)} \, dx_k + \int\limits_{\overline{P_{13} P_{14}}} w_m^{(ij)} \, dx_k - \left(\int\limits_{\overline{P_3 P_4}} w_p^{(ij)} \, dx_k + \int\limits_{\overline{P_4 P_5}} w_e^{(ij)} \, dx_k \right)$$

$$= \int\limits_{x_{k0e}}^{\sqrt{R_2^2 - x_{ij}^2}} w_e^{(ij)} \, dx_k + \int\limits_{\sqrt{R_2^2 - x_{ij}^2}}^{d/2} w_m^{(ij)} \, dx_k$$

$$- \left(\int\limits_{0}^{\sqrt{R_1^2 - x_{ij}^2}} w_p^{(ij)} \, dx_k + \int\limits_{\sqrt{R_1^2 - x_{ij}^2}}^{x_{k0e}} w_e^{(ij)} \, dx_k \right) \geq 0,$$

$$x_{ij} \in \langle x_{ije}, R_1 \rangle, \quad x_{ije} \in \langle 0, R_1 \rangle, \quad x_{ijm} \in \left(R_1, \frac{d}{2c_\varphi} \right\rangle, \tag{8.73}$$

$$W_{cp}^{(ij)} = \int\limits_{\overline{P_9 P_{15}}} w_m^{(ij)} \, dx_k - \left(\int\limits_{\overline{P_6 P_7}} w_p^{(ij)} \, dx_k + \int\limits_{\overline{P_7 P_8}} w_e^{(ij)} \, dx_k + \int\limits_{\overline{P_8 P_9}} w_m^{(ij)} \, dx_k \right)$$

$$= \int\limits_{x_{k0m}}^{d/2} w_m^{(ij)} \, dx_k$$

$$- \left(\int\limits_{0}^{\sqrt{R_1^2 - x_{ij}^2}} w_p^{(ij)} \, dx_k + \int\limits_{\sqrt{R_1^2 - x_{ij}^2}}^{\sqrt{R_2^2 - x_{ij}^2}} w_e^{(ij)} \, dx_k + \int\limits_{\sqrt{R_2^2 - x_{ij}^2}}^{x_{k0m}} w_m^{(ij)} \, dx_k \right) \geq 0,$$

$$x_{ij} \in \langle x_{ije}, R_1 \rangle, \quad x_{ije} \in \langle 0, R_1 \rangle, \quad x_{ijm} \in \left(R_1, \frac{d}{2c_\varphi} \right\rangle. \tag{8.74}$$

If $x_{ije} \in (R_1, R_2)$ and $x_{ijm} \in (R_2, d/(2c_\varphi))$ (see Eqs. (2.32), (2.33)), then $W_{cp}^{(ij)} = W_{cp}^{(ij)}(x_{ij}, \varphi, R_1, t, v)$ for $x_{ij} \in \langle 0, x_{ijp} \rangle$ and $x_{ij} \in \langle x_{ijp}, R_1 \rangle$ is given by Eqs. (8.72) and (8.73), respectively.

Additionally, $W_{cp}^{(ij)} = W_{cp}^{(ij)}(x_{ij}, \varphi, R_1, t, v)$ is define for such interval $x_{ij} \in \langle x_{ijp1}, x_{ijp0} \rangle \subset \langle 0, R_1 \rangle$ or $x_{ij} \in \langle x_{ijp0}, x_{ijp2} \rangle \subset \langle 0, R_1 \rangle$ for which the condition $W_{cp}^{(ij)}(x_{ij}, \varphi, R_1, t, v) \geq 0$ is valid, where x_{ijp0} is determined by the condition $W_{cp}^{(ij)}(x_{ij}, \varphi, R_1, t, v) = 0$.

If the conditions

$$R_{1cp\,max}^{(ij)} < R_{1cp\,max}^{(jk)}, \quad R_{1cp\,max}^{(ij)} < R_{1cp\,max}^{(ki)}, \quad R_{1cp\,max}^{(ij)} < R_{1ce\,max}^{(ij)} \tag{8.75}$$

Related phenomena 139

are simultaneously valid for the critical particle radii $R_{1cp\,max}^{(jk)}$, $R_{1cp\,max}^{(ki)}$, $R_{1ce\,max}^{(ij)}$ and $R_{1cp\,max}^{(ij)}$ determined by the methods 1 and 2, respectively, then the crack is formed in the plane $x_i x_j$ in the spherical particle

- in spite of the fact that the conditions in Item 2, Section 8.2.4.2 are valid,

- and in spite of the fact that the crack would be expected to be formed in the cell matrix regarding these conditions in Item 2, Section 8.2.4.2.

In this case, the 'minimal resistance' is represented by the critical particle radius $R_{1cp\,max}^{(ij)}$ as a minimal value of the set $\left\{ R_{1cp\,max}^{(ij)}, R_{1cp\,max}^{(jk)}, R_{1cp\,max}^{(ki)}, R_{1ce\,max}^{(ij)} \right\}$.

Spherical envelope. Let the condition $x_{ije} \in \langle 0, R_1 \rangle$ or $x_{ije} \in (R_1, R_2)$ is valid which is related to $x_{ijm} \in (R_2, d/(2c_\varphi))$ (see Eqs. (2.32), (2.33); Fig. 8.24c, 8.25b) or $x_{ijm} = d/(2c_\varphi)$ (see Fig. 8.24d, 8.25c).

Let the critical radii $R_{\eta ce,\,max}^{(jk)} = R_{\eta ce,\,max}^{(jk)}(t, v)$, $R_{\eta ce,\,max}^{(ki)} = R_{\eta ce,\,max}^{(ki)}(t, v)$ ($\eta = 1, 2$) analysed in Items (a)–(c), Section 8.2.4.2.A are determined by the curve integrals $W_{ce}^{(jk)} = W_{ce}^{(jk)}(x_{jk}, \varphi, R_1, t, v)$ for $\varphi = \angle(x_j, x_{jk})$, $x_{jk} \subset x_j x_k$ and $W_{ce}^{(ki)} = W_{ce}^{(ki)}(x_{ki}, \varphi, R_1, t, v)$ for $\varphi = \angle(x_k, x_{ki})$, $x_{ki} \subset x_k x_i$, respectively, where $W_{ce}^{(jk)}$, $W_{ce}^{(ki)}$ are determined by the **method 1**, and $R_2 = R_1 + t$ (see Eq. (2.5)). Additionally, a condition given by Eq. (8.17) for $q = e$ is considered for the determination of $R_{\eta ce,\,max}^{(jk)}$, $R_{\eta ce,\,max}^{(ki)}$.

Similarly, let the critical radius $R_{\eta cq,\,max}^{(ij)} = R_{\eta cq,\,max}^{(ij)}(t, v)$ ($q = p, m$) is determined by the curve integral $W_{cq}^{(ij)} = W_{cq}^{(ij)}(x_{ij}, \varphi, R_1, t, v)$ for $\varphi = \angle(x_i, x_{ij})$, $x_{ij} \subset x_i x_j$, where $W_{cq}^{(ij)}$ is also determined by the **method 1**, and a condition given by Eq. (8.17) for $q = p, m$ is considered for the determination of $R_{\eta cq,\,max}^{(ij)}$.

Finally, let the critical radius $R_{\eta ce,\,max}^{(ij)} = R_{\eta ce,\,max}^{(ij)}(t, v)$ is determined by the curve integral $W_{ce}^{(ij)} = W_{ce}^{(ij)}(x_{ij}, \varphi, R_1, t, v)$ for $\varphi = \angle(x_i, x_{ij})$, $x_{ij} \subset x_i x_j$, where $W_{ce}^{(ij)}$ is determined by the **method 2**, and a condition given by Eq. (8.17) for $q = e$ is considered for the determination of $R_{\eta ce,\,max}^{(ij)}$.

If $x_{ije} \in \langle 0, R_1 \rangle$ and $x_{ijm} \in (R_2, d/(2c_\varphi))$ (see Eqs. (2.32), (2.33)), then the curve integral $W_{ce}^{(ij)} = W_{ce}^{(ij)}(x_{ij}, \varphi, R_1, t, v)$ is determined by the **method 2** along the abscissa $\overline{P_1 P_2 P_3 P_4}$ for $x_{ij} \in \langle R_1, R_2 \rangle$ in Fig. 8.34.

With regard to Eq. (8.3), we get

$$
W_{ce}^{(ij)} = \int_{\overline{P_3 P_4}} w_m^{(ij)}\, dx_k - \left(\int_{\overline{P_1 P_2}} w_e^{(ij)}\, dx_k + \int_{\overline{P_2 P_3}} w_m^{(ij)}\, dx_k \right)
$$

$$
= \int_{x_{k0m}}^{d/2} w_m^{(ij)}\, dx_k - \left(\int_0^{\sqrt{R_2^2 - x_{ij}^2}} w_e^{(ij)}\, dx_k + \int_{\sqrt{R_2^2 - x_{ij}^2}}^{x_{k0m}} w_m^{(ij)}\, dx_k \right) \geq 0,
$$

$$x_{ij} \in \langle R_1, R_2 \rangle, \quad x_{ije} \in \langle 0, R_1 \rangle, \quad x_{ijm} \in \left(R_1, \frac{d}{2c_\varphi} \right). \quad (8.76)$$

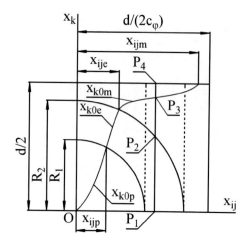

Figure 8.34: The abscissa $\overline{P_1 P_2 P_3 P_4}$ for $x_{ij} \in \langle R_1, R_2 \rangle$ along which the curve integral $W_{ce}^{(ij)} = W_{ce}^{(ij)}(x_{ij}, \varphi, R_1, t, v)$ (see Eqs. (8.3), (8.76)) for $x_{ije} \in \langle 0, R_1 \rangle$ and $x_{ijm} \in (R_2, d/(2c_\varphi))$ is determined by the method 2 due to the crack formation in the spherical particle in spite of the fact that the conditions in Item 2, Section 8.2.4.2 are valid. This determination is also considered for $x_{ijm} = d/(2c_\varphi)$ (see Eqs. (2.32), (2.33); Fig. 8.24d), although Fig. 8.34 is related to the condition $x_{ijm} \in (R_2, d/(2c_\varphi))$ (see Fig. 8.24c).

If $x_{ije} \in (R_1, R_2)$ and $x_{ijm} \in (R_2, d/(2c_\varphi)\rangle$ (see Eqs. (2.32), (2.33); Fig. 8.25b,c), then $W_{ce}^{(ij)} = W_{ce}^{(ij)}(x_{ij}, \varphi, R_1, t, v)$ for $x_{ij} \in \langle x_{ije}, R_2 \rangle$ is given by Eq. (8.76). Consequently, $W_{ce}^{(ij)} = W_{ce}^{(ij)}(x_{ij}, \varphi, R_1, t, v)$ is determined along the abscissa $\overline{P_1 P_2 P_3 P_4}$ for $x_{ij} \in \langle R_1, x_{ije} \rangle$ in Fig. 8.35.

With regard to Eq. (8.3), we get

$$W_{ce}^{(ij)} = \int_{\overline{P_2 P_3}} w_e^{(ij)} \, dx_k + \int_{\overline{P_3 P_4}} w_m^{(ij)} \, dx_k - \int_{\overline{P_1 P_2}} w_e^{(ij)} \, dx_k$$

$$= \int_{x_{k0e}}^{\sqrt{R_2^2 - x_{ij}^2}} w_e^{(ij)} \, dx_k + \int_{\sqrt{R_2^2 - x_{ij}^2}}^{d/2} w_m^{(ij)} \, dx_k - \int_0^{x_{k0e}} w_e^{(ij)} \, dx_k \geq 0,$$

$$x_{ij} \in \langle R_1, x_{ije} \rangle, \quad x_{ije} \in (R_1, R_2), \quad x_{ijm} \in \left(R_1, \frac{d}{2c_\varphi} \right). \quad (8.77)$$

If the conditions

$$R_{1ce\,max}^{(ij)} < R_{1ce\,max}^{(jk)}, \quad R_{1ce\,max}^{(ij)} < R_{1ce\,max}^{(ki)},$$
$$R_{1ce\,max}^{(ij)} < R_{1cp\,max}^{(ij)}, \quad R_{1ce\,max}^{(ij)} < R_{1cm\,max}^{(ij)} \quad (8.78)$$

are simultaneously valid for the critical particle radii $R_{1ce\,max}^{(jk)}$, $R_{1ce\,max}^{(ki)}$, $R_{1cp\,max}^{(ij)}$, $R_{1cm\,max}^{(ij)}$ and $R_{1ce\,max}^{(ij)}$ determined by the methods 1 and 2, respectively, then the crack is formed in the plane $x_i x_j$ in the spherical envelope

- in spite of the fact that the conditions in Item 2, Section 8.2.4.2 are valid,

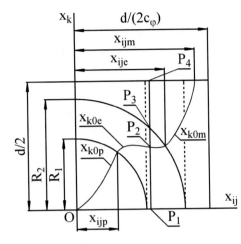

Figure 8.35: The abscissa $\overline{P_1P_2P_3P_4}$ for $x_{ij} \in \langle R_1, x_{ije}\rangle$ along which the curve integral $W_{ce}^{(ij)} = W_{ce}^{(ij)}(x_{ij}, \varphi, R_1, t, v)$ (see Eqs. (8.3), (8.76)) for $x_{ije} \in (R_1, R_2)$ and $x_{ijm} \in (R_2, d/(2c_\varphi))$ is determined by the method 2 due to the crack formation in the spherical particle in spite of the fact that the conditions in Item 2, Section 8.2.4.2 are valid. This determination is also considered for $x_{ijm} = d/(2c_\varphi)$ (see Eqs. (2.32), (2.33); Fig. 8.25c), although Fig. 8.35 is related to the condition $x_{ijm} \in (R_2, d/(2c_\varphi))$ (see Fig. 8.25b).

- and in spite of the fact that the crack would be expected to be formed in the spherical particle and/or the cell matrix regarding these conditions in Item 2, Section 8.2.4.2.

In this case, the 'minimal resistance' is represented by the critical particle radius $R_{1ce\,max}^{(ij)}$ as a minimal value of the set $\left\{R_{1cp\,max}^{(ij)}, R_{1ce\,max}^{(ij)}, R_{1ce\,max}^{(jk)}, R_{1ce\,max}^{(ki)}, R_{1cm\,max}^{(ij)}\right\}$.

Cell matrix. Let the condition $x_{ijm} = d/(2c_\varphi)$ (see Eqs. (2.32), (2.33)) is valid disregarding the condition $x_{ije} \in \langle 0, R_1\rangle$ (see Fig. 8.24d) or $x_{ije} \in (R_1, R_2)$ (see Fig. 8.25c).

Let the critical radii $R_{\eta cm,\,max}^{(jk)} = R_{\eta cm,\,max}^{(jk)}(t, v)$, $R_{\eta cm,\,max}^{(ki)} = R_{\eta cm,\,max}^{(ki)}(t, v)$ ($\eta = 1,2$) analysed in Items (a)–(c), Section 8.2.4.2.A are determined by the curve integrals $W_{cm}^{(jk)} = W_{cm}^{(jk)}(x_{jk}, \varphi, R_1, t, v)$ for $\varphi = \angle(x_j, x_{jk})$, $x_{jk} \subset x_j x_k$ and $W_{cm}^{(ki)} = W_{cm}^{(ki)}(x_{ki}, \varphi, R_1, t, v)$ for $\varphi = \angle(x_k, x_{ki})$, $x_{ki} \subset x_k x_i$, respectively, where $W_{cm}^{(jk)}$, $W_{cm}^{(ki)}$ are determined by the **method 1**, and $R_2 = R_1 + t$ (see Eq. (2.5)). Additionally, a condition given by Eq. (8.17) for $q = m$ is considered for the determination of $R_{\eta cm,\,max}^{(jk)}$, $R_{\eta cm,\,max}^{(ki)}$.

Similarly, let the critical radius $R_{\eta ce,\,max}^{(ij)} = R_{\eta ce,\,max}^{(ij)}(t, v)$ is determined by the curve integral $W_{ce}^{(ij)} = W_{ce}^{(ij)}(x_{ij}, \varphi, R_1, t, v)$ for $\varphi = \angle(x_i, x_{ij})$, $x_{ij} \subset x_i x_j$, where $W_{ce}^{(ij)}$ is also determined by the **method 1**, and a condition given by Eq. (8.17) for $q = e$ is considered for the determination of $R_{\eta ce,\,max}^{(ij)}$.

Finally, let the critical radius $R_{\eta cm,\,max}^{(ij)} = R_{\eta cm,\,max}^{(ij)}(t, v)$ is determined by the curve integral $W_{cm}^{(ij)} = W_{cm}^{(ij)}(x_{ij}, \varphi, R_1, t, v)$ for $\varphi = \angle(x_i, x_{ij})$, $x_{ij} \subset x_i x_j$, where $W_{cm}^{(ij)}$ is determined by the **method 2**, and a condition given by Eq. (8.17) for $q = m$ is considered for the determination of $R_{\eta cm,\,max}^{(ij)}$.

If $x_{ijm} = d/(2c_\varphi)$ (see Eqs. (2.32), (2.33)), then the curve integral $W_{cm}^{(ij)} = W_{cm}^{(ij)}(x_{ij}, \varphi, R_1, t, v)$ is determined by the **method 2** along the abscissa $\overline{P_1P_2P_3}$ for $x_{ij} \in \langle R_2, d/(2c_\varphi)\rangle$ in Fig. 8.36.

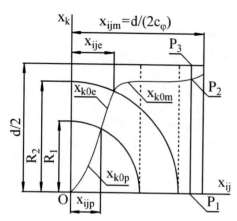

Figure 8.36: The abscissa $\overline{P_1 P_2 P_3}$ for $x_{ij} \in \langle R_2, d/(2c_\varphi) \rangle$ (see Eqs. (2.32), (2.33)) along which the curve integral $W_{cm}^{(ij)} = W_{cm}^{(ij)}(x_{ij}, \varphi, R_1, t, v)$ (see Eqs. (8.3), (8.79)) for $x_{ije} \in \langle 0, R_1 \rangle$ and $x_{ijm} = d/(2c_\varphi)$ is determined by the method 2 due to the crack formation in the spherical particle in spite of the fact that the conditions in Item 2, Section 8.2.4.2 are valid. This determination is also considered for $x_{ije} \in (R_1, R_2)$ (see 8.25c), although Fig. 8.36 is related to the condition $x_{ije} \in \langle 0, R_1 \rangle$ (see Fig. 8.24d).

With regard to Eq. (8.3), we get

$$W_{cm}^{(ij)} = \int_{\overline{P_2 P_3}} w_m^{(ij)} dx_k - \int_{\overline{P_1 P_2}} w_m^{(ij)} dx_k = \int_{x_{k0m}}^{d/2} w_m^{(ij)} dx_k - \int_0^{x_{k0m}} w_m^{(ij)} dx_k,$$

$$x_{ij} \in \left\langle R_1, \frac{d}{2c_\varphi} \right\rangle; \quad x_{ije} \in \langle 0, R_1 \rangle \text{ or } x_{ije} \in (R_1, R_2); \quad x_{ijm} = \frac{d}{2c_\varphi}. \quad (8.79)$$

If the conditions

$$R_{1cm\,max}^{(ij)} < R_{1cm\,max}^{(jk)}, \quad R_{1cm\,max}^{(ij)} < R_{1cm\,max}^{(ki)}, \quad R_{1cm\,max}^{(ij)} < R_{1ce\,max}^{(ij)} \quad (8.80)$$

are simultaneously valid for the critical particle radii $R_{1cm\,max}^{(jk)}$, $R_{1cm\,max}^{(ki)}$, $R_{1ce\,max}^{(ij)}$ and $R_{1cm\,max}^{(ij)}$ determined by the methods 1 and 2, respectively, then the crack is formed in the plane $x_i x_j$ in the spherical particle

- in spite of the fact that the conditions in Item 2, Section 8.2.4.2 are valid,

- and in spite of the fact that the crack would be expected to be formed in the cell matrix regarding these conditions in Item 2, Section 8.2.4.2.

In this case, the 'minimal resistance' is represented by the critical particle radius $R_{1cm\,max}^{(ij)}$ as a minimal value of the set $\left\{ R_{1cm\,max}^{(ij)}, R_{1cm\,max}^{(jk)}, R_{1cm\,max}^{(ki)}, R_{1ce\,max}^{(ij)} \right\}$.

8.2.5 Transformations concerning cracking in planes $x_i x_j$, $x_j x_k$, $x_k x_i$

With regard to the cracking in planes $x_i x_j$, $x_j x_k$, $x_k x_i$, i.e. in the planes $x_1 x_2$, $x_2 x_3$, $x_3 x_1$, the following analyses and transformations concerning

- the coefficient a_{11}, a_{12}, a_{13}, a_{31}, a_{32}, a_{33} (see Eqs. (9.2)–(9.4), (9.8)–(9.10)),

Related phenomena 143

- the elastic moduli $s_{11q}, s_{12q}, \ldots, s_{56q}, s_{66q}$ (see Eqs. (9.1), (9.13)–(9.34))

- and the thermal expansion coefficien α_{iq} ($i = 1,2,3$) (see Eqs. (9.11), (9.13))

are required to be considered.

With regard to the coefficient $a_{11}, a_{12}, a_{13}, a_{31}, a_{32}, a_{33}$ included in Eqs. (9.1), (9.11), (9.14)–(9.34), we get:

1. In case of the cracking in the plane $x_i x_j$, the coefficient $a_{11}, a_{12}, a_{13}, a_{31}, a_{32}, a_{33}$ have the forms

$$a_{11} = \frac{x_{ij} \cos \varphi}{\sqrt{x_{ij}^2 + x_k^2}}, \tag{8.81}$$

$$a_{12} = \frac{x_{ij} \sin \varphi}{\sqrt{x_{ij}^2 + x_k^2}}, \tag{8.82}$$

$$a_{13} = \frac{x_k}{\sqrt{x_{ij}^2 + x_k^2}}, \tag{8.83}$$

$$a_{31} = -\frac{x_k \cos \varphi}{\sqrt{x_{ij}^2 + x_k^2}}, \tag{8.84}$$

$$a_{32} = -\frac{x_k \sin \varphi}{\sqrt{x_{ij}^2 + x_k^2}}, \tag{8.85}$$

$$a_{33} = -\frac{x_{ij}}{\sqrt{x_{ij}^2 + x_k^2}}. \tag{8.86}$$

2. In case of the cracking in the plane $x_j x_k$, the variables x_{ij} and x_k in Eq. (8.81)–(8.86) are replaced by x_{jk} and x_i, respectively.

3. In case of the cracking in the plane $x_k x_i$, the variables x_{ij} and x_k in Eq. (8.81)–(8.86) are replaced by x_{ki} and x_j, respectively.

Consequently, the elastic moduli $s_{11q}, s_{12q}, \ldots, s_{56q}, s_{66q}$ (see Eq. (9.13)) included in Eqs. (9.1), (9.14)–(9.34) and the thermal expansion coefficien α_{iq} ($i = 1,2,3$) (see Eq. (9.13)) included in Eq. (9.11) are replaced by $s_{11q}^{(y)}, s_{12q}^{(y)}, \ldots, s_{56q}^{(y)}, s_{66q}^{(y)}$ and $\alpha_{iq}^{(y)}$, respectively, where $y = ij, jk, ki$. In case of the cracking in the planes $x_i x_j$, $x_j x_k$ and $x_k x_i$, we get $y = ij$, $y = jk$ and $y = ki$, respectively.

Transformations of $s_{11q}^{(y)}, s_{12q}^{(y)}, \ldots, s_{56q}^{(y)}, s_{66q}^{(y)}$ and $\alpha_{iq}^{(y)}$ due to the cracking in the planes $x_1 x_2$, $x_2 x_3$, $x_3 x_1$ are as follows. Let the cracking in the plane $x_i x_j$ is related to the Cartesian system $\left(O x_1^{(y)} x_2^{(y)} x_3^{(y)} \right)$, where $y \equiv ij = 12, 23, 31$. Relationships between $x_i^{(y_1)}$, $x_j^{(y_2)}$ ($i, j = 1,2,3$; $i \neq j$; $y_1, y_2 = 12, 23, 31$; $y_1 \neq y_2$) are derived as

$$x_1^{(12)} = x_3^{(23)} = x_2^{(31)}, \quad x_2^{(12)} = x_1^{(23)} = x_3^{(31)}, \quad x_3^{(12)} = x_2^{(23)} = x_1^{(31)}. \tag{8.87}$$

The elastic modulus $s_{rstuq}^{(y)}$ ($\equiv s_{rstuq}$) and the thermal expansion coefficien $\alpha_{iq}^{(y)}$ ($\equiv \alpha_{iq}$) ($i, r, s, t, u = 1,2,3$; $y = 12, 23, 31$) are related to the axes $x_r^{(y)}$, $x_s^{(y)}$, $x_t^{(y)}$, $x_u^{(y)}$ and $x_i^{(y)}$, respectively.

The transformations $s_{ijklq}^{(y_1)} \rightarrow s_{stuvq}^{(y_2)}$ and $\alpha_{iq}^{(y_1)} \rightarrow \alpha_{jq}^{(y_2)}$ ($i, j, k, l, s, t, u, v = 1,2,3$; $y_1, y_2 = 12,23,31$; $y_1 \neq y_2$) can be thus determined using the relationships given by Eq. (8.87). Strictly speaking, if the numerical values $s_{ijklq}^{(y_1)}$, $\alpha_{iq}^{(y_1)}$ are known, then the numerical values $s_{stuvq}^{(y_2)}$, $\alpha_{jq}^{(y_2)}$ can be determined by Eq. (8.87). Finally, with regard to the subscripts $ijkl$, $stuv$, the subscript transformations given by Eq. (9.12) are considered. As an example, considering Eqs. (8.87), (9.12), the elastic modulus $s_{45q}^{(23)}$ in the Cartesian system $\left(Ox_1^{(23)}x_2^{(23)}x_3^{(23)}\right)$ which is derived regarding the Cartesian system $\left(Ox_1^{(31)}x_2^{(31)}x_3^{(31)}\right)$ has the form

$$s_{45q}^{(23)} = s_{23\,13q}^{(23)} = s_{12\,32q}^{(31)} = s_{12\,23q}^{(31)} = s_{64q}^{(31)} = s_{46q}^{(31)}. \tag{8.88}$$

Accordingly, the transformations $s_{ijklq}^{(y_1)} \rightarrow s_{stuvq}^{(y_2)}$ and $\alpha_{iq}^{(y_1)} \rightarrow \alpha_{jq}^{(y_2)}$ are derived as

$$
\begin{aligned}
s_{11q}^{(12)} &= s_{33q}^{(23)} = s_{22q}^{(31)} & s_{12q}^{(12)} &= s_{13q}^{(23)} = s_{23q}^{(31)}, & s_{13q}^{(12)} &= s_{23q}^{(23)} = s_{12q}^{(31)}, \\
s_{14q}^{(12)} &= s_{36q}^{(23)} = s_{25q}^{(31)}, & s_{15q}^{(12)} &= s_{34q}^{(23)} = s_{26q}^{(31)}, & s_{16q}^{(12)} &= s_{35q}^{(23)} = s_{24q}^{(31)}, \\
s_{22q}^{(12)} &= s_{11q}^{(23)} = s_{33q}^{(31)}, & s_{23q}^{(12)} &= s_{12q}^{(23)} = s_{13q}^{(31)}, & s_{24q}^{(12)} &= s_{16q}^{(23)} = s_{35q}^{(31)}, \\
s_{25q}^{(12)} &= s_{14q}^{(23)} = s_{36q}^{(31)}, & s_{26q}^{(12)} &= s_{15q}^{(23)} = s_{34q}^{(31)}, & s_{33q}^{(12)} &= s_{22q}^{(23)} = s_{11q}^{(31)}, \\
s_{34q}^{(12)} &= s_{26q}^{(23)} = s_{15q}^{(31)}, & s_{35q}^{(12)} &= s_{24q}^{(23)} = s_{16q}^{(31)}, & s_{36q}^{(12)} &= s_{25q}^{(23)} = s_{14q}^{(31)}, \\
s_{44q}^{(12)} &= s_{66q}^{(23)} = s_{55q}^{(31)}, & s_{45q}^{(12)} &= s_{46q}^{(23)} = s_{56q}^{(31)}, & s_{46q}^{(12)} &= s_{56q}^{(23)} = s_{45q}^{(31)}, \\
s_{55q}^{(12)} &= s_{44q}^{(23)} = s_{66q}^{(31)}, & s_{56q}^{(12)} &= s_{45q}^{(23)} = s_{46q}^{(31)}, & s_{66q}^{(12)} &= s_{55q}^{(23)} = s_{44q}^{(31)}, \tag{8.89}
\end{aligned}
$$

$$\alpha_{1q}^{(12)} = \alpha_{3q}^{(23)} = \alpha_{2q}^{(31)}, \quad \alpha_{2q}^{(12)} = \alpha_{1q}^{(23)} = \alpha_{3q}^{(31)}, \quad \alpha_{3q}^{(12)} = \alpha_{2q}^{(23)} = \alpha_{1q}^{(31)}. \tag{8.90}$$

8.3 Analytical model of energy barrier

8.3.1 General analysis

The analytical model of an energy barrier along the axes x_i, x_j, x_k ($i, j, k = 1,2,3$; $i \neq j \neq k$) (see Fig. 8.37) presented in Section 8.3 includes

- the definitio of the energy barriers $W_{bi} = W_{bi}(x_i)$, $W_{bj} = W_{bj}(x_j)$ and $W_{bk} = W_{bk}(x_k)$ along the axes x_i, x_j and x_k, respectively, where W_{bi}, W_{bj}, W_{bk} as 'surface' energy density represent energy gradients (see Eqs. (8.92)–(8.94)),

- the determination of the integration intervals $x_i \in \langle x_{i1q}, x_{i2q} \rangle$, $x_j \in \langle x_{j1q}, x_{j2q} \rangle$, $x_k \in \langle x_{k1q}, x_{k2q} \rangle$ for the spherical particle ($q = p$), the spherical envelope ($q = e$) and the cell matrix ($q = m$) which represent components of the multi-particle-matrix and multi-particle-envelope-matrix systems define in Section 2.1 (see Fig. 2.1a).

The definitio of the energy barriers $W_{bi} = W_{bi}(x_i)$, $W_{bj} = W_{bj}(x_j)$, $W_{bk} = W_{bk}(x_k)$ is as follows.

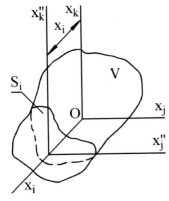

Figure 8.37: The solid continuum with a general shape with the volume V, and with the surface S_i in a position given by $x_i \in \langle x_{i1}, x_{i2} \rangle$, where the difference $x_{i2} - x_{i1}$ represents length of the solid continuum along the axis x_i. The surface S_i with a normal represented by the axis x_i is a cross-section of the solid continuum. The cross-section is given by the plane $x_j'' x_k''$ which is parallel to the plane $x_j x_k$.

Let a solid continuum with a general shape as presented in Fig. 8.37 is considered. Let W is energy accumulated in the volume V of this solid continuum, and then we get

$$W = \int_V w \, dV = \int_{x_{i1}}^{x_{i2}} \int_{x_{j1}}^{x_{j2}} \int_{x_{k1}}^{x_{k2}} w \, dx_i \, dx_j \, dx_k$$

$$= \int_{x_{i1}}^{x_{i2}} \int_{S_i} w \, dx_i \, dS_i = \int_{x_{j1}}^{x_{j2}} \int_{S_j} w \, dx_j \, dS_j = \int_{x_{k1}}^{x_{k2}} \int_{S_k} w \, dx_k \, dS_k, \quad (8.91)$$

where $w = w(x_i, x_j, x_k)$ is energy density induced by stress-deformation field acting in the solid continuum. $x_{i1}, x_{i2}; x_{j1}, x_{j2}; x_{k1}, x_{k2}$ are integration boundaries related to the variables $x_i; x_j; x_k$, respectively. The surfaces S_i, S_j, S_k in positions given by $x_i \in \langle x_{1i}, x_{2i} \rangle$, $x_j \in \langle x_{1j}, x_{2j} \rangle$, $x_k \in \langle x_{1k}, x_{2k} \rangle$ represent cross-sections of the solid continuum, where these surfaces are perpendicular to the axes x_i, x_j, x_k, respectively. As an example (see Fig. 8.37), the surface S_i with the normal x_i is given by the plane $x_j'' x_k''$ which is parallel to

Consequently, $W_{bi} = W_{bi}(x_i)$, $W_{bj} = W_{bj}(x_j)$, $W_{bk} = W_{bk}(x_k)$ as surface integrals related to S_i, S_j, S_k, respectively, have the forms

$$W_{bi} = W_{bi}(x_i) = \frac{\partial W}{\partial x_i} = \int_{S_i} w \, dS_i = \int_{x_{j1}}^{x_{j2}} \int_{x_{k1}}^{x_{k2}} w \, dx_j \, dx_k, \quad (8.92)$$

$$W_{bj} = W_{bj}(x_j) = \frac{\partial W}{\partial x_j} = \int_{S_j} w \, dS_j = \int_{x_{i1}}^{x_{i2}} \int_{x_{k1}}^{x_{k2}} w \, dx_i \, dx_k, \quad (8.93)$$

$$W_{bk} = W_{bk}(x_k) = \frac{\partial W}{\partial x_k} = \int_{S_k} w \, dS_k = \int_{x_{i1}}^{x_{i2}} \int_{x_{j1}}^{x_{j2}} w \, dx_i \, dx_j. \qquad (8.94)$$

Accordingly, $W_{bi} = W_{bi}(x_i)$, $W_{bj} = W_{bj}(x_j)$, $W_{bk} = W_{bk}(x_k)$ thus represent 'surface' energy density related to S_i, S_j, S_k, respectively.

If the energy density $w = w(x_i, x_j, x_k) = w(r, \varphi, \nu)$ is derived by the spherical coordinates (r, φ, ν), then the following relationships between the coordinates (x_i, x_j, x_k) and (r, φ, ν) are considered:

1. If $\varphi = \angle(x_i, x_{ij})$, $\nu = \angle(x_1', x_k)$ (see Fig. 3.1), $x_{ij} \subset x_i x_j$ (see Fig. 8.1), then we get
$$x_i = r \cos \varphi \sin \nu, \quad x_j = r \sin \varphi \sin \nu, \quad x_k = r \cos \nu. \qquad (8.95)$$

2. If $\varphi = \angle(x_j, x_{jk})$, $\nu = \angle(x_1', x_i)$, $x_{jk} \subset x_j x_k$, then we get
$$x_i = r \cos \nu, \quad x_j = r \cos \varphi \sin \nu, \quad x_k = r \sin \varphi \sin \nu. \qquad (8.96)$$

3. If $\varphi = \angle(x_k, x_{ki})$, $\nu = \angle(x_1', x_j)$, $x_{ki} \subset x_k x_i$, then we get
$$x_i = r \sin \varphi \sin \nu, \quad x_j = r \cos \nu, \quad x_k = r \cos \varphi \sin \nu. \qquad (8.97)$$

In general, energy barriers influenc motion of dislocations, magnetic domain walls, etc., where a magnetic domain wall represents a front between magnetic domains with and without ordered magnetic moments [11, p. 63].

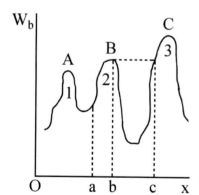

Figure 8.38: The energy barriers *1*, *2*, *3* with maximal values in the points A, B, C, respectively. The energy barrier *2* exhibits the maximum in the position $x = b$.

As an example, a planar magnetic domain wall in a magnetic material is shifted along the axis x from the position $x = a$ to the position $x = b$ due to the change $\Delta H_{ba} = H_b - H_a$ of the magnetic fiel intensity H (see Fig 8.38). The point B with the position $x = b$ represents a maximum of the energy barrier *2* as presented in Fig 8.38. Consequently, this magnetic domain wall moves along the axis x to the position $x = c$ at the constant intensity $H_b = H_c$. The wall is then stopped by the energy barrier *3* which is higher than the energy barrier *2*. The position change $\Delta x_{cb} = c - b$ of the wall results in the

Related phenomena

change ΔM_{cb} of the magnetic moment M of the magnetic material. Accordingly, ΔM_{ba} related to $\Delta x_{cb} = c - b$ results in a discrete change of the magnetization hysteresis loop at zero change of H, i.e. for $\Delta H_{cb} = H_c - H_b = 0$. This discrete change at the constant intensity $H_b = H_c$ induces a voltage impulse in a pickup coil, where the voltage impulse is known as the Barkhausen jump. Additionally, maximal values of the energy barriers *1, 2, 3* are reasons of the coercivity of magnetic materials including magnetic multi-component materials [11, p. 63].

Sections 8.3.2, 8.3.3 presents the determination of the surface integrals $W_{bi} = W_{bi}(x_i)$, $W_{bj} = W_{bj}(x_j)$, $W_{bk} = W_{bk}(x_k)$ in multi-particle-matrix and multi-particle-envelope-matrix systems.

Let the multi-particle-matrix and multi-particle-envelope-matrix systems consist of cubic cells with the dimension d which is equal to the inter-particle distance (see Fig. 2.1a). The surface integrals $W_{bi} = W_{bi}(x_i)$, $W_{bj} = W_{bj}(x_j)$, $W_{bk} = W_{bk}(x_k)$ are thus sufficien to be determined within one eighth of the cubic cell, i.e. for $x_i \in \langle 0, d/2 \rangle$, $x_j \in \langle 0, d/2 \rangle$ and $x_k \in \langle 0, d/2 \rangle$. This is a consequence of symmetry of these systems due to the matrix infinit and the periodic distribution of the spherical particles and the spherical envelopes (see Section 2.1).

8.3.2 Multi-particle-matrix system

The surface integral $W_{bi} = W_{bi}(x_i)$ for $x_i \in \langle 0, R_1 \rangle$ and $x_i \in \langle R_1, d/2 \rangle$ has the forms

$$W_{bi} = W_{bi}(x_i)$$

$$= 4 \left(\int_{x_{j1p}}^{x_{j2p}} \int_{x_{k1p}}^{x_{k2p}} w_p \, dx_j \, dx_k + \int_{x_{j1m}}^{x_{j2m}} \int_{x_{k1m}}^{x_{k2m}} w_m \, dx_j \, dx_k \right),$$

$$x_i \in \langle 0, R_1 \rangle \, ;$$

$$\langle x_{j1p}, x_{j2p} \rangle = \left\langle 0, \sqrt{R_1^2 - x_i^2} \right\rangle, \quad \langle x_{k1p}, x_{k2p} \rangle = \left\langle 0, \sqrt{R_1^2 - \left(x_i^2 + x_j^2 \right)} \right\rangle ;$$

$$\langle x_{j1m}, x_{j2m} \rangle = \left\langle 0, \frac{d}{2} \right\rangle,$$

$$\langle x_{k1m}, x_{k2m} \rangle = \left\langle \sqrt{R_1^2 - \left(x_i^2 + x_j^2 \right)}, \frac{d}{2} \right\rangle$$

$$\text{for} \quad \langle x_{j1m}, x_{j2m} \rangle = \left\langle 0, \sqrt{R_1^2 - x_i^2} \right\rangle,$$

$$\langle x_{k1m}, x_{k2m} \rangle = \left\langle 0, \frac{d}{2} \right\rangle \quad \text{for} \quad \langle x_{j1m}, x_{j2m} \rangle = \left\langle \sqrt{R_1^2 - x_i^2}, \frac{d}{2} \right\rangle, \tag{8.98}$$

$$W_{bi} = W_{bi}(x_i) = 4 \int_{x_{j1m}}^{x_{j2m}} \int_{x_{k1m}}^{x_{k2m}} w_m \, dx_j \, dx_k,$$

$$x_i \in \left\langle R_1, \frac{d}{2} \right\rangle; \quad \langle x_{j1m}, x_{j2m} \rangle = \langle x_{k1m}, x_{k2m} \rangle = \left\langle 0, \frac{d}{2} \right\rangle, \qquad (8.99)$$

where the intervals $x_j \in \langle x_{j1q}, x_{j2q} \rangle$, $x_k \in \langle x_{k1q}, x_{k2q} \rangle$ for the energy density $w_p = w_p(x_i, x_j, x_k)$ $(q = p, m)$ are related to the surface S_{iq} which represents a cross-section within the spherical particle $(q = p)$ and the cell matrix $(q = m)$. The cross-section S_{iq} with a normal represented by the axis x_i is given by the plane $x_j'' x_k''$ (see Fig. 8.37) in a position for $x_i \in \langle 0, d/2 \rangle$, where the plane $x_j'' x_k''$ is parallel to the plane $x_j x_k$.

The surfaces integrals $W_{bj} = W_{bj}(x_j)$ and $W_{bk} = W_{bk}(x_k)$ (see Eqs. (8.93), (8.94)) are given by Eqs. (8.98), (8.99), where the transformations $i \to j$, $j \to i$, $k \to k$ and $i \to k$, $j \to j$, $i \to k$ of the subscripts i, j, k are considered in Eqs. (8.98), (8.99), respectively.

8.3.3 Multi-particle-envelope-matrix system

The surface integral $W_{bi} = W_{bi}(x_i)$ for $x_i \in \langle R_2, d/2 \rangle$ is given by Eq. (8.99). Consequently, $W_{bi} = W_{bi}(x_i)$ for $x_i \in \langle 0, R_1 \rangle$ and $x_i \in \langle R_1, R_2 \rangle$ has the forms

$$W_{bi} = W_{bi}(x_i)$$

$$= 4 \left(\int_{x_{j1p}}^{x_{j2p}} \int_{x_{k1p}}^{x_{k2p}} w_p \, dx_j \, dx_k + \int_{x_{j1e}}^{x_{j2e}} \int_{x_{k1e}}^{x_{k2e}} w_e \, dx_j \, dx_k \right.$$

$$\left. + \int_{x_{j1m}}^{x_{j2m}} \int_{x_{k1m}}^{x_{k2m}} w_m \, dx_j \, dx_k \right), \quad x_i \in \langle 0, R_1 \rangle;$$

$$\langle x_{j1p}, x_{j2p} \rangle = \left\langle 0, \sqrt{R_1^2 - x_i^2} \right\rangle, \quad \langle x_{k1p}, x_{k2p} \rangle = \left\langle 0, \sqrt{R_1^2 - \left(x_i^2 + x_j^2 \right)} \right\rangle;$$

$$\langle x_{j1e}, x_{j2e} \rangle = \left\langle 0, \sqrt{R_2^2 - x_i^2} \right\rangle,$$

$$\langle x_{k1e}, x_{k2e} \rangle = \left\langle \sqrt{R_1^2 - \left(x_i^2 + x_j^2 \right)}, \sqrt{R_2^2 - \left(x_i^2 + x_j^2 \right)} \right\rangle$$

$$\text{for} \quad \langle x_{j1e}, x_{j2e} \rangle = \left\langle 0, \sqrt{R_1^2 - x_i^2} \right\rangle,$$

$$\langle x_{k1e}, x_{k2e} \rangle = \left\langle 0, \sqrt{R_2^2 - \left(x_i^2 + x_j^2 \right)} \right\rangle$$

$$\text{for} \quad \langle x_{j1e}, x_{j2e} \rangle = \left\langle \sqrt{R_1^2 - x_i^2}, \sqrt{R_2^2 - x_i^2} \right\rangle;$$

$$\langle x_{j1m}, x_{j2m} \rangle = \left\langle 0, \frac{d}{2} \right\rangle,$$

$$\langle x_{k1m}, x_{k2m} \rangle = \left\langle \sqrt{R_2^2 - \left(x_i^2 + x_j^2\right)}, \frac{d}{2} \right\rangle$$

$$\text{for} \quad \langle x_{j1m}, x_{j2m} \rangle = \left\langle 0, \sqrt{R_2^2 - x_i^2} \right\rangle,$$

$$\langle x_{k1m}, x_{k2m} \rangle = \left\langle 0, \frac{d}{2} \right\rangle \quad \text{for} \quad \langle x_{j1m}, x_{j2m} \rangle = \left\langle \sqrt{R_2^2 - x_i^2}, \frac{d}{2} \right\rangle, \tag{8.100}$$

$$W_{bi} = W_{bi}(x_i)$$

$$= 4 \left(\int_{x_{j1e}}^{x_{j2e}} \int_{x_{k1e}}^{x_{k2e}} w_e \, dx_j \, dx_k + \int_{x_{j1m}}^{x_{j2m}} \int_{x_{k1m}}^{x_{k2m}} w_m \, dx_j \, dx_k \right),$$

$$x_i \in \langle R_1, R_2 \rangle ;$$

$$\langle x_{j1e}, x_{j2e} \rangle = \left\langle 0, \sqrt{R_2^2 - x_i^2} \right\rangle, \quad \langle x_{k1e}, x_{k2e} \rangle = \left\langle 0, \sqrt{R_2^2 - \left(x_i^2 + x_j^2\right)} \right\rangle ;$$

$$\langle x_{j1m}, x_{j2m} \rangle = \left\langle 0, \frac{d}{2} \right\rangle,$$

$$\langle x_{k1m}, x_{k2m} \rangle = \left\langle \sqrt{R_2^2 - \left(x_i^2 + x_j^2\right)}, \frac{d}{2} \right\rangle$$

$$\text{for} \quad \langle x_{j1m}, x_{j2m} \rangle = \left\langle 0, \sqrt{R_2^2 - x_i^2} \right\rangle,$$

$$\langle x_{k1m}, x_{k2m} \rangle = \left\langle 0, \frac{d}{2} \right\rangle \quad \text{for} \quad \langle x_{j1m}, x_{j2m} \rangle = \left\langle \sqrt{R_2^2 - x_i^2}, \frac{d}{2} \right\rangle, \tag{8.101}$$

where the intervals $x_j \in \langle x_{j1q}, x_{j2q} \rangle$, $x_k \in \langle x_{k1q}, x_{k2q} \rangle$ for the energy density $w_p = w_p(x_i, x_j, x_k)$ $(q = p,e,m)$ are related to the surface S_{iq} which represents a cross-section within the spherical particle $(q = p)$, the spherical envelope $(q = e)$ and the cell matrix $(q = m$. The cross-section S_{iq} with a normal represented by the axis x_i is given by the plane $x_j'' x_k''$ (see Fig. 8.37) in a position for $x_i \in \langle 0, d/2 \rangle$, where the plane $x_j'' x_k''$ is parallel to the plane $x_j x_k$.

The surfaces integrals $W_{bj} = W_{bj}(x_j)$ and $W_{bk} = W_{bk}(x_k)$ (see Eqs. (8.93), (8.94)) are given by Eqs. (8.100), (8.101), where the transformations $i \to j, j \to i, k \to k$ and $i \to k, j \to j, i \to k$ of the subscripts i, j, k are considered in Eqs. (8.100), (8.101), respectively.

8.4 Analytical model of strengthening

8.4.1 General analysis

The analytical model of the micro-strengthening $\sigma_{si} = \sigma_{si}(x_i)$, $\sigma_{sj} = \sigma_{si}(x_j)$, $\sigma_{sk} = \sigma_{si}(x_k)$ and the macro-strengthening $\overline{\sigma}_{si}, \overline{\sigma}_{sj}, \overline{\sigma}_{sk}$ along the axes x_i, x_j, x_k ($i, j, k = 1,2,3$; $i \neq j \neq k$) (see Fig. 8.37) presented in Section 8.4 is based on the following energy analysis.

Let a solid continuum with a general shape and with the volume V as presented in Fig. 8.37 is considered. Let the surface S_i in a position by $x_i \in \langle x_{1i}, x_{2i} \rangle$ is a cross-sections of this solid continuum, where $S_i(x_i)$ is an area of S_i. The cross-section S_i with a normal represented by the axis x_i is given by the plane $x_j'' x_k''$ (see Fig. 8.37), where the plane $x_j'' x_k''$ is parallel to the plane $x_j x_k$. The surface S_i is described by the coordinates $x_j \in \langle x_{j1}, x_{j2} \rangle$ and $x_k \in \langle x_{k1}, x_{k2} \rangle$.

Let $\sigma_{si} = \sigma_{si}(x_i)$ represents a normal stress on the surface S_i, i.e. the normal stress $\sigma_{si} = \sigma_{si}(x_i)$ which is required to be determined within this analysis acts along the axis x_i. Additionally, let $\sigma_{si} = \sigma_{si}(x_i)$ is constant regarding each point of the surface S_i, i.e. $\sigma_{si} = \sigma_{si}(x_i) \neq f(x_j, x_k)$ is not a function of $x_j \in \langle x_{j1}, x_{j2} \rangle$ and $x_k \in \langle x_{k1}, x_{k2} \rangle$.

Consequently, let the stress $\sigma_{si} = \sigma_{si}(x_i)$ induces the elastic strain $\varepsilon_{sii} = \varepsilon_{sii}(x_i) = s_{ii}\sigma_{si}(x_i)$ along the axis x_i. Accordingly, the 'surface' elastic energy density $W_{\sigma_{si}} = W_{\sigma_{si}}(x_i)$ accumulated on the surface area $S_i = S_i(x_i)$ has the form

$$W_{\sigma_{si}} = W_{\sigma_{si}}(x_i) = \frac{\sigma_{si}\,\varepsilon_{sii}\,S_i}{2} = \frac{s_{ii}\,S_i\,\sigma_{si}^2}{2}, \quad s_{ii} = \frac{1}{E_i}, \tag{8.102}$$

where E_i is the Young's modulus along the axis x_i.

Let the stress $\sigma_i = \sigma_i(x_i, x_j, x_k)$ of stress-deformation field in the solid continuum acts along the axis x_i. Let $w_i = w_i(x_i, x_j, x_k)$ is energy density induced by the stress $\sigma_i = \sigma_i(x_i, x_j, x_k)$. Accordingly, the 'surface' energy density $W_i = W_i(x_i)$ accumulated on the surface area $S_i = S_i(x_i)$ is derived as

$$W_i = W_i(x_i) = \int_{S_i} w_i \, dS_i. \tag{8.103}$$

With regard to a sign of the stress $\sigma_{si} = \sigma_{si}(x_i)$, the determination of the integral in Eq. (8.103) is required to consider the following conditions:

1. If $\sigma_i(x_i, x_j, x_k) < 0$ at a point with the coordinates (x_i, x_j, x_k) on the surface S_i, then the energy density $w_i = w_i(x_i, x_j, x_k)$ at this point is considered to be $-w_i$.

2. If $\sigma_i(x_i, x_j, x_k) > 0$ at a point with the coordinates (x_i, x_j, x_k) on the surface S_i, then the energy density $w_i = w_i(x_i, x_j, x_k)$ at this point is considered to be $+w_i$.

The micro-strengthening $\sigma_{si} = \sigma_{si}(x_i)$ along the axis x_i is determined by the condition $W_{\sigma_{si}} = W_i$, and then we get

$$\sigma_{si} = \sigma_{si}(x_i) = \pm \sqrt{\frac{2\,|W_i|}{s_{ii}\,S_i}}, \quad s_{ii} = \frac{1}{E_i}, \tag{8.104}$$

where the signs '+' and '-' in Eq. (8.104) are considered for $W_i > 0$ and $W_i < 0$, respectively.

The macro-strengthening $\overline{\sigma}_{si}$ along the axis x_i represents a mean value of $\sigma_{si} = \sigma_{si}(x_i)$ regarding the interval $x_i \in \langle x_{1i}, x_{2i} \rangle$, and then we get

$$\overline{\sigma}_{si} = \frac{1}{x_{2i} - x_{1i}} \int_{x_{1i}}^{x_{2i}} \sigma_{si}\,dx_i = \frac{1}{x_{2i} - x_{1i}} \sqrt{\frac{2}{s_{ii}}} \int_{x_{1i}}^{x_{2i}} \pm \sqrt{\frac{|W_i|}{S_i}}\,dx_i,$$

$$s_{ii} = \frac{1}{E_i}, \tag{8.105}$$

where the following conditions are considered:

3. If $W_i(x_i) > 0$ in a position given by the coordinate x_i, then the sign '+' in Eq. (8.105) is considered.

4. If $W_i(x_i) < 0$ in a position given by the coordinate x_i, then the sign '-' in Eq. (8.105) is considered.

Finally, with regard to a sign of the macro-strengthening $\overline{\sigma}_{si}$, the following analysis is considered:

5. If $\overline{\sigma}_{si} > 0$, then the macro-strengthening $\overline{\sigma}_{si}$ represents a 'resistance' against compressive mechanical loading.

6. If $\overline{\sigma}_{si} < 0$, then the macro-strengthening $\overline{\sigma}_{si}$ represents a 'resistance' against tensile mechanical loading.

With regard to the micro-/macro-strengthening along the axes x_j and x_k, the transformations $i \rightarrow j, j \rightarrow i, k \rightarrow k$ and $i \rightarrow k, j \rightarrow j, i \rightarrow k$ of the subscripts i, j, k are considered in Eqs. (8.102)–(8.105), respectively.

8.4.2 Strengthening in multi-particle-(envelope)-matrix system

Let the multi-particle-matrix and multi-particle-envelope-matrix systems consist of cubic cells with the dimension d which is equal to the inter-particle distance (see Fig. 2.1a). The micro- and macro-strengthening in these systems is sufficien to be determined within one eighth of the cubic cell, i.e. for $x_i \in \langle 0, d/2 \rangle$, $x_j \in \langle 0, d/2 \rangle$ and $x_k \in \langle 0, d/2 \rangle$. This is a consequence of symmetry of these systems due to the matrix infinit and the periodical distribution of the spherical particles and the spherical envelopes (see Section 2.1).

8.4.2.1   Multi-particle-matrix system

If $x_i \in \langle 0, R_1 \rangle$, then the areas $S_{ip} = S_{ip}(x_i)$ and $S_{im1} = S_{im1}(x_i)$ of the surfaces S_{ip} and S_{im1} in the spherical particle and the cell matrix within one eighth of the cubic cell, respectively, are derived as

$$S_{ip} = \frac{\pi \left(R_1^2 - x_i^2 \right)}{4}, \quad x_i \in \langle 0, R_1 \rangle, \tag{8.106}$$

$$S_{im1} = \frac{d^2}{4} - S_{ip} = \frac{d^2 - \pi \left(R_1^2 - x_i^2 \right)}{4}, \quad x_i \in \langle 0, R_1 \rangle. \tag{8.107}$$

Consequently, the 'surface' elastic energy density $W_{\sigma_{si}1} = W_{\sigma_{si}1}(x_i)$ for $x_i \in \langle 0, R_1 \rangle$ (see Eq. (8.102)) has the form

$$W_{\sigma_{si}1} = \frac{\sigma_{si}^2 \left(s_{iip} S_{ip} + s_{iim} S_{im1} \right)}{2}, \quad s_{iiq} = \frac{1}{E_{iq}}, \quad x_i \in \langle 0, R_1 \rangle, \quad q = p, m, \tag{8.108}$$

where E_{iq} is the Young's modulus along the axis x_i in the spherical particle ($q = p$) and the cell matrix ($q = m$).

The 'surface' energy density $W_{i1} = W_{i1}(x_i)$ for $x_i \in \langle 0, R_1 \rangle$ (see Eq. (8.103)) accumulated on the surfaces S_{ip} and S_{im1} which are described by $x_j \in \langle x_{j1p}, x_{j2p} \rangle$, $x_k \in \langle x_{k1p}, x_{k2p} \rangle$ and $x_j \in \langle x_{j1m}, x_{j2m} \rangle$, $x_k \in \langle x_{k1m}, x_{k2m} \rangle$, respectively, is derived as

$$W_{i1} = \int\limits_{S_{ip}} w_{ip} \, dS_i + \int\limits_{S_{im1}} w_{im} \, dS_i$$

$$= \int\limits_{x_{j1p}}^{x_{j2p}} \int\limits_{x_{k1p}}^{x_{k2p}} w_{ip} \, dx_j \, dx_k + \int\limits_{x_{j1m}}^{x_{j2m}} \int\limits_{x_{k1m}}^{x_{k2m}} w_{im} \, dx_j \, dx_k,$$

$$x_i \in \langle 0, R_1 \rangle \,;$$

$$\langle x_{j1p}, x_{j2p} \rangle = \left\langle 0, \sqrt{R_1^2 - x_i^2} \right\rangle, \quad \langle x_{k1p}, x_{k2p} \rangle = \left\langle 0, \sqrt{R_1^2 - \left(x_i^2 + x_j^2 \right)} \right\rangle \,;$$

$$\langle x_{j1m}, x_{j2m} \rangle = \left\langle 0, \frac{d}{2} \right\rangle,$$

$$\langle x_{k1m}, x_{k2m} \rangle = \left\langle \sqrt{R_1^2 - \left(x_i^2 + x_j^2 \right)}, \frac{d}{2} \right\rangle$$

$$\text{for} \quad \langle x_{j1m}, x_{j2m} \rangle = \left\langle 0, \sqrt{R_1^2 - x_i^2} \right\rangle,$$

$$\langle x_{k1m}, x_{k2m} \rangle = \left\langle 0, \frac{d}{2} \right\rangle \quad \text{for} \quad \langle x_{j1m}, x_{j2m} \rangle = \left\langle \sqrt{R_1^2 - x_i^2}, \frac{d}{2} \right\rangle, \tag{8.109}$$

where the determination of the integrals in Eq. (8.109) is required to consider the conditions in Items 1, 1, Section 8.4.1.

Related phenomena 153

The micro-strengthening $\sigma_{si1} = \sigma_{si1}(x_i)$ for $x_i \in \langle 0, R_1 \rangle$ along the axis x_i (see Eq. (8.104)) which is determined by the condition $W_{\sigma_{si1}} = W_{i1}$ has the form

$$\sigma_{si1} = \pm \sqrt{\frac{2|W_{i1}|}{s_{iip}S_{ip} + s_{iim}S_{im1}}}, \quad s_{iiq} = \frac{1}{E_{iq}}, \quad x_i \in \langle 0, R_1 \rangle, \quad q = p, m, \quad (8.110)$$

where the signs '+' and '-' in Eq. (8.110) are considered for $W_{i1} > 0$ and $W_{i1} < 0$, respectively.

If $x_i \in \langle R_1, d/2 \rangle$, then $S_{im2} = d^2/4$ is an surface area in the the cell matrix within one eighth of the cubic cell. Consequently, the 'surface' elastic energy density $W_{\sigma_{si2}} = W_{\sigma_{si2}}(x_i)$ for $x_i \in \langle R_1, d/2 \rangle$ (see Eq. (8.102)) has the form

$$W_{\sigma_{si2}} = \frac{s_{iim}\, d^2\, \sigma_{si}^2}{8}, \quad s_{iim} = \frac{1}{E_{im}}, \quad x_i \in \left\langle R_1, \frac{d}{2} \right\rangle. \quad (8.111)$$

The 'surface' energy density $W_{i2} = W_{i2}(x_i)$ for $x_i \in \langle R_1, d/2 \rangle$ (see Eq. (8.103)) accumulated on the surface S_{im2} which is described by $x_j \in \langle 0, d/2 \rangle$, $x_k \in \langle 0, d/2 \rangle$ is derived as

$$W_{i2} = \int\limits_{S_{im2}} w_{im}\, dS_i = \int\limits_0^{d/2}\int\limits_0^{d/2} w_{im}\, dx_j\, dx_k, \quad x_i \in \left\langle R_1, \frac{d}{2} \right\rangle, \quad (8.112)$$

where the determination of the integral in Eq. (8.112) is required to consider the conditions in Items 1, 1, Section 8.4.1.

The micro-strengthening $\sigma_{si2} = \sigma_{si2}(x_i)$ for $x_i \in \langle R_1, d/2 \rangle$ along the axis x_i (see Eq. (8.104)) which is determined by the condition $W_{\sigma_{si2}} = W_{i2}$ has the form

$$\sigma_{si2} = \pm \frac{2}{d}\sqrt{\frac{2|W_{i2}|}{s_{iim}}}, \quad s_{iim} = \frac{1}{E_{im}}, \quad x_i \in \left\langle R_1, \frac{d}{2} \right\rangle, \quad (8.113)$$

where the signs '+' and '-' in Eq. (8.113) are considered for $W_{i2} > 0$ and $W_{i2} < 0$, respectively. Additionally, we get $[\sigma_{si1}(x_i)]_{x_i=R_1} = [\sigma_{si2}(x_i)]_{x_i=R_1}$ due to $[S_{im1}(x_i)]_{x_i=R_1} = S_{im2}$ and $[W_{i1}(x_i)]_{x_i=R_1} = [W_{i2}(x_i)]_{x_i=R_1}$.

The macro-strengthening $\overline{\sigma}_{si}$ along the axis x_i (see Eq. (8.105)) which represents a mean value of $\sigma_{si} = \sigma_{si1}(x_i) + \sigma_{si2}(x_i)$ regarding the interval $x_i \in \langle x_{1i}, x_{2i} \rangle = \langle 0, d/2 \rangle$ is derived as

$$\overline{\sigma}_{si} = \frac{2}{d}\left(\int\limits_0^{R_1} \sigma_{si1}\, dx_i + \int\limits_{R_1}^{d/2} \sigma_{si2}\, dx_i \right)$$

$$= \frac{2\sqrt{2}}{d}\left(\int\limits_0^{R_1} \pm\sqrt{\frac{|W_{i1}|}{s_{iip}S_{ip} + s_{iim}S_{im1}}}\, dx_i + \frac{2}{d \times \sqrt{s_{iim}}} \int\limits_{R_1}^{d/2} \pm\sqrt{|W_{i2}|}\, dx_i \right),$$

$$s_{iiq} = \frac{1}{E_{iq}}, \quad q = p, m, \quad (8.114)$$

154   Ladislav Ceniga

where the conditions and the analysis in Items 3–6, Section 8.4.1 are considered.

8.4.2.2   Multi-particle-envelope-matrix system

If $x_i \in \langle 0, R_1 \rangle$, then the area $S_{ip} = S_{ip}(x_i)$ of the surface S_{ip} in the spherical particle is given by Eq. (8.106). The areas $S_{ie1} = S_{ie1}(x_i)$ and $S_{im3} = S_{im3}(x_i)$ of the surfaces S_{ie1} and S_{im3} for $x_i \in \langle 0, R_1 \rangle$ in the spherical envelope and the cell matrix within one eighth of the cubic cell, respectively, are derived as

$$S_{ie1} = \frac{\pi\left(R_2^2 - x_i^2\right)}{4} - S_{ip} = \frac{\pi\left(R_2^2 - R_1^2\right)}{4}, \quad x_i \in \langle 0, R_1 \rangle, \tag{8.115}$$

$$S_{im3} = \frac{d^2 - \pi\left(R_2^2 - x_i^2\right)}{4}, \quad x_i \in \langle 0, R_1 \rangle. \tag{8.116}$$

Consequently, the 'surface' elastic energy density $W_{\sigma si3} = W_{\sigma si3}(x_i)$ for $x_i \in \langle 0, R_1 \rangle$ (see Eq. (8.102)) has the form

$$W_{\sigma si3} = \frac{\sigma_{si}^2\left(s_{iip}\, S_{ip} + s_{iie}\, S_{ie1} + s_{iim}\, S_{im3}\right)}{2}, \quad s_{iiq} = \frac{1}{E_{iq}},$$
$$x_i \in \langle 0, R_1 \rangle, \quad q = p, e, m, \tag{8.117}$$

where E_{iq} is the Young's modulus along the axis x_i in the spherical particle ($q = p$), the spherical envelope ($q = e$) and the cell matrix ($q = m$).

The 'surface' energy density $W_{i3} = W_{i3}(x_i)$ for $x_i \in \langle 0, R_1 \rangle$ (see Eq. (8.103)) accumulated on the surfaces S_{ip}; S_{ie1} and S_{im3} which are described by $x_j \in \langle x_{j1p}, x_{j2p} \rangle$, $x_k \in \langle x_{k1p}, x_{k2p} \rangle$; $x_j \in \langle x_{j1e}, x_{j2e} \rangle$, $x_k \in \langle x_{k1e}, x_{k2e} \rangle$ and $x_j \in \langle x_{j1m}, x_{j2m} \rangle$, $x_k \in \langle x_{k1m}, x_{k2m} \rangle$, respectively, is derived as

$$W_{i3} = \int_{S_{ip}} w_{ip}\, dS_i + \int_{S_{ie1}} w_{ie}\, dS_i + \int_{S_{im3}} w_{im}\, dS_i$$

$$= \int_{x_{j1p}}^{x_{j2p}} \int_{x_{k1p}}^{x_{k2p}} w_{ip}\, dx_j\, dx_k + \int_{x_{j1e}}^{x_{j2e}} \int_{x_{k1e}}^{x_{k2e}} w_{ie}\, dx_j\, dx_k$$

$$+ \int_{x_{j1m}}^{x_{j2m}} \int_{x_{k1m}}^{x_{k2m}} w_{im}\, dx_j\, dx_k,$$

$$x_i \in \langle 0, R_1 \rangle\, ;$$

$$\langle x_{j1p}, x_{j2p} \rangle = \left\langle 0, \sqrt{R_1^2 - x_i^2} \right\rangle, \quad \langle x_{k1p}, x_{k2p} \rangle = \left\langle 0, \sqrt{R_1^2 - \left(x_i^2 + x_j^2\right)} \right\rangle\, ;$$

$$\langle x_{j1e}, x_{j2e} \rangle = \left\langle 0, \sqrt{R_2^2 - x_i^2} \right\rangle,$$

$$\langle x_{k1e}, x_{k2e}\rangle = \left\langle \sqrt{R_1^2 - \left(x_i^2 + x_j^2\right)}, \sqrt{R_2^2 - \left(x_i^2 + x_j^2\right)} \right\rangle$$

$$\text{for} \quad \langle x_{j1e}, x_{j2e}\rangle = \left\langle 0, \sqrt{R_1^2 - x_i^2} \right\rangle,$$

$$\langle x_{k1e}, x_{k2e}\rangle = \left\langle 0, \sqrt{R_2^2 - \left(x_i^2 + x_j^2\right)} \right\rangle$$

$$\text{for} \quad \langle x_{j1e}, x_{j2e}\rangle = \left\langle \sqrt{R_1^2 - x_i^2}, \sqrt{R_2^2 - x_i^2} \right\rangle;$$

$$\langle x_{j1m}, x_{j2m}\rangle = \left\langle 0, \frac{d}{2} \right\rangle,$$

$$\langle x_{k1m}, x_{k2m}\rangle = \left\langle \sqrt{R_2^2 - \left(x_i^2 + x_j^2\right)}, \frac{d}{2} \right\rangle$$

$$\text{for} \quad \langle x_{j1m}, x_{j2m}\rangle = \left\langle 0, \sqrt{R_2^2 - x_i^2} \right\rangle,$$

$$\langle x_{k1m}, x_{k2m}\rangle = \left\langle 0, \frac{d}{2} \right\rangle \quad \text{for} \quad \langle x_{j1m}, x_{j2m}\rangle = \left\langle \sqrt{R_2^2 - x_i^2}, \frac{d}{2} \right\rangle, \tag{8.118}$$

where the determination of the integrals in Eq. (8.118) is required to consider the conditions in Items 1, 1, Section 8.4.1.

The micro-strengthening $\sigma_{si3} = \sigma_{si3}(x_i)$ for $x_i \in \langle 0, R_1\rangle$ along the axis x_i (see Eq. (8.104)) which is determined by the condition $W_{\sigma_{si3}} = W_{i3}$ has the form

$$\sigma_{si3} = \pm \sqrt{\frac{2\,|W_{i3}|}{s_{iip}\,S_{ip} + s_{iie}\,S_{ie1} + s_{iim}\,S_{im3}}}, \quad s_{iiq} = \frac{1}{E_{iq}},$$
$$x_i \in \langle 0, R_1\rangle, \quad q = p, e, m, \tag{8.119}$$

where the signs '+' and '-' in Eq. (8.119) are considered for $W_{i3} > 0$ and $W_{i3} < 0$, respectively.

If $x_i \in \langle R_1, R_2\rangle$, then the area $S_{im3} = S_{im3}(x_i)$ of the surface S_{im3} in the cell matrix, given by Eq. (8.116), is considered. The area $S_{ie2} = S_{ie2}(x_i)$ of the surface S_{ie2} for $x_i \in \langle R_1, R_2\rangle$ in the spherical envelope within one eighth of the cubic cell is derived as

$$S_{ie2} = \frac{\pi\left(R_2^2 - x_i^2\right)}{4}, \quad x_i \in \langle R_1, R_2\rangle. \tag{8.120}$$

Consequently, the 'surface' elastic energy density $W_{\sigma_{si4}} = W_{\sigma_{si4}}(x_i)$ for $x_i \in \langle R_1, R_2\rangle$ (see Eq. (8.102)) has the form

$$W_{\sigma_{si4}} = \frac{\sigma_{si}^2\left(s_{iie}\,S_{ie2} + s_{iim}\,S_{im3}\right)}{2}, \quad s_{iiq} = \frac{1}{E_{iq}}, \quad x_i \in \langle 0, R_1\rangle, \quad q = e, m. \tag{8.121}$$

The 'surface' energy density $W_{i4} = W_{i4}(x_i)$ for $x_i \in \langle R_1, R_2 \rangle$ (see Eq. (8.103)) accumulated on the surfaces S_{ie2} and S_{im3} which are described by $x_j \in \langle x_{j1e}, x_{j2e} \rangle$, $x_k \in \langle x_{k1e}, x_{k2e} \rangle$ and $x_j \in \langle x_{j1m}, x_{j2m} \rangle$, $x_k \in \langle x_{k1m}, x_{k2m} \rangle$, respectively, is derived as

$$
W_{i4} = \int_{S_{ie2}} w_{ie}\, dS_i + \int_{S_{im3}} w_{im}\, dS_i
$$

$$
= \int_{x_{j1e}}^{x_{j2e}} \int_{x_{k1e}}^{x_{k2e}} w_{ie}\, dx_j\, dx_k + \int_{x_{j1m}}^{x_{j2m}} \int_{x_{k1m}}^{x_{k2m}} w_{im}\, dx_j\, dx_k,
$$

$$
x_i \in \langle R_1, R_2 \rangle\,;
$$

$$
\langle x_{j1e}, x_{j2e} \rangle = \left\langle 0, \sqrt{R_2^2 - x_i^2} \right\rangle, \quad \langle x_{k1e}, x_{k2e} \rangle = \left\langle 0, \sqrt{R_2^2 - \left(x_i^2 + x_j^2 \right)} \right\rangle;
$$

$$
\langle x_{j1m}, x_{j2m} \rangle = \left\langle 0, \frac{d}{2} \right\rangle,
$$

$$
\langle x_{k1m}, x_{k2m} \rangle = \left\langle \sqrt{R_2^2 - \left(x_i^2 + x_j^2 \right)}, \frac{d}{2} \right\rangle
$$

$$
\text{for} \quad \langle x_{j1m}, x_{j2m} \rangle = \left\langle 0, \sqrt{R_2^2 - x_i^2} \right\rangle,
$$

$$
\langle x_{k1m}, x_{k2m} \rangle = \left\langle 0, \frac{d}{2} \right\rangle \quad \text{for} \quad \langle x_{j1m}, x_{j2m} \rangle = \left\langle \sqrt{R_2^2 - x_i^2}, \frac{d}{2} \right\rangle, \quad (8.122)
$$

where the determination of the integrals in Eq. (8.122) is required to consider the conditions in Items 1, 1, Section 8.4.1.

The micro-strengthening $\sigma_{si3} = \sigma_{si4}(x_i)$ for $x_i \in \langle R_1, R_2 \rangle$ along the axis x_i (see Eq. (8.104)) which is determined by the condition $W_{\sigma_{si4}} = W_{i4}$ has the form

$$
\sigma_{si4} = \pm \sqrt{\frac{2\,|W_{i4}|}{s_{iie}\, S_{ie2} + s_{iim}\, S_{im3}}}, \quad s_{iiq} = \frac{1}{E_{iq}}, \quad x_i \in \langle R_1, R_2 \rangle, \quad q = e, m, \quad (8.123)
$$

where the signs '+' and '-' in Eq. (8.123) are considered for $W_{i4} > 0$ and $W_{i4} < 0$, respectively.

If $x_i \in \langle R_2, d/2 \rangle$, then $S_{im2} = d^2/4$, and $W_{\sigma_{si}} = W_{\sigma_{si}}(x_i)$, $W_i = W_i(x_i)$, $\sigma_{si} = \sigma_{si}(x_i)$ (see Eqs. (8.102)–(8.104)) for $x_i \in \langle R_2, d/2 \rangle$ are given by Eqs. (8.111)–(8.113).

The macro-strengthening $\overline{\sigma}_{si}$ along the axis x_i (see Eq. (8.105)) which represents a mean value of $\sigma_{si} = \sigma_{si3}(x_i) + \sigma_{si4}(x_i) + \sigma_{si2}(x_i)$ regarding the interval $x_i \in \langle x_{1i}, x_{2i} \rangle = \langle 0, d/2 \rangle$ is derived as

$$
\overline{\sigma}_{si} = \frac{2}{d} \left(\int_0^{R_1} \sigma_{si3}\, dx_i + \int_{R_1}^{R_2} \sigma_{si4}\, dx_i + \int_{R_2}^{d/2} \sigma_{si2}\, dx_i \right)
$$

$$= \frac{2\sqrt{2}}{d} \left(\int_0^{R_1} \pm \sqrt{\frac{|W_{i3}|}{s_{iip}\,S_{ip} + s_{iie}\,S_{ie1} + s_{iim}\,S_{im3}}}\, dx_i \right.$$

$$\left. + \int_{R_1}^{R_2} \pm \sqrt{\frac{|W_{i4}|}{s_{iie}\,S_{ie2} + s_{iim}\,S_{im3}}}\, dx_i + \frac{2}{d \times \sqrt{s_{iim}}} \int_{R_2}^{d/2} \pm \sqrt{|W_{i2}|}\, dx_i \right),$$

$$s_{iiq} = \frac{1}{E_{iq}}, \quad q = p, e, m, \tag{8.124}$$

where the conditions and the analysis in Items 3–6, Section 8.4.1 are considered.

8.5 Analytical-computational and analytical-experimental-computational methods of lifetime prediction

The analytical-computational and analytical-experimental-computational methods of the lifetime prediction presented in Sections 8.5.2 and 8.5.3, respectively, are applicable to the three-component material define in Item 3, Section 2.1, which consists of grains with and without a continuous component on a grain surface. The grains with the continuous component and the grains without the continuous component are identical or different from microstructural point of view, and thus exhibit identical or different thermal expansion co-efficients

With regard to analytical modelling of the thermal stresses, this three-component material is replaced by the multi-particle-envelope-matrix system as analysed in Section 2.1.

Let the multi-particle-envelope-matrix system consists of cubic cells with the dimension d which is equal to the inter-particle distance (see Fig. 2.1a). Consequently, as analysed in Section 2.1, we get:

1. The grain with the continuous component on the grain surface and with the thermal expansion coefficien β_p corresponds to the spherical particle of the multi-particle-envelope-matrix system, where $v \in (0, v_{max})$ is a volume fraction of the spherical particle with the radius R_1, and the coefficien v_{max} (see Eq. (2.7)) is a function of R_1, t.

2. The grain without the continuous component on the grain surface and with the thermal expansion coefficien β_m corresponds to the infinit matrix of the multi-particle-envelope-matrix system, where $\beta_m = \beta_p$ or $\beta_m \neq \beta_p$.

3. The continuous component with the thickness t and with the thermal expansion coefficien β_e corresponds to the spherical envelope of the multi-particle-envelope-matrix system, where $\beta_e \neq \beta_m = \beta_p$ or $\beta_e \neq \beta_m \neq \beta_p$.

Finally, the parameters R_1, t, v of the multi-particle-envelope-matrix system represent microstructural parameters of the three-component material define in Item 3, Section 2.1. The analytical-computational and analytical-experimental-computational methods

158   Ladislav Ceniga

of the lifetime prediction consider dependences of the radial stresses $\overline{p_1} = \overline{p_1}(R_1, t, v)$ and $\overline{p_2} = \overline{p_2}(R_1, t, v)$ on these microstructural parameters.

The mean values $\overline{p_1} = \overline{p_1}(R_1, t, v)$ and $\overline{p_2} = \overline{p_2}(R_1, t, v)$ of the φ, ν-dependent radial stresses $p_1 = p_1(\varphi, \nu, R_1, t, v)$ and $p_2 = p_2(\varphi, \nu, R_1, t, v)$ acting at the particle-envelope and matrix-envelope boundaries of the multi-particle-envelope-matrix system, respectively, are define as [8, p. 225]

$$\overline{p_i} = \overline{p_i}(R_1, t, v) = \left(\frac{2}{\pi}\right)^2 \int_0^{\pi/2} \int_0^{\pi/2} p_i \, d\varphi \, d\nu, \quad i = 1, 2. \tag{8.125}$$

The integration is sufficien to be determined within one eighth of the cubic cell, i.e. for $\varphi \in \langle 0, \pi/2 \rangle$ and $\nu \in \langle 0, \pi/2 \rangle$. This is a consequence of symmetry of the multi-particle-envelope-matrix system due to the matrix infinit and the periodical distribution of the spherical particles and the spherical envelopes (see Fig. 2.1a).

On the one hand, the analyses in Sections 8.5.1–8.5.3 concern a stress-strain state which is induced by the thermal stresses. On the other hand, these analyses are also applicable to a stress-strain state with radial; tangential; and shear stresses in the cell matrix which are derived as $\sigma'_{11m} = -p_2 f_{11m}$; $\sigma'_{22m} = -p_2 f_{22m}$, $\sigma'_{33m} = -p_2 f_{33m}$; and $\sigma'_{12m} = -p_2 f_{12m}$, $\sigma'_{13m} = -p_2 f_{13m}$, respectively, where f_m ($\equiv f_{11m}, f_{22m}, f_{33m}, f_{12m}, f_{13m}$) is a function of a position in the cell matrix.

8.5.1   Resistive and contributory effects of thermal stresses

An analysis of the resistive and contributory effects of the thermal stresses which considers dependences of the radial stresses $\overline{p_1} = \overline{p_1}(R_1, t, v)$ and $\overline{p_2} = \overline{p_2}(R_1, t, v)$ on the microstructural parameters R_1, t, v is as follows.

The compressive or tensile radial stress $\overline{p_1}(R_1, v) > 0$ or $\overline{p_1}(R_1, v) < 0$ for $v \in (0, \pi/6)$ (see Eq. (2.7)) acting at the particle-matrix boundary of the multi-particle-matrix system corresponds to the condition $\beta_m > \beta_p$ or $\beta_m < \beta_p$ (see Eqs. (3.100)–(3.102)), respectively. The mean value $\overline{p_1} = \overline{p_1}(R_1, v)$ of the φ, ν-dependent radial stress $p_1 = p_1(\varphi, \nu, R_1, v)$ is determined by Eq. (8.125).

In contrast to the multi-particle-matrix system, the radial stresses $\overline{p_1}(R_1, v, t) \gtrless 0$ and $\overline{p_2}(R_1, v, t) \gtrless 0$, both for $v \in (0, v_{max})$ (see Eq. (2.7)), acting at the particle-envelope and matrix-envelope boundaries can or need not correspond to the conditions $\beta_e \gtrless \beta_p$ and $\beta_m \gtrless \beta_e$ (see Eqs. (3.100)–(3.102)), respectively. The fact that $\overline{p_1}(R_1, v, t) \gtrless 0$ and $\overline{p_2}(R_1, v, t) \gtrless 0$ need not correspond to the conditions $\beta_e \gtrless \beta_p$ and $\beta_m \gtrless \beta_e$, respectively, results from a dependence of $\overline{p_1} = \overline{p_1}(R_1, v, t)^1$, $\overline{p_2} = \overline{p_2}(R_1, v, t)$

---

[1]On the condition $\beta_p \neq \beta_e = \beta_m$ (see Eqs. (3.100)–(3.102)) for the multi-particle-envelope-matrix system, the radial stress p_1 given by Eq. (5.106), and the mean value $\overline{p_1}$ are functions of the coefficient $\beta_p, \beta_e, \xi_{61p}$ (see Eq. (3.88) for ($q = p$)), ϱ_{1e}^{me} (see Eq. (5.77)). These coefficient are not functions of the microstructural parameters R_1, t, v of the three-component materials define in Items 1, 3, 4 in Section 2.1. Accordingly, with regard to Eq. (5.106), the radial stress p_1 acting at the particle-envelope boundary, and the mean value $\overline{p_1}$ are

Related phenomena   159

(see Eqs. (5.140), (5.141), (5.172)) on the microstructural parameters R_1, t, v of the three-component materials define in Items 1, 3, 4 in Section 2.1. This dependence results from the function $f_c = f_c(R_1, t, v)$ (see Eqs. (2.6), (2.7), (2.30)–(2.33), (2.36)) which is included in Eqs. (5.140), (5.141), (5.172).

Consequently, one of the radial stresses $\overline{p_1} = \overline{p_1}(R_1, t, v)$, $\overline{p_2} = \overline{p_2}(R_1, t, v)$ or both $\overline{p_1} = \overline{p_1}(R_1, t, v)$, $\overline{p_2} = \overline{p_2}(R_1, t, v)$ can exhibit zero values at R_{1c}, t_c. Accordingly, R_{1c}, t_c represent such critical values of the microstructural parameters R_1, t, v at which one of the radial stresses $\overline{p_1}$, $\overline{p_2}$ or both $\overline{p_1}$, $\overline{p_2}$ are transformed from compressive to tensile or vice versa.

The compressive radial stress $\overline{p_i}(R_1, t, v) > 0$ $(i = 1,2)$ results in compressive radial and tensile tangential thermal stresses, $\sigma'_{1q}(R_1, t, v) < 0$ and $\sigma'_{2q}(R_1, t, v) > 0$, $\sigma'_{3q}(R_1, t, v) > 0$ (see Fig. 3.1), respectively, which thus represent 'resistance' against the compressive radial stress $\overline{\sigma_{ri}}(R_1, t, v) < 0$ $(i = 1,2)$, where the radial stress $\overline{\sigma_{ri}} = \overline{\sigma_{ri}}(R_1, t, v)$ is explained below. With regard to Fig. 3.1, the stresses $\sigma'_{1q} = \sigma_{rq}$, $\sigma'_{2q} = \sigma_{\varphi q}$, $\sigma'_{3q} = \sigma_{\nu q}$ act along the axes $x'_1 = x_r$, $x'_2 = x_\varphi$, $x'_3 = x_\nu$, respectively.

Similarly, the tensile radial stress $\overline{p_i}(R_1, t, v) < 0$ $(i = 1,2)$ results in tensile radial and compressive tangential thermal stresses, $\sigma'_{1q}(R_1, t, v) > 0$ and $\sigma'_{2q}(R_1, t, v) < 0$, $\sigma'_{3q}(R_1, t, v) < 0$, respectively, which thus represent 'resistance' against the tensile radial stress $\overline{\sigma_{ri}}(R_1, t, v) > 0$. Additionally, we get:

- In case of $\overline{p_1}(R_1, t, v) \gtrless 0$, this analysis concerning the radial and tangential thermal stresses is considered for the thermal stresses in the spherical particle, i.e. $q = p$.

- In case of $\overline{p_2}(R_1, t, v) \gtrless 0$, this analysis concerning the radial and tangential thermal stresses is considered for the thermal stresses in the cell matrix, i.e. $q = m$.

Finally, due to $\beta_e \neq \beta_m = \beta_p$, the thermal stresses in the spherical particle and the cell matrix of the multi-particle-envelope-matrix system exhibit mutually opposite signs, i.e. $\sigma_{\eta p} \times \sigma_{\eta m} < 0$ for $\eta = r, \varphi, \nu$.

Let the three-component material is loaded by mechanical loading represented by the stresses $\sigma_{1\,mech}$, $\sigma_{2\,mech}$, $\sigma_{3\,mech}$ acting along the axes x_1, x_2, x_3 (see Fig. 3.1). Let $\sigma_{r1} = \sigma_{r1}(\varphi, \nu, R_1, t, v)$ and $\sigma_{r2} = \sigma_{r2}(\varphi, \nu, R_1, t, v)$ represent 'responses' of the mechanical loading $\sigma_{1\,mech}$, $\sigma_{2\,mech}$, $\sigma_{3\,mech}$, where the radial stresses $\sigma_{r1} = \sigma_{r1}(\varphi, \nu, R_1, t, v)$ and $\sigma_{r2} = \sigma_{r2}(\varphi, \nu, R_1, t, v)$ act at the particle-envelope and matrix-envelope boundaries, respectively. The radial stresses $\overline{\sigma_{r1}} = \overline{\sigma_{r1}}(R_1, t, v)$ and $\overline{\sigma_{r2}} = \overline{\sigma_{r2}}(R_1, t, v)$ thus represent mean values of the φ, ν-dependent radial stresses $\sigma_{r1} = \sigma_{r1}(\varphi, \nu, R_1, t, v)$ and $\sigma_{r2} = \sigma_{r2}(\varphi, \nu, R_1, t, v)$, respectively. Finally, the radial stress $\sigma_{ri} = \sigma_{ri}(\varphi, \nu, R_1, t, v)$

---

paradoxically independent on R_1, t, v, i.e. $p_1 \neq f(R_1, t, v)$, $\overline{p_1} \neq f(R_1, t, v)$. The radial displacements u_{1p}, u_{1e} and then the thermal stresses in the spherical particle and spherical envelope are not functions of R_1, t, v. This is also valid for $(u_{1m})_{r=R_2} = (u_{1e})_{r=R_2} \neq f(R_1, t, v)$, where the condition $(u_{1m})_{r=R_2} = (u_{1e})_{r=R_2}$ results from the condition $\beta_e = \beta_p$. Finally, the thermal stresses in the cell matrix given by Eqs. (5.92)–(5.96) can be dependent on R_1, t, v for $r \in \langle R_2, r_c \rangle$ (see Eqs. (2.29)–(2.33), (2.35), (2.36)). This dependence results from the function $f_c = f_c(R_1, t, v)$ (see Eqs. (2.6), (2.7), (2.30)–(2.33), (2.36)) which is included in Eqs. (5.92)–(5.96).

($i=1,2$) can be determined by e.g. the Eshelby's model. The Eshelby's model and its development [1]–[3] defin the disturbance of an applied stress-fiel in a solid continuum, where the applied stress-fiel is disturbed due to the presence of inclusions in a solid continuum.

The critical microstructural parameters R_{1c}, t_c for $v \in (0, v_{max})$ (see Eq. (2.7)) thus represent such critical values at which the radial/tangential thermal stresses in the cell matrix change from positive/negative to negative/positive, and then this resistive effect of the thermal stresses in the cell matrix is transformed to a contributory effect or vice versa.

Let the conditions $\overline{p_i}(R_1, t, v) = 0$ and $|\overline{p_i}(R_1, t, v)| = |\overline{\sigma_{ri}}(R_1, t, v)|$ ($i=1,2$) are required to be valid for each value of the parameter $v \in (0, v_{max})$ (see Eq. (2.7)). Consequently, the conditions $\overline{p_i}(R_1, t, v) = 0$ and $|\overline{p_i}(R_1, t, v)| = |\overline{\sigma_{ri}}(R_1, t, v)|$ result in the $R_{1c} - t_c$ dependences, $R_{1Bc} = f_i(t_{Bc})$ and $R_{1Ac} = f_i(t_{Ac}, \overline{\sigma_{ri}})$, between the variables R_1, t of the functions $\overline{p_i} = \overline{p_i}(R_1, t, v)$, $\overline{\sigma_{ri}} = \overline{\sigma_{ri}}(R_1, t, v)$, respectively.

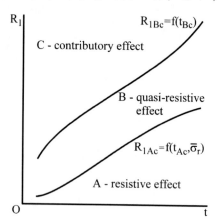

Figure 8.39: A schematic illustration of the dependences $R_{1Ac} = f(t_{Ac}, \overline{\sigma_r})$ and $R_{1Bc} = f(t_{Bc})$ which result from the conditions $|\overline{p_i}(R_1, t, v)| = |\overline{\sigma_{ri}}(R_1, t, v)|$ and $\overline{p_i}(R_1, t, v) = 0$, respectively, where $f(t_{Ac}, \overline{\sigma_r}) \equiv f_i(t_{Ac}, \overline{\sigma_{ri}})$, $f(t_{Bc}) \equiv f_i(t_{Bc})$. These conditions are required to be valid for each value of the parameter $v \in (0, v_{max})$ (see Eq. (2.7)).

Fig. 8.39 shows a schematic illustration of the dependences $R_{1Ac} = f(t_{Ac}, \overline{\sigma_r})$ and $R_{1Bc} = f(t_{Bc})$ for $v \in (0, v_{max})$ (see Eq. (2.7)) which defin the areas A, B, C, where $f(t_{Ac}, \overline{\sigma_r}) = f_i(t_{Ac}, \overline{\sigma_{ri}})$, $f(t_{Bc}) = f_i(t_{Ac}, \overline{\sigma_{ri}})$ ($i=1,2$).

The area A is characterized by such coordinates (R_1, t) of the microstructural parameters R_1, t for which the thermal stresses exhibit the resistive effect against the mechanical loading, and additionally $|\overline{p_i}(R_1, t, v)| \geq |\overline{\sigma_{ri}}(R_1, t, v)|$, and $\overline{p_i}(R_1, t, v) \times \overline{\sigma_{ri}}(R_1, t, v) < 0$.

The area B is characterized by such coordinates (R_1, t) of the microstructural parameters R_1, t for which the thermal stresses exhibit the quasi-resistive effect against the mechanical loading, i.e. $|\overline{p_i}(R_1, t, v)| \in \langle 0, |\overline{\sigma_{ri}}(R_1, t, v)|\rangle$, and $\overline{p_i}(R_1, t, v) \times \overline{\sigma_{ri}}(R_1, t, v) \leq 0$.

The area C is characterized by such coordinates (R_1, t) of the microstructural parameters R_1, t for which the thermal stresses exhibit the contributory effect regarding the mechanical loading, and $\overline{p_i}(R_1, t, v) \times \overline{\sigma_{ri}}(R_1, t, v) > 0$.

With regard to Fig. 8.39, material scientists are able to design and develop microstructure of the three-component material with such microstructural parameters R_1, t, v which result in the thermal-stress induced resistance against required mechanical loading.

Additionally, let the microstructure which is tailored regarding the required mechanical loading is time-dependent. If the time-dependent development of the microstructure is determined analytically or experimentally, then the time τ_c when the three-component material exhibits the critical microstructural parameters R_{1c}, t_c for each value of $v \in (0, v_{max})$ (see Eq. (2.7)) can be determined.

8.5.2   Analytical-computational method

The transformation of the resistive effect to contributory effect of the thermal stresses against mechanical loading is considered for the determination of a method of the lifetime prediction. The lifetime prediction method is applicable to e.g. structural steels which are used within high-temperature applications. The structural steel represents multi-component materials which exhibit time-dependent development of microstructure.

The analytical-computational method represents a connection of analytical and computational techniques which result from the 'resistance-contribution' transformation and from a computational simulation of the time-dependent microstructure development at the temperature T. The temperature T exhibits a function of a parameter within the analytical-computational method.

An initial state of microstructure of the structural steel at the beginning of the time-temperature exploitation is characterized by the presence of grains with aperiodically distributed precipitates of more or less define shape in the grain matrix. In general, the grain matrix is represented by the ferritic matrix with the substitutive atoms Mn, Cr, Mo, W. The precipitates are represented by the carbonitride MX (M = Nb,V,Ti; X = C,N), various carbide types of Cr, Mo, V, W, and inter-metallic phases (Laves and sigma phases) [13]–[17]. Consequently, a fina state of the microstructure of the structural steel at the end of the time-temperature exploitation regarding the lifetime is characterized by the presence of grains with negligible amount of the precipitates in the matrix, and without or with an envelope on the grain surface. The envelope is represented the carbides $M_{23}C_6$, M_6C, and/or the inter-metallic phases [13]–[17].

The microstructure of the fina state of the structural steel thus corresponds to the microstructure which is define in Items 1–3, Section 8.5.

Let $T < T_r$, where T ($\equiv T_e$) and T_r is exploitation and relaxation temperature, respectively, and additionally T_r is dependent on the melting point T_m of the structural steel (see Items 4, 5, Section 3.4.1). With regard to values of the thermal expansion coefficient of the matrix and the envelope, β_m and β_e (see Eq. (3.100)–(3.102)) [13]–[17], respectively, the structural steel is thus acted by the thermal stresses in a range of 8–10 MPa which are dependent on $\Delta\beta \times \Delta T$, where $\Delta T = T_r - T$ ($\equiv T_r - T_e$). On the contrary, if $T > T_r$

($\equiv T_e > T_r$), then the thermal stresses are released by plastic deformation (see Items 4, 5, Section 3.4.1).

With regard to $T \approx 550 - 580°C$, $T_r \approx 600°C$, $\alpha_m \approx 10^{-5}$, $\alpha_m \approx (3-5) \times 10^{-6}$ [4, p. 162], the range of 8–10 MPa is a considerable value in comparison to a value of 20 MPa of mechanical loading which is applied to constructions made of the structural steel. Accordingly, the thermal stresses which are elastic regarding the yield stress $\sigma_y = 200 - 800$ MPa [4, p. 261] are assumed to influenc the lifetime significantl .

The lifetime prediction method determines the critical time τ_{Ac} and τ_{Bc} when the microstructure exhibits the critical parameters R_{1Ac}, t_{1Ac} and R_{1Bc}, t_{1Bc}, respectively, where $\tau_{Ac} < \tau_{Bc}$. The determination of τ_{Ac}, τ_{Bc} is as follows.

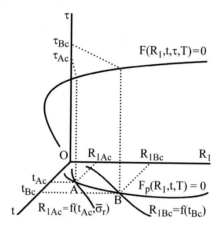

Figure 8.40: The function $F(R_1, t, \tau, T) = 0$ which is determined by a computational simulation of the time-temperature-dependent development of the microstructure. The radius R_1 of the grains which are covered by the envelope on their surfaces, the thickness t of the component on the grain surface, and the time τ represent variables of the function $F(R_1, t, \tau, T) = 0$ at the exploitation temperature $T (\equiv T_e)$ which is a parameter of the function $F(R_1, t, \tau, T) = 0$, where $T < T_r$ and T_r is relaxation temperature (see Items 4, 5, Section 3.4.1). The function $F_p(R_1, t, T) = 0$ represents a projection of $F(R_1, t, \tau, T) = 0$ into the plane $R_1 - t$. The functions $R_{1Ac} = f(t_{Ac}, \overline{\sigma_r})$, $R_{1Bc} = f(t_{Bc})$ (see Fig. 8.39) are described in Section 8.5.1, where $f(t_{Ac}, \overline{\sigma_r}) \equiv f_i(t_{Ac}, \overline{\sigma_{ri}})$, $f(t_{Bc}) \equiv f_i(t_{Bc})$ ($i = 1, 2$).

With regard to Fig. 8.40, let the Cartesian system $(ORt\tau)$ with the same functions $R_{1Ac} = f(t_{Ac}, \overline{\sigma_r})$ and $R_{1Bc} = f(t_{Bc})$ in the plane $R - t$ as those in Fig. 8.39 is considered, where $f(t_{Ac}, \overline{\sigma_r}) \equiv f_i(t_{Ac}, \overline{\sigma_{ri}})$, $f(t_{Bc}) \equiv f_i(t_{Bc})$.

Let $F(R_1, t, \tau, T) = 0$ represents a function which is determined by a computational simulation of the time-temperature-dependent development of the microstructure. The grain radius R_1, the thickness t of the component on the grain surface, and the time τ represent variables of the function $F(R_1, t, \tau, T) = 0$ at the exploitation temperature $T (\equiv T_e)$ which is a parameter of the function $F(R_1, t, \tau, T) = 0$, where $T < T_r$. The $F(R_1, t, \tau, T) = 0$

thus define a time dependence of both R_1 and t at the exploitation temperature T. Consequently, let the function $F_p(R_1, t, T) = 0$ represents a projection of $F(R_1, t, \tau, T) = 0$ into the plane $R_1 - t$.

Additionally, as mentioned above, the volume fraction $v \in (0, v_{max})$ (see Eq. (2.7)) represents a fraction of the grains with the radius R_1 which are covered by the envelope on their surfaces. On the one hand, v also represents a microstructural parameters of the structural steel. On the other hand, the time dependences $R_1 - \tau$ and $t - \tau$ at the exploitation temperature T are sufficien to be determined by the computational simulation which is consequently considered for the determination of the lifetime prediction. Let a suitable computational simulation considers the time dependence $v - \tau$ as well, in spite of the fact that $v - \tau$ is not necessary to be known. Accordingly, the functions $F(R_1, t, \tau, T) = 0$ and $F_p(R_1, t, T) = 0$ are replaced by $F(R_1, t, v, \tau, T) = 0$ and $F_p(R_1, t, v, T) = 0$, respectively.

As shown in Fig. 8.40, the points A and B with the coordinates (R_{1Ac}, t_{Ac}) and (R_{1Bc}, t_{Bc}) represent intersections of $F_p(R_1, t, T) = 0$ with $R_{1Ac} = f(t_{Ac}, \overline{\sigma}_r)$ and $R_{1Bc} = f(t_{Bc})$, respectively, where $f(t_{Ac}, \overline{\sigma}_r) \equiv f_i(t_{Ac}, \overline{\sigma}_{ri})$, $f(t_{Bc}) \equiv f_i(t_{Bc})$ $(i = 1, 2)$.

The substitution of $R_1 = f(t_{Ac}, \overline{\sigma}_r)$ and $R_1 = f(t_{Bc})$ to the function $F_p(R_1, t, T) = 0$ results in the conditions $F_p(f(t_{Ac}, \overline{\sigma}_r), t_{Ac}, T) = 0$ and $F_p(f(t_{Bc}), t_{Bc}, T) = 0$, respectively. The conditions $F_p(f(t_{Ac}, \overline{\sigma}_r), t_{Ac}, T) = 0$ and $F_p(f(t_{Bc}), t_{Bc}, T) = 0$ with the variables t_{Ac} and t_{Bc} are considered for the determination of the t-coordinates t_{Ac} and t_{Bc} of the point A and B in the plane $R_1 - t$, respectively.

Consequently, the substitution of the coordinates t_{Ac} and t_{Bc} to $R_{1Ac} = f(t_{Ac}, \overline{\sigma}_r)$ and $R_{1Bc} = f(t_{Bc})$ results in the determination of the R_1-coordinates R_{1Ac} and R_{1Bc} of the points A and B, respectively.

The substitution of the coordinates (R_{1Ac}, t_{Ac}) and (R_{1Bc}, t_{Bc}) to $F(R_1, t, \tau, T) = 0$ results in the condition $F(R_{1Ac}, t_{Ac}, \tau_{Ac}, T) = 0$ and $F(R_{1Bc}, t_{Bc}, \tau, T) = 0$, respectively. The conditions $F(R_{1Ac}, t_{Ac}, \tau, T) = 0$ and $F(R_{1Bc}, t_{Bc}, \tau_{Bc}, T) = 0$ with the variables τ_{Ac} and τ_{Bc} are considered for the determination of the time τ_{Ac} and τ_{Bc}, respectively. Additionally, τ_{Ac} is a function of $\overline{\sigma}_r$, i.e. $\tau_{Ac} = \tau_{Ac}(\overline{\sigma}_r)$.

Consequently, the intervals $\tau < \tau_{Ac}(\overline{\sigma}_r)$, $\tau \in \langle \tau_{Ac}(\overline{\sigma}_r), \tau_{Bc} \rangle$, $\tau > \tau_{Bc}$ define the time τ which is related to the resistive, quasi-resistive, contributory effects of the thermal stresses against the mechanical loading, respectively. The intervals $\tau < \tau_{Ac}(\overline{\sigma}_r)$, $\tau \in \langle \tau_{Ac}(\overline{\sigma}_r), \tau_{Bc} \rangle$, $\tau > \tau_{Bc}$ determine non-critical, quasi-critical, critical time periods of the exploitation, respectively.

The critical time $\tau_{Ac} = \tau_{Ac}(\overline{\sigma}_r)$ and τ_{Bc} is assumed to represent contribution to the total lifetime related to the thermal-stress induced resistive effects of a three-component material.

Finally, $\tau_{Ac} = \tau_{Ac}(\overline{\sigma}_r)$, τ_{Bc} are determined for $f(t_{Ac}, \overline{\sigma}_r) \equiv f_1(t_{Ac}, \overline{\sigma}_{r1})$, $f(t_{Bc}) \equiv f_1(t_{Bc})$ or $f(t_{Ac}, \overline{\sigma}_r) \equiv f_2(t_{Ac}, \overline{\sigma}_{r2})$, $f(t_{Bc}) \equiv f_2(t_{Bc})$, and then $\tau_{Ac}(\overline{\sigma}_r) \rightarrow \tau_{Ac1}(\overline{\sigma}_{r1})$, $\tau_{Bc} \rightarrow \tau_{Bc1}$ or $\tau_{Ac}(\overline{\sigma}_r) \rightarrow \tau_{Ac2}(\overline{\sigma}_{r2})$, $\tau_{Bc} \rightarrow \tau_{Bc2}$, respectively. Consequently, τ_{Acmin} and τ_{Bcmin} which represent minimal values of the sets $\{\tau_{Ac1}(\overline{\sigma}_{r1}), \tau_{Ac2}(\overline{\sigma}_{r2})\}$

164   Ladislav Ceniga

and $\{\tau_{Bc1}, \tau_{Bc2}\}$ are considered to represent the lifetime regarding the 'resistance–quasi-resistance' and 'quasi-resistance–contribution' transformations (see Fig. 8.39), respectively.

8.5.3 Analytical-computational-experimental method

The analytical-computational-experimental method represents a connection of the analytical model of the thermal stresses with computational simulation of time-dependent microstructure development as well as with experimental results. The temperature T also exhibits a function of a parameter within this analytical-computational-experimental method. The analysis of the analytical-computational-experimental method is as follows.

The computational simulation is represented by a time-dependent development of the thickness t. Let the time-dependent development is represented by the function $t = g(\tau)$, where $\tau = f(t)$ is an inverse function of $t = g(\tau)$. Let $\log \tau = f(\log t)$ is derived by the Taylor series in terms of $\log t$ regarding the time $\tau \in (0, 10^6)$ in hours. Consequently, the function $\log \tau = f(\log t)$ has the form

$$\log \tau = \sum_{i=0}^{n} c_i^{(\tau)} (\log t)^i. \tag{8.126}$$

The experimental results are represented by a time-dependent development of the radius R_1 of the grains which are covered by the component on their surfaces. Let the $R_1 - \tau$ dependence for $\tau \in (0, 10^6)$ is derived as

$$\log R_1 = \sum_{j=0}^{m} c_j^{(R_1)} (\log \tau)^j. \tag{8.127}$$

With regard to Eqs. (8.126), (8.127), we get

$$\log R_1 = \sum_{j=0}^{m} c_j^{(R_1)} \left(\sum_{i=0}^{n} c_i^{(\tau)} (\log t)^i \right)^j. \tag{8.128}$$

The analytical model of the thermal stresses is represented by the functions $R_{1Ac} = f(t_{Ac}, \overline{\sigma}_r)$ and $R_{1Bc} = f(t_{Bc})$ (see Fig. 8.39), strictly speaking by the functions $\log R_{1Ac} = \log[f(t_{Ac}, \overline{\sigma}_r)]$ and $\log R_{1Bc} = \log[f(t_{Bc})]$, where $f(t_{Ac}, \overline{\sigma}_r) \equiv f_i(t_{Ac}, \overline{\sigma}_{ri})$, $f(t_{Bc}) \equiv f_i(t_{Bc})$.

With regard to Eq. (8.128), and considering $t \to t_{Ac}$, we get

$$\log[f(t_{Ac}, \overline{\sigma}_r)] - \sum_{j=0}^{m} c_j^{(R_1)} \left(\sum_{i=0}^{n} c_i^{(\tau)} (\log t_{Ac})^i \right)^j = 0. \tag{8.129}$$

Similarly, considering $t \to t_{Bc}$, we get

$$\log[f(t_{Bc})] - \sum_{j=0}^{m} c_j^{(R_1)} \left(\sum_{i=0}^{n} c_i^{(\tau)} (\log t_{Bc})^i \right)^j = 0. \tag{8.130}$$

Consequently, the roots $t_{Ac} = t_{Ac}(\overline{\sigma}_r)$ and t_{Bc} of Eqs. (8.129) and (8.130) are determined by a numerical method.

Substituting $t_{Ac} = t_{Ac}(\overline{\sigma}_r)$ and t_{Bc} to Eq. (8.126), the critical time $\tau_{Ac} = \tau_{Ac}(\overline{\sigma}_r)$ and τ_{Bc} has the form

$$\tau_{Qc} = \exp\left[(\ln 10)\sum_{i=0}^{n} c_i^{(\tau)}(\log t_{Qc})^i\right], \quad Q = A, B. \tag{8.131}$$

Finally, Equations (8.126) and (8.127) can be also determined from experimental results and a computational simulation, respectively.

Finally, with regard to $f(t_{Ac}, \overline{\sigma}_r) \equiv f_1(t_{Ac}, \overline{\sigma}_{r1})$, $f(t_{Bc}) \equiv f_1(t_{Bc})$ or $f(t_{Ac}, \overline{\sigma}_r) \equiv f_2(t_{Ac}, \overline{\sigma}_{r2})$, $f(t_{Bc}) \equiv f_2(t_{Bc})$, we get $\tau_{Ac}(\overline{\sigma}_r) \rightarrow \tau_{Ac1}(\overline{\sigma}_{r1})$, $\tau_{Bc} \rightarrow \tau_{Bc1}$ or $\tau_{Ac}(\overline{\sigma}_r) \rightarrow \tau_{Ac2}(\overline{\sigma}_{r2})$, $\tau_{Bc} \rightarrow \tau_{Bc2}$, respectively. Consequently, $\tau_{Ac\,min}$ and $\tau_{Bc\,min}$ which represent minimal values of the sets $\{\tau_{Ac1}(\overline{\sigma}_{r1}), \tau_{Ac2}(\overline{\sigma}_{r2})\}$ and $\{\tau_{Bc1}, \tau_{Bc2}\}$ are considered to represent the lifetime regarding the 'resistance–quasi-resistance' and 'quasi-resistance–contribution' transformations (see Fig. 8.39), respectively.

Chapter 9

Appendix

9.1   Elastic modulus and thermal expansion coefficien

The elastic modulus s'_{ijkl} ($i, j, k, l = 1,2,3$) in the Cartesian system with the axes x'_1, x'_2, x'_3 derived by $s_{11}, s_{12}, \ldots, s_{56}, s_{66}$ in the Cartesian system $(Ox_1x_2x_3)$ (see Fig. 3.1) has the form [6, p. 20,21]

$$s'_{ijkl} = \sum_{r,s,t,u=1}^{3} a_{ir}\, a_{js}\, a_{kt}\, a_{lu}\, s_{rstu} = \sum_{r=1}^{3} a_{ir} a_{jr} a_{kr} a_{lr} s_{rr}$$

$$+ \sum_{r=1}^{3} [\delta_{1r}\, (a_{i2}a_{j3}a_{k2}a_{l3} + a_{i2}a_{j3}a_{k3}a_{l2} + a_{i3}a_{j2}a_{k2}a_{l3} + a_{i3}a_{j2}a_{k3}a_{l2})$$

$$+ \delta_{2r}\, (a_{i1}a_{j3}a_{k1}a_{l3} + a_{i1}a_{j3}a_{k3}a_{l1} + a_{i3}a_{j1}a_{k1}a_{l3} + a_{i3}a_{j1}a_{k3}a_{l1})$$

$$+ \delta_{3r}\, (a_{i1}a_{j2}a_{k1}a_{l2} + a_{i1}a_{j2}a_{k2}a_{l1} + a_{i2}a_{j1}a_{k1}a_{l2} + a_{i2}a_{j1}a_{k2}a_{l1})]\, s_{3+r3+r}$$

$$+ \sum_{r,s=1}^{3} [\delta_{1s}\, (a_{ir}a_{jr}a_{k2}a_{l3} + a_{ir}a_{jr}a_{k3}a_{l2} + a_{i2}a_{j3}a_{kr}a_{lr} + a_{i3}a_{j2}a_{kr}a_{lr})$$

$$+ \delta_{2s}\, (a_{ir}a_{jr}a_{k1}a_{l3} + a_{ir}a_{jr}a_{k3}a_{l1} + a_{i1}a_{j3}a_{kr}a_{lr} + a_{i3}a_{j1}a_{kr}a_{lr})$$

$$+ \delta_{3s}\, (a_{ir}a_{jr}a_{k1}a_{l2} + a_{ir}a_{jr}a_{k2}a_{l1} + a_{i1}a_{j2}a_{kr}a_{lr} + a_{i2}a_{j1}a_{kr}a_{lr})]\, s_{r3+s}$$

$$+ \sum_{r,s=1;r\neq s}^{3} (a_{ir}a_{jr}a_{ks}a_{ls} + a_{is}a_{js}a_{kr}a_{lr})\, s_{rs}$$

$$+ \sum_{r=1}^{3} \sum_{s;s>r} [\delta_{1r}\delta_{2s}\, (a_{i2}a_{j3}a_{k1}a_{l3} + a_{i2}a_{j3}a_{k3}a_{l1} + a_{i3}a_{j2}a_{k1}a_{l3} + a_{i3}a_{j2}a_{k3}a_{l1}$$

$$+ a_{i1}a_{j3}a_{k2}a_{l3} + a_{i1}a_{j3}a_{k3}a_{l2} + a_{i3}a_{j1}a_{k2}a_{l3} + a_{i3}a_{j1}a_{k3}a_{l2})$$

$$+ \delta_{1r}\delta_{3s}\, (a_{i2}a_{j3}a_{k1}a_{l2} + a_{i2}a_{j3}a_{k2}a_{l1} + a_{i3}a_{j2}a_{k1}a_{l2} + a_{i3}a_{j2}a_{k2}a_{l1}$$

$$+ a_{i1}a_{j2}a_{k2}a_{l3} + a_{i1}a_{j2}a_{k3}a_{l2} + a_{i2}a_{j1}a_{k2}a_{l3} + a_{i2}a_{j1}a_{k3}a_{l2})$$

$$+ \delta_{2r}\delta_{3s}\, (a_{i1}a_{j3}a_{k1}a_{l2} + a_{i1}a_{j3}a_{k2}a_{l1} + a_{i3}a_{j1}a_{k1}a_{l2} + a_{i3}a_{j1}a_{k2}a_{l1}$$

$$+ a_{i1}a_{j2}a_{k1}a_{l3} + a_{i1}a_{j2}a_{k3}a_{l1} + a_{i2}a_{j1}a_{k1}a_{l3} + a_{i2}a_{j1}a_{k3}a_{l1})] \, s_{3+r3+s},$$
$$i, j, k, l = 1, 2, 3. \tag{9.1}$$

The coefficien $a_{vw} = \cos\left[\angle\left(x'_v, x_w\right)\right]$ (v, w = 1,2,3) representing a direction cosine of an angle of the axes x'_v, x_w (see Fig. 3.1) is derived as

$$a_{11} = \cos\varphi\sin\nu, \tag{9.2}$$

$$a_{12} = \sin\varphi\sin\nu, \tag{9.3}$$

$$a_{13} = \cos\nu, \tag{9.4}$$

$$a_{21} = -\sin\varphi, \tag{9.5}$$

$$a_{22} = -\cos\varphi, \tag{9.6}$$

$$a_{23} = 0, \tag{9.7}$$

$$a_{31} = -\cos\varphi\cos\nu, \tag{9.8}$$

$$a_{32} = -\sin\varphi\cos\nu, \tag{9.9}$$

$$a_{33} = -\sin\nu. \tag{9.10}$$

The thermal expansion coefficien α'_1 along the axis x'_1 (see Fig. 3.1) has the form [18]

$$\alpha'_1 = a_{11}^2\alpha_1 + a_{12}^2\alpha_2 + a_{13}^2\alpha_3, \tag{9.11}$$

where α_i is a thermal expansion coefficien along the axis x_i (i = 1,2,3)).

The formulae for the elastic modulus s'_{ijkl} ($i, j, k, l = 1, \ldots, 3$) and the thermal expansion coefficien α'_1 are valid for uniaxial- and triaxial-anisotropic elastic solid continuum which consists of uniaxial- and triaxial-anisotropic crystal lattices. The only difference concerning the uniaxial and triaxial anisotropy is represented by the elastic modulus $s_{rs} = s_{sr}$ ($r, s = 1, \ldots, 6$) and the thermal expansion coefficien α_i (i = 1, 2,3). The elastic modulus s_{rs} and the thermal expansion coefficien α_i included in s'_{ijkl} and α'_1, respectively, are different for the uniaxial-anisotropic and triaxial-anisotropic crystal lattices [6, p. 28,29], [18].

9.1.1 Transformations concerning subscripts ij and q

The transformations concerning the subscript $ij \rightarrow n$ (i, j = 1,2,3; n = 1, \ldots, 6), valid also for the subscript kl (k, l = 1,2,3), are derived as [6, p. 20]

$$ij \equiv ji,$$
$$i = j \quad \rightarrow \quad ij \equiv i;$$
$$i \neq j \quad \rightarrow \quad ij = 12 \equiv 6, \quad ij = 13 \equiv 5, \quad ij = 23 \equiv 4. \tag{9.12}$$

Consequently, the transformations concerning the subscript $q = p,e,m$ have the forms

$$s_{ij} \rightarrow s_{ijq}, \quad s'_{ij} \rightarrow s'_{ijq}, \quad \alpha_k \rightarrow \alpha_{kq}, \quad \alpha'_1 \rightarrow \alpha'_{1q},$$

$$i, j = 1, \ldots, 6; \quad k = 1, 2, 3, \tag{9.13}$$

where the subscripts $q = p$, $q = e$, and $q = m$ are considered for the spherical particle, the spherical envelope, and the infinit matrix, respectively.

With regard to the determination of the thermal stresses along with the thermal-stress induced phenomena (see Chapter 8), considering the elastic modulus s'_{ijkl} ($i, j, k, l = 1,2,3$) in the Cartesian system $(Px'_1 x'_2 x'_3)$ (Fig. 3.1) and the transformations given by Eqs. (9.1) and (9.12), respectively, the elastic modulus $s_{ijq} = s_{jiq}$ ($i, j = 1, \ldots, 6$) is required to be derived in the forms

$$
\begin{aligned}
s'_{11} ={}& s_{11} (a_{11})^4 + 2s_{12} (a_{11}a_{12})^2 + 2s_{13} (a_{11}a_{13})^2 + 4s_{14} (a_{11})^2 a_{12}a_{13} \\
&+ 4s_{15} (a_{11})^3 a_{13} + 4s_{16} (a_{11})^3 a_{12} + s_{22} (a_{12})^4 + 2s_{23} (a_{12}a_{13})^2 \\
&+ 4s_{24} (a_{12})^3 a_{13} + 4s_{25}a_{11} (a_{12})^2 a_{13} + 4s_{26}a_{11} (a_{12})^3 + s_{33} (a_{13})^4 \\
&+ 4s_{34}a_{12} (a_{13})^3 + 4s_{35}a_{11} (a_{13})^3 + 4s_{36}a_{11}a_{12} (a_{13})^2 + 4s_{44} (a_{12}a_{13})^2 \\
&+ 8s_{45}a_{11}a_{12} (a_{13})^2 + 8s_{46}a_{11} (a_{12})^2 a_{13} + 4s_{55} (a_{11}a_{13})^2 + 8s_{56} (a_{11})^2 a_{12}a_{13} \\
&+ 4s_{66} (a_{11}a_{12})^2 ,
\end{aligned}
\tag{9.14}
$$

$$
\begin{aligned}
s'_{12} ={}& s_{11} (a_{11}a_{21})^2 + s_{12} \left[(a_{11}a_{22})^2 + (a_{12}a_{21})^2 \right] + s_{13} \left[(a_{11}a_{23})^2 + (a_{13}a_{21})^2 \right] \\
&+ 2s_{14} \left[(a_{11})^2 a_{22}a_{23} + a_{12}a_{13} (a_{21})^2 \right] + 2s_{15} \left[(a_{11})^2 a_{21}a_{23} + a_{11}a_{13} (a_{21})^2 \right] \\
&+ 2s_{16} \left[(a_{11})^2 a_{21}a_{22} + a_{11}a_{12} (a_{21})^2 \right] + s_{22} (a_{12}a_{22})^2 \\
&+ s_{23} \left[(a_{12}a_{23})^2 + (a_{13}a_{22})^2 \right] + 2s_{24} \left[(a_{12})^2 a_{22}a_{23} + a_{12}a_{13} (a_{22})^2 \right] \\
&+ 2s_{25} \left[(a_{12})^2 a_{21}a_{23} + a_{11}a_{13} (a_{22})^2 \right] + 2s_{26} \left[(a_{12})^2 a_{21}a_{22} + a_{11}a_{12} (a_{22})^2 \right] \\
&+ s_{33} (a_{13}a_{23})^2 + 2s_{34} \left[(a_{13})^2 a_{22}a_{23} + a_{12}a_{13} (a_{23})^2 \right] \\
&+ 2s_{35} \left[(a_{13})^2 a_{21}a_{23} + a_{11}a_{13} (a_{23})^2 \right] + 2s_{36} \left[(a_{13})^2 a_{21}a_{22} + a_{11}a_{12} (a_{23})^2 \right] \\
&+ 4s_{44}a_{12}a_{13}a_{22}a_{23} + 4s_{45} (a_{12}a_{13}a_{21}a_{23} + a_{11}a_{13}a_{22}a_{23}) \\
&+ 4s_{46} (a_{12}a_{13}a_{21}a_{22} + a_{11}a_{12}a_{22}a_{23}) + 4s_{55}a_{11}a_{13}a_{21}a_{23} \\
&+ 4s_{56} (a_{11}a_{13}a_{21}a_{22} + a_{11}a_{12}a_{21}a_{23}) + 4s_{66}a_{11}a_{12}a_{21}a_{22} ,
\end{aligned}
\tag{9.15}
$$

$$
\begin{aligned}
s'_{13} ={}& s_{11} (a_{11}a_{31})^2 + s_{12} \left[(a_{11}a_{32})^2 + (a_{12}a_{31})^2 \right] + s_{13} \left[(a_{11}a_{33})^2 + (a_{13}a_{31})^2 \right] \\
&+ 2s_{14} \left[(a_{11})^2 a_{32}a_{33} + a_{12}a_{13} (a_{31})^2 \right] + 2s_{15} \left[(a_{11})^2 a_{31}a_{33} + a_{11}a_{13} (a_{31})^2 \right] \\
&+ 2s_{16} \left[(a_{11})^2 a_{31}a_{32} + a_{11}a_{12} (a_{31})^2 \right] + s_{22} (a_{12}a_{32})^2 \\
&+ s_{23} \left[(a_{12}a_{33})^2 + (a_{13}a_{32})^2 \right] + 2s_{24} \left[(a_{12})^2 a_{32}a_{33} + a_{12}a_{13} (a_{32})^2 \right] \\
&+ 2s_{25} \left[(a_{12})^2 a_{31}a_{33} + a_{11}a_{13} (a_{32})^2 \right] + 2s_{26} \left[(a_{12})^2 a_{31}a_{32} + a_{11}a_{12} (a_{32})^2 \right]
\end{aligned}
$$

$$+ s_{33} (a_{13}a_{33})^2 + 2s_{34} \left[(a_{13})^2 a_{32}a_{33} + a_{12}a_{13} (a_{33})^2\right]$$

$$+ 2s_{35} \left[(a_{13})^2 a_{31}a_{33} + a_{11}a_{13} (a_{33})^2\right] + 2s_{36} \left[(a_{13})^2 a_{31}a_{32} + a_{11}a_{12} (a_{33})^2\right]$$

$$+ 4s_{44}a_{12}a_{13}a_{32}a_{33} + 4s_{45} (a_{12}a_{13}a_{31}a_{33} + a_{11}a_{13}a_{32}a_{33})$$

$$+ 4s_{46} (a_{12}a_{13}a_{31}a_{32} + a_{11}a_{12}a_{32}a_{33}) + 4s_{55}a_{11}a_{13}a_{31}a_{33}$$

$$+ 4s_{56} (a_{11}a_{13}a_{31}a_{32} + a_{11}a_{12}a_{31}a_{33}) + 4s_{66}a_{11}a_{12}a_{31}a_{32}, \tag{9.16}$$

$$s'_{14} = s_{11} (a_{11})^2 a_{21}a_{31} + s_{12} \left[(a_{11})^2 a_{22}a_{32} + (a_{12})^2 a_{21}a_{31}\right]$$

$$+ s_{13} \left[(a_{11})^2 a_{23}a_{33} + (a_{13})^2 a_{21}a_{31}\right]$$

$$+ s_{14} \left[(a_{11})^2 (a_{22}a_{33} + a_{23}a_{32}) + 2a_{12}a_{13}a_{21}a_{31}\right]$$

$$+ s_{15} \left[(a_{11})^2 (a_{21}a_{33} + a_{23}a_{31}) + 2a_{11}a_{13}a_{21}a_{31}\right]$$

$$+ s_{16} \left[(a_{11})^2 (a_{21}a_{32} + a_{22}a_{31}) + 2a_{11}a_{12}a_{21}a_{31}\right] + s_{22} (a_{12})^2 a_{22}a_{32}$$

$$+ s_{23} \left[(a_{12})^2 a_{23}a_{33} + (a_{13})^2 a_{22}a_{32}\right]$$

$$+ s_{24} \left[(a_{12})^2 (a_{22}a_{33} + a_{23}a_{32}) + 2a_{12}a_{13}a_{22}a_{32}\right]$$

$$+ s_{25} \left[(a_{12})^2 (a_{21}a_{33} + a_{23}a_{31}) + 2a_{11}a_{13}a_{22}a_{32}\right]$$

$$+ s_{26} \left[(a_{12})^2 (a_{21}a_{32} + a_{22}a_{31}) + 2a_{11}a_{12}a_{22}a_{32}\right] + s_{33} (a_{13})^2 a_{23}a_{33}$$

$$+ s_{34} \left[(a_{13})^2 (a_{22}a_{33} + a_{23}a_{32}) + 2a_{12}a_{13}a_{23}a_{33}\right]$$

$$+ s_{35} \left[(a_{13})^2 (a_{21}a_{33} + a_{23}a_{31}) + 2a_{11}a_{13}a_{23}a_{33}\right]$$

$$+ s_{36} \left[(a_{13})^2 (a_{21}a_{32} + a_{22}a_{31}) + 2a_{11}a_{12}a_{23}a_{33}\right]$$

$$+ 2s_{44}a_{12}a_{13} (a_{22}a_{33} + a_{23}a_{32})$$

$$+ 2s_{45}a_{13} \left[a_{12} (a_{21}a_{33} + a_{23}a_{31}) + a_{11} (a_{22}a_{33} + a_{23}a_{32})\right]$$

$$+ 2s_{46}a_{12} \left[a_{13} (a_{21}a_{32} + a_{22}a_{31}) + a_{11} (a_{22}a_{33} + a_{23}a_{32})\right]$$

$$+ 2s_{55}a_{11}a_{13} (a_{21}a_{33} + a_{23}a_{31})$$

$$+ 2s_{56}a_{11} \left[a_{12} (a_{21}a_{33} + a_{23}a_{31}) + a_{13} (a_{21}a_{32} + a_{22}a_{31})\right]$$

$$+ 2s_{66}a_{11}a_{12} (a_{21}a_{32} + a_{22}a_{31}), \tag{9.17}$$

$$s'_{15} = s_{11} (a_{11})^3 a_{31} + s_{12}a_{11}a_{12} (a_{11}a_{32} + a_{12}a_{31}) + s_{13}a_{11}a_{13} (a_{11}a_{33} + a_{13}a_{31})$$

$$+ s_{14}a_{11} \left[a_{11} (a_{12}a_{33} + a_{13}a_{32}) + 2a_{12}a_{13}a_{31}\right] + s_{15} (a_{11})^2 (a_{11}a_{33} + 3a_{13}a_{31})$$

$$+ s_{16} (a_{11})^2 (a_{11}a_{32} + 3a_{12}a_{31}) + s_{22} (a_{12})^3 a_{32} + s_{23}a_{12}a_{13} (a_{12}a_{33} + a_{13}a_{32})$$

$$+ s_{24} (a_{12})^2 (a_{12}a_{33} + 3a_{13}a_{32}) + s_{25}a_{12} \left[a_{12} (a_{11}a_{33} + a_{13}a_{31}) + 2a_{11}a_{13}a_{32}\right]$$

$$+ s_{26} (a_{12})^2 (a_{12}a_{31} + 3a_{11}a_{32}) + s_{33}a_{13}a_{13}a_{13}a_{33}$$
$$+ s_{34} (a_{13})^2 (a_{13}a_{32} + 3a_{12}a_{33}) + s_{35} (a_{13})^2 (a_{13}a_{31} + 3a_{11}a_{33})$$
$$+ s_{36}a_{13} [a_{13} (a_{11}a_{32} + a_{12}a_{31}) + 2a_{11}a_{12}a_{33}] + 2s_{44}a_{12}a_{13} (a_{12}a_{33} + a_{13}a_{32})$$
$$+ 2s_{45}a_{13} [a_{13} (a_{12}a_{31} + a_{11}a_{32}) + 2a_{11}a_{12}a_{33}]$$
$$+ 2s_{46}a_{12} [a_{12} (a_{13}a_{31} + a_{11}a_{33}) + 2a_{11}a_{13}a_{32}] + 2s_{55}a_{11}a_{13} (a_{11}a_{33} + a_{13}a_{31})$$
$$+ 2s_{56}a_{11} [a_{11} (a_{13}a_{32} + a_{12}a_{33}) + 2a_{12}a_{13}a_{31}]$$
$$+2s_{66}a_{11}a_{12} (a_{11}a_{32} + a_{12}a_{31}), \tag{9.18}$$

$$s'_{16} = s_{11} (a_{11})^3 a_{21} + s_{12}a_{11}a_{12} (a_{11}a_{22} + a_{12}a_{21}) + s_{13}a_{11}a_{13} (a_{11}a_{23} + a_{13}a_{21})$$
$$+ s_{14}a_{11} [a_{11} (a_{12}a_{23} + a_{13}a_{22}) + 2a_{12}a_{13}a_{21}] + s_{15} (a_{11})^2 (a_{11}a_{23} + 3a_{13}a_{21})$$
$$+ s_{16} (a_{11})^2 (a_{11}a_{22} + 3a_{12}a_{21}) + s_{22} (a_{12})^3 a_{22} + s_{23}a_{12}a_{13} (a_{12}a_{23} + a_{13}a_{22})$$
$$+ s_{24} (a_{12})^2 (a_{12}a_{23} + 3a_{13}a_{22}) + s_{25}a_{12} [a_{12} (a_{11}a_{23} + a_{13}a_{21}) + 2a_{11}a_{13}a_{22}]$$
$$+ s_{26} (a_{12})^2 (a_{12}a_{21} + 3a_{11}a_{22}) + s_{33} (a_{13})^2 a_{23} + s_{34} (a_{13})^2 (a_{13}a_{22} + 3a_{12}a_{23})$$
$$+ s_{35} (a_{13})^2 (a_{13}a_{21} + 3a_{11}a_{23}) + s_{36}a_{13} [a_{13} (a_{11}a_{22} + a_{12}a_{21}) + 2a_{11}a_{12}a_{23}]$$
$$+ 2s_{44}a_{12}a_{13} (a_{12}a_{23} + a_{13}a_{22}) + 2s_{45}a_{13} [a_{13} (a_{12}a_{21} + a_{11}a_{22}) + 2a_{11}a_{12}a_{23}]$$
$$+ 2s_{46}a_{12} [a_{12} (a_{11}a_{23} + a_{13}a_{21}) + 2a_{11}a_{13}a_{22}] + 2s_{55}a_{11}a_{13} (a_{11}a_{23} + a_{13}a_{21})$$
$$+ 2s_{56}a_{11} [a_{11} (a_{13}a_{22} + a_{12}a_{23}) + 2a_{12}a_{13}a_{21}]$$
$$+2s_{66}a_{11}a_{12} (a_{11}a_{22} + a_{12}a_{21}), \tag{9.19}$$

$$s'_{22} = s_{11} (a_{21})^4 + 2s_{12} (a_{21}a_{22})^2 + 2s_{13} (a_{21}a_{23})^2 + 4s_{14} (a_{21})^2 a_{22}a_{23}$$
$$+ 4s_{15} (a_{21})^3 a_{23} + 4s_{16} (a_{21})^3 a_{22} + s_{22} (a_{22})^4 + 2s_{23} (a_{22}a_{23})^2$$
$$+ 4s_{24} (a_{22})^3 a_{23} + 4s_{25}a_{21} (a_{22})^2 a_{23} + 4s_{26}a_{21} (a_{22})^3 + s_{33} (a_{23})^4$$
$$+ 4s_{34}a_{22} (a_{23})^3 + 4s_{35}a_{21} (a_{23})^3 + 4s_{36}a_{21}a_{22} (a_{23})^2 + 4s_{44} (a_{22}a_{23})^2$$
$$+ 8s_{45}a_{21}a_{22} (a_{23})^2 + 8s_{46}a_{21} (a_{22})^2 a_{23} + 4s_{55} (a_{21}a_{23})^2 + 8s_{56} (a_{21})^2 a_{22}a_{23}$$
$$+ 4s_{66} (a_{21}a_{22})^2, \tag{9.20}$$

$$s'_{23} = s_{11} (a_{21}a_{31})^2 + s_{12} \left[(a_{21}a_{32})^2 + (a_{22}a_{31})^2 \right] + s_{13} \left[(a_{21}a_{33})^2 + (a_{23}a_{31})^2 \right]$$
$$+ 2s_{14} \left[(a_{21})^2 a_{32}a_{33} + a_{22}a_{23} (a_{31})^2 \right] + 2s_{15} \left[(a_{21})^2 a_{31}a_{33} + a_{21}a_{23} (a_{31})^2 \right]$$
$$+ 2s_{16} \left[(a_{21})^2 a_{31}a_{32} + a_{21}a_{22} (a_{31})^2 \right] + s_{22} (a_{22}a_{32})^2$$
$$+ s_{23} \left[(a_{22}a_{33})^2 + (a_{23}a_{32})^2 \right] + 2s_{24} \left[(a_{22})^2 a_{32}a_{33} + a_{22}a_{23} (a_{32})^2 \right]$$
$$+ 2s_{25} \left[(a_{22})^2 a_{31}a_{33} + a_{21}a_{23} (a_{32})^2 \right] + 2s_{26} \left[(a_{22})^2 a_{31}a_{32} + a_{21}a_{22} (a_{32})^2 \right]$$
$$+ s_{33} (a_{23}a_{33})^2 + 2s_{34} \left[(a_{23})^2 a_{32}a_{33} + a_{22}a_{23} (a_{33})^2 \right]$$

$$+ 2s_{35} \left[(a_{23})^2 a_{31} a_{33} + a_{21} a_{23} (a_{33})^2 \right] + 2s_{36} \left[(a_{23})^2 a_{31} a_{32} + a_{21} a_{22} (a_{33})^2 \right]$$
$$+ 4s_{44} a_{22} a_{23} a_{32} a_{33} + 4s_{45} a_{23} a_{33} (a_{21} a_{32} + a_{22} a_{31})$$
$$+ 4s_{46} a_{22} a_{32} (a_{21} a_{33} + a_{23} a_{31}) + 4s_{55} a_{21} a_{23} a_{31} a_{33}$$
$$+ 4s_{56} a_{21} a_{31} (a_{22} a_{33} + a_{23} a_{32}) + 4s_{66} a_{21} a_{22} a_{31} a_{32}, \tag{9.21}$$

$$s'_{24} = s_{11} (a_{21})^3 a_{31} + s_{12} a_{21} a_{22} (a_{21} a_{32} + a_{22} a_{31}) + s_{13} a_{21} a_{23} (a_{21} a_{33} + a_{23} a_{31})$$
$$+ s_{14} a_{21} \left[a_{21} (a_{22} a_{33} + a_{23} a_{32}) + 2 a_{22} a_{23} a_{31} \right] + s_{15} (a_{21})^2 (a_{21} a_{33} + 3 a_{23} a_{31})$$
$$+ s_{16} (a_{21} a_{21})^2 (a_{21} a_{32} + 3 a_{22} a_{31}) + s_{22} (a_{22})^3 a_{32} + s_{23} a_{22} a_{23} (a_{22} a_{33} + a_{23} a_{32})$$
$$+ s_{24} (a_{22})^2 (a_{22} a_{33} + 3 a_{23} a_{32}) + s_{25} a_{22} \left[a_{22} (a_{21} a_{33} + a_{23} a_{31}) + 2 a_{21} a_{23} a_{32} \right]$$
$$+ s_{26} (a_{22})^2 (a_{22} a_{31} + 3 a_{21} a_{32}) + s_{33} (a_{23})^3 a_{33} + s_{34} (a_{23})^2 (a_{23} a_{32} + 3 a_{22} a_{33})$$
$$+ s_{35} (a_{23})^2 (a_{23} a_{31} + 3 a_{21} a_{33}) + s_{36} a_{23} \left[a_{23} (a_{21} a_{32} + a_{22} a_{31}) + 2 a_{21} a_{22} a_{33} \right]$$
$$+ 2 s_{44} a_{22} a_{23} (a_{22} a_{33} + a_{23} a_{32}) + 2 s_{45} a_{23} \left[a_{23} (a_{21} a_{32} + a_{22} a_{31}) + 2 a_{21} a_{22} a_{33} \right]$$
$$+ 2 s_{46} a_{22} \left[a_{22} (a_{23} a_{31} + a_{21} a_{33}) + 2 a_{21} a_{23} a_{32} \right] + 2 s_{55} a_{21} a_{23} (a_{21} a_{33} + a_{23} a_{31})$$
$$+ 2 s_{56} a_{21} \left[a_{21} (a_{23} a_{32} + a_{22} a_{33}) + 2 a_{22} a_{23} a_{31} \right]$$
$$+ 2 s_{66} a_{21} a_{22} (a_{21} a_{32} + a_{22} a_{31}), \tag{9.22}$$

$$s'_{25} = s_{11} a_{11} (a_{21})^2 a_{31} + s_{12} \left[a_{11} (a_{22})^2 a_{31} + a_{12} (a_{21})^2 a_{32} \right]$$
$$+ s_{13} \left[a_{13} (a_{21})^2 a_{33} + a_{11} (a_{23})^2 a_{31} \right]$$
$$+ s_{14} \left[(a_{21})^2 (a_{12} a_{33} + a_{13} a_{32}) + 2 a_{11} a_{22} a_{23} a_{31} \right]$$
$$+ s_{15} \left[(a_{21})^2 (a_{11} a_{33} + a_{13} a_{31}) + 2 a_{11} a_{21} a_{23} a_{31} \right]$$
$$+ s_{16} \left[(a_{21})^2 (a_{11} a_{32} + a_{12} a_{31}) + 2 a_{11} a_{21} a_{22} a_{31} \right] + s_{22} a_{12} (a_{22})^2 a_{32}$$
$$+ s_{23} \left[a_{12} (a_{23})^2 a_{32} + a_{13} (a_{22})^2 a_{33} \right]$$
$$+ s_{24} \left[(a_{22})^2 (a_{12} a_{33} + a_{13} a_{32}) + 2 a_{12} a_{22} a_{23} a_{32} \right]$$
$$+ s_{25} \left[(a_{22})^2 (a_{11} a_{33} + a_{13} a_{31}) + 2 a_{12} a_{21} a_{23} a_{32} \right]$$
$$+ s_{26} \left[(a_{22})^2 (a_{11} a_{32} + a_{12} a_{31}) + 2 a_{12} a_{21} a_{22} a_{32} \right] + s_{33} a_{13} (a_{23})^2 a_{33}$$
$$+ s_{34} \left[(a_{23})^2 (a_{12} a_{33} + a_{13} a_{32}) + 2 a_{13} a_{22} a_{23} a_{33} \right]$$
$$+ s_{35} \left[(a_{23})^2 (a_{11} a_{33} + a_{13} a_{31}) + 2 a_{13} a_{21} a_{23} a_{33} \right]$$
$$+ s_{36} \left[(a_{23})^2 (a_{11} a_{32} + a_{12} a_{31}) + 2 a_{13} a_{21} a_{22} a_{33} \right]$$
$$+ 2 s_{44} a_{22} a_{23} (a_{12} a_{33} + a_{13} a_{32})$$
$$+ 2 s_{45} a_{23} \left[a_{13} (a_{22} a_{31} + a_{21} a_{32}) + a_{33} (a_{11} a_{22} + a_{12} a_{21}) \right]$$

$$+\ 2s_{46}\left[a_{12}a_{22}\left(a_{23}a_{31}+a_{21}a_{33}\right)+a_{22}a_{32}\left(a_{11}a_{23}+a_{13}a_{21}\right)\right]$$
$$+\ 2s_{55}a_{21}a_{23}\left(a_{11}a_{33}+a_{13}a_{31}\right)$$
$$+\ 2s_{56}\left[a_{11}a_{21}\left(a_{23}a_{32}+a_{22}a_{33}\right)+a_{21}a_{31}\left(a_{12}a_{23}+a_{13}a_{22}\right)\right]$$
$$+\ 2s_{66}a_{21}a_{22}\left(a_{11}a_{32}+a_{12}a_{31}\right),\tag{9.23}$$

$$\begin{aligned}
s'_{26} &= s_{11}a_{11}\left(a_{21}\right)^{3}+s_{12}a_{21}a_{22}\left(a_{11}a_{22}+a_{12}a_{21}\right)\\
&+ s_{13}a_{21}a_{23}\left(a_{11}a_{23}+a_{13}a_{21}\right)\\
&+ s_{14}a_{21}\left[a_{21}\left(a_{12}a_{23}+a_{13}a_{22}\right)+2a_{11}a_{22}a_{23}\right]\\
&+ s_{15}\left(a_{21}\right)^{2}\left(3a_{11}a_{23}+a_{13}a_{21}\right)+s_{16}\left(a_{21}\right)^{2}\left(a_{12}a_{21}+3a_{11}a_{22}\right)\\
&+ s_{22}a_{22}a_{22}a_{12}a_{22}+s_{23}a_{22}a_{23}\left(a_{12}a_{23}+a_{13}a_{22}\right)\\
&+ s_{24}\left(a_{22}\right)^{2}\left(a_{13}a_{22}+3a_{12}a_{23}\right)\\
&+ s_{25}a_{22}\left[a_{22}\left(a_{11}a_{23}+a_{13}a_{21}\right)+2a_{12}a_{21}a_{23}\right]\\
&+ s_{26}\left(a_{22}\right)^{2}\left(a_{11}a_{22}+3a_{12}a_{21}\right)+s_{33}\left(a_{23}\right)^{3}a_{13}\\
&+ s_{34}\left(a_{23}\right)^{2}\left(a_{12}a_{23}+3a_{13}a_{22}\right)+s_{35}\left(a_{23}\right)^{2}\left(a_{11}a_{23}+3a_{13}a_{21}\right)\\
&+ s_{36}a_{23}\left[a_{23}\left(a_{11}a_{22}+a_{12}a_{21}\right)+2a_{13}a_{21}a_{22}\right]\\
&+ 2s_{44}a_{22}a_{23}\left(a_{12}a_{23}+a_{13}a_{22}\right)\\
&+ 2s_{45}a_{23}\left[a_{23}\left(a_{11}a_{22}+a_{12}a_{21}\right)+2a_{13}a_{21}a_{22}\right]\\
&+ 2s_{46}a_{22}\left[a_{22}\left(a_{11}a_{23}+a_{13}a_{21}\right)+2a_{12}a_{21}a_{23}\right]\\
&+ 2s_{55}a_{21}a_{23}\left(a_{11}a_{23}+a_{13}a_{21}\right)\\
&+ 2s_{56}a_{21}\left[a_{21}\left(a_{12}a_{23}+a_{13}a_{22}\right)+2a_{11}a_{22}a_{23}\right]\\
&+ 2s_{66}a_{21}a_{22}\left(a_{11}a_{22}+a_{12}a_{21}\right),
\end{aligned}\tag{9.24}$$

$$\begin{aligned}
s'_{33} &= s_{11}\left(a_{31}\right)^{4}+2s_{12}\left(a_{31}a_{32}\right)^{2}+2s_{13}\left(a_{31}a_{33}\right)^{2}+4s_{14}\left(a_{31}\right)^{2}a_{32}a_{33}\\
&+ 4s_{15}\left(a_{31}\right)^{3}a_{33}+4s_{16}\left(a_{31}\right)^{3}a_{32}+s_{22}\left(a_{32}\right)^{4}+2s_{23}\left(a_{32}a_{33}\right)^{2}\\
&+ 4s_{24}\left(a_{32}\right)^{3}a_{33}+4s_{25}a_{31}\left(a_{32}\right)^{2}a_{33}+4s_{26}a_{31}\left(a_{32}\right)^{3}+s_{33}\left(a_{33}\right)^{4}\\
&+ 4s_{34}a_{32}\left(a_{33}\right)^{3}+4s_{35}a_{31}\left(a_{33}\right)^{3}+4s_{36}a_{31}a_{32}\left(a_{33}\right)^{2}+4s_{44}\left(a_{32}a_{33}\right)^{2}\\
&+ 8s_{45}a_{31}a_{32}\left(a_{33}\right)^{2}+8s_{46}a_{31}\left(a_{32}\right)^{2}a_{33}+4s_{55}\left(a_{31}a_{33}\right)^{2}\\
&+ 8s_{56}\left(a_{31}\right)^{2}a_{32}a_{33}+4s_{66}\left(a_{31}a_{32}\right)^{2},
\end{aligned}\tag{9.25}$$

$$\begin{aligned}
s'_{34} &= s_{11}a_{21}\left(a_{31}\right)^{3}+s_{12}a_{31}a_{32}\left(a_{21}a_{32}+a_{22}a_{31}\right)+s_{13}a_{31}a_{33}\left(a_{21}a_{33}+a_{23}a_{31}\right)\\
&+ s_{14}a_{31}\left[a_{31}\left(a_{22}a_{33}+a_{23}a_{32}\right)+2a_{21}a_{32}a_{33}\right]+s_{15}\left(a_{31}\right)^{2}\left(a_{23}a_{31}+3a_{21}a_{33}\right)\\
&+ s_{16}\left(a_{31}\right)^{2}\left(a_{22}a_{31}+3a_{21}a_{32}\right)+s_{22}a_{22}\left(a_{32}\right)^{3}+s_{23}a_{32}a_{33}\left(a_{22}a_{33}+a_{23}a_{32}\right)\\
&+ s_{24}\left(a_{32}\right)^{2}\left(a_{23}a_{32}+3a_{22}a_{33}\right)+s_{25}a_{32}\left[a_{32}\left(a_{21}a_{33}+a_{23}a_{31}\right)+2a_{22}a_{31}a_{33}\right]\\
&+ s_{26}\left(a_{32}\right)^{2}\left(a_{21}a_{32}+3a_{22}a_{31}\right)+s_{33}a_{23}\left(a_{33}\right)^{3}+s_{34}\left(a_{33}\right)^{2}\left(a_{22}a_{33}+3a_{23}a_{32}\right)\\
&+ s_{35}\left(a_{33}\right)^{2}\left(a_{21}a_{33}+3a_{23}a_{31}\right)+s_{36}a_{33}\left[a_{33}\left(a_{21}a_{32}+a_{22}a_{31}\right)+2a_{23}a_{31}a_{32}\right]
\end{aligned}$$

$$+ 2s_{44}a_{32}a_{33}\left(a_{22}a_{33} + a_{23}a_{32}\right) + 2s_{45}a_{33}\left[a_{33}\left(a_{21}a_{32} + a_{22}a_{31}\right) + 2a_{23}a_{31}a_{32}\right]$$
$$+ 2s_{46}a_{32}\left[a_{32}\left(a_{21}a_{33} + a_{23}a_{31}\right) + 2a_{22}a_{31}a_{33}\right] + 2s_{55}a_{31}a_{33}\left(a_{21}a_{33} + a_{23}a_{31}\right)$$
$$+ 2s_{56}a_{31}\left[a_{31}\left(a_{22}a_{33} + a_{23}a_{32}\right) + 2a_{21}a_{32}a_{33}\right]$$
$$+ 2s_{66}a_{31}a_{32}\left(a_{21}a_{32} + a_{22}a_{31}\right), \tag{9.26}$$

$$s'_{35} = s_{11}a_{11}\left(a_{31}\right)^3 + s_{12}a_{31}a_{32}\left(a_{11}a_{32} + a_{12}a_{31}\right) + s_{13}a_{31}a_{33}\left(a_{11}a_{33} + a_{13}a_{31}\right)$$
$$+ s_{14}a_{31}\left[a_{31}\left(a_{12}a_{33} + a_{13}a_{32}\right) + 2a_{11}a_{32}a_{33}\right] + s_{15}\left(a_{31}\right)^2\left(a_{13}a_{31} + 3a_{11}a_{33}\right)$$
$$+ s_{16}\left(a_{31}\right)^2\left(a_{12}a_{31} + 3a_{11}a_{32}\right) + s_{22}a_{12}\left(a_{32}\right)^3 + s_{23}a_{32}a_{33}\left(a_{13}a_{32} + a_{12}a_{33}\right)$$
$$+ s_{24}\left(a_{32}\right)^2\left(a_{13}a_{32} + 3a_{12}a_{33}\right) + s_{25}a_{32}\left[a_{32}\left(a_{11}a_{33} + a_{13}a_{31}\right) + 2a_{12}a_{31}a_{33}\right]$$
$$+ s_{26}\left(a_{32}\right)^2\left(a_{11}a_{32} + 3a_{12}a_{31}\right) + s_{33}a_{33}a_{33}a_{13}a_{33}$$
$$+ s_{34}\left(a_{33}\right)^2\left(a_{12}a_{33} + 3a_{13}a_{32}\right) + s_{35}\left(a_{33}\right)^2\left(a_{11}a_{33} + 3a_{13}a_{31}\right)$$
$$+ s_{36}a_{33}\left[a_{33}\left(a_{11}a_{32} + a_{12}a_{31}\right) + 2a_{13}a_{31}a_{32}\right] + 2s_{44}a_{32}a_{33}\left(a_{12}a_{33} + a_{13}a_{32}\right)$$
$$+ 2s_{45}a_{33}\left[a_{33}\left(a_{11}a_{32} + a_{12}a_{31}\right) + 2a_{13}a_{31}a_{32}\right]$$
$$+ 2s_{46}a_{32}\left[a_{32}\left(a_{11}a_{33} + a_{13}a_{31}\right) + 2a_{12}a_{31}a_{33}\right] + 2s_{55}a_{31}a_{33}\left(a_{11}a_{33} + a_{13}a_{31}\right)$$
$$+ 2s_{56}a_{31}\left[a_{31}\left(a_{12}a_{33} + a_{13}a_{32}\right) + 2a_{11}a_{32}a_{33}\right]$$
$$+ 2s_{66}a_{31}a_{32}\left(a_{11}a_{32} + a_{12}a_{31}\right), \tag{9.27}$$

$$s'_{36} = s_{11}a_{11}a_{21}\left(a_{31}\right)^2 + s_{12}\left[a_{11}a_{21}\left(a_{32}\right)^2 + a_{12}a_{22}\left(a_{31}\right)^2\right]$$
$$+ s_{13}\left[a_{11}a_{21}\left(a_{33}\right)^2 + a_{13}a_{23}\left(a_{31}\right)^2\right]$$
$$+ s_{14}\left[\left(a_{31}\right)^2\left(a_{12}a_{23} + a_{13}a_{22}\right) + 2a_{11}a_{21}a_{32}a_{33}\right]$$
$$+ s_{15}a_{31}\left[a_{31}\left(a_{11}a_{23} + a_{13}a_{21}\right) + 2a_{11}a_{21}a_{33}\right]$$
$$+ s_{16}a_{31}\left[a_{31}\left(a_{11}a_{22} + a_{12}a_{21}\right) + 2a_{11}a_{21}a_{32}\right] + s_{22}a_{12}a_{22}\left(a_{32}\right)^2$$
$$+ s_{23}\left[a_{12}a_{22}\left(a_{33}\right)^2 + a_{13}a_{23}\left(a_{32}\right)^2\right]$$
$$+ s_{24}a_{32}\left[a_{32}\left(a_{12}a_{23} + a_{13}a_{22}\right) + 2a_{12}a_{22}a_{33}\right]$$
$$+ s_{25}\left[\left(a_{32}\right)^2\left(a_{11}a_{23} + a_{13}a_{21}\right) + 2a_{12}a_{22}a_{31}a_{33}\right]$$
$$+ s_{26}a_{32}\left[a_{32}\left(a_{11}a_{22} + a_{12}a_{21}\right) + 2a_{12}a_{22}a_{31}\right] + s_{33}a_{13}a_{23}\left(a_{33}\right)^2$$
$$+ s_{34}a_{33}\left[a_{33}\left(a_{12}a_{23} + a_{13}a_{22}\right) + 2a_{13}a_{23}a_{32}\right]$$
$$+ s_{35}a_{33}\left[a_{33}\left(a_{11}a_{23} + a_{13}a_{21}\right) + 2a_{13}a_{23}a_{31}\right]$$
$$+ s_{36}\left[\left(a_{33}\right)^2\left(a_{11}a_{22} + a_{12}a_{21}\right) + 2a_{13}a_{23}a_{31}a_{32}\right]$$
$$+ s_{44}a_{33}\left[a_{32}\left(a_{12}a_{23} + a_{12}a_{23}\right) + 2a_{13}a_{22}a_{32}\right]$$
$$+ 2s_{45}a_{33}\left[a_{31}\left(a_{12}a_{23} + a_{13}a_{22}\right) + a_{32}\left(a_{11}a_{23} + a_{13}a_{21}\right)\right]$$
$$+ 2s_{46}a_{32}\left[a_{31}\left(a_{12}a_{23} + a_{13}a_{22}\right) + a_{33}\left(a_{11}a_{22} + a_{12}a_{21}\right)\right]$$
$$+ 2s_{55}a_{31}a_{33}\left(a_{11}a_{23} + a_{13}a_{21}\right)$$

$$+ 2s_{56}a_{31}\left[a_{32}\left(a_{11}a_{23}+a_{13}a_{21}\right)+a_{33}\left(a_{11}a_{22}+a_{12}a_{21}\right)\right]$$
$$+ 2s_{66}a_{31}a_{32}\left(a_{11}a_{22}+a_{12}a_{21}\right), \tag{9.28}$$

$$s'_{44} = s_{11}\left(a_{21}a_{31}\right)^2 + 2s_{12}a_{21}a_{22}a_{31}a_{32} + 2s_{13}a_{21}a_{23}a_{31}a_{33}$$
$$+ 2s_{14}a_{21}a_{31}\left(a_{22}a_{33}+a_{23}a_{32}\right) + 2s_{15}a_{21}a_{31}\left(a_{21}a_{33}+a_{23}a_{31}\right)$$
$$+ 2s_{16}a_{21}a_{31}\left(a_{21}a_{32}+a_{22}a_{31}\right) + s_{22}\left(a_{22}a_{32}\right)^2 + 2s_{23}a_{22}a_{23}a_{32}a_{33}$$
$$+ 2s_{24}a_{22}a_{32}\left(a_{22}a_{33}+a_{23}a_{32}\right) + 2s_{25}a_{22}a_{32}\left(a_{21}a_{33}+a_{23}a_{31}\right)$$
$$+ 2s_{26}a_{22}a_{32}\left(a_{21}a_{32}+a_{22}a_{31}\right) + s_{33}\left(a_{23}a_{33}\right)^2 + 2s_{34}a_{23}a_{33}\left(a_{22}a_{33}+a_{23}a_{32}\right)$$
$$+ 2s_{35}a_{23}a_{33}\left(a_{21}a_{33}+a_{23}a_{31}\right) + 2s_{36}a_{23}a_{33}\left(a_{21}a_{32}+a_{22}a_{31}\right)$$
$$+ s_{44}\left(a_{22}a_{33}+a_{23}a_{32}\right)^2$$
$$+ 2s_{45}\left[\left(a_{23}\right)^2 a_{31}a_{32}+a_{21}a_{22}\left(a_{33}\right)^2+a_{23}a_{33}\left(a_{21}a_{32}+a_{22}a_{31}\right)\right]$$
$$+ 2s_{46}\left[a_{21}a_{23}\left(a_{32}\right)^2+\left(a_{22}\right)^2 a_{31}a_{33}+a_{22}a_{32}\left(a_{21}a_{33}+a_{23}a_{31}\right)\right]$$
$$+ s_{55}\left(a_{21}a_{33}+a_{23}a_{31}\right)^2$$
$$+ 2s_{56}\left[\left(a_{21}\right)^2 a_{32}a_{33}+a_{22}a_{23}\left(a_{31}\right)^2+a_{21}a_{31}\left(a_{23}a_{32}+a_{22}a_{33}\right)\right]$$
$$+ s_{66}\left(a_{21}a_{32}+a_{22}a_{31}\right)^2, \tag{9.29}$$

$$s'_{45} = s_{11}a_{11}a_{21}\left(a_{31}\right)^2 + s_{12}a_{31}a_{32}\left(a_{11}a_{22}+a_{12}a_{21}\right)$$
$$+ s_{13}a_{31}a_{33}\left(a_{13}a_{21}+a_{11}a_{23}\right)$$
$$+ s_{14}a_{31}\left[a_{32}\left(a_{13}a_{21}+a_{11}a_{23}\right)+a_{33}\left(a_{12}a_{21}+a_{11}a_{22}\right)\right]$$
$$+ s_{15}a_{31}\left[a_{31}\left(a_{13}a_{21}+a_{11}a_{23}\right)+a_{33}\left(a_{11}a_{21}+a_{11}a_{21}\right)\right]$$
$$+ s_{16}a_{31}\left[a_{31}\left(a_{12}a_{21}+a_{11}a_{22}\right)+2a_{11}a_{21}a_{32}\right] + s_{22}a_{12}a_{22}\left(a_{32}\right)^2$$
$$+ s_{23}a_{32}a_{33}\left(a_{13}a_{22}+a_{12}a_{23}\right)$$
$$+ s_{24}a_{32}\left[a_{32}\left(a_{12}a_{23}+a_{13}a_{22}\right)+2a_{12}a_{22}a_{33}\right]$$
$$+ s_{25}a_{32}\left[a_{12}\left(a_{21}a_{33}+a_{23}a_{31}\right)+a_{22}\left(a_{11}a_{33}+a_{13}a_{31}\right)\right]$$
$$+ s_{26}a_{32}\left[a_{32}\left(a_{11}a_{22}+a_{12}a_{21}\right)+2a_{12}a_{22}a_{31}\right] + s_{33}a_{13}a_{23}\left(a_{33}\right)^2$$
$$+ s_{34}a_{33}\left[a_{33}\left(a_{12}a_{23}+a_{13}a_{22}\right)+2a_{13}a_{23}a_{32}\right]$$
$$+ s_{35}a_{33}\left[a_{33}\left(a_{11}a_{23}+a_{13}a_{21}\right)+2a_{13}a_{23}a_{31}\right]$$
$$+ s_{36}a_{33}\left[a_{13}\left(a_{21}a_{32}+a_{22}a_{31}\right)+a_{23}\left(a_{11}a_{32}+a_{12}a_{31}\right)\right]$$
$$+ s_{44}\left[a_{12}a_{22}\left(a_{33}\right)^2+a_{13}a_{23}\left(a_{32}\right)^2+a_{32}a_{33}\left(a_{12}a_{23}+a_{13}a_{22}\right)\right]$$
$$+ s_{45}\left[\left(a_{33}\right)^2\left(a_{11}a_{22}+a_{12}a_{21}\right)+a_{31}a_{33}\left(a_{12}a_{23}+a_{13}a_{22}\right)\right.$$
$$\left.+ a_{32}a_{33}\left(a_{11}a_{23}+a_{13}a_{21}\right)+2a_{13}a_{23}a_{31}a_{32}\right]$$
$$+ s_{46}\left[\left(a_{32}\right)^2\left(a_{11}a_{23}+a_{13}a_{21}\right)+a_{12}a_{32}\left(a_{23}a_{31}+a_{21}a_{33}\right)\right.$$
$$\left.+ a_{22}a_{32}\left(a_{11}a_{33}+a_{13}a_{31}\right)+2a_{12}a_{22}a_{31}a_{33}\right]$$

$$+ s_{55} \left[a_{11}a_{33} \left(a_{21}a_{33} + a_{23}a_{31} \right) + a_{13}a_{31} \left(a_{21}a_{33} + a_{23}a_{31} \right) \right]$$
$$+ s_{56} \left[\left(a_{31} \right)^2 \left(a_{12}a_{23} + a_{13}a_{22} \right) + a_{11}a_{31} \left(a_{22}a_{33} + a_{23}a_{32} \right) \right.$$
$$\left. + a_{21}a_{31} \left(a_{12}a_{33} + a_{13}a_{32} \right) + 2a_{11}a_{21}a_{32}a_{33} \right]$$
$$+ s_{66} \left[a_{11}a_{21} \left(a_{32} \right)^2 + a_{12}a_{22} \left(a_{31} \right)^2 + a_{31}a_{32} \left(a_{12}a_{21} + a_{11}a_{22} \right) \right], \quad (9.30)$$

$$s'_{46} = s_{11}a_{11} \left(a_{21}a_{21} \right)^2 a_{31} + s_{12}a_{21}a_{22} \left(a_{12}a_{31} + a_{11}a_{32} \right)$$
$$+ s_{13}a_{21}a_{23} \left(a_{13}a_{31} + a_{11}a_{33} \right)$$
$$+ s_{14}a_{21} \left[a_{11} \left(a_{22}a_{33} + a_{23}a_{32} \right) + a_{31} \left(a_{12}a_{23} + a_{13}a_{22} \right) \right]$$
$$+ s_{15}a_{21} \left[a_{21} \left(a_{13}a_{31} + a_{11}a_{33} \right) + 2a_{11}a_{23}a_{31} \right]$$
$$+ s_{16}a_{21} \left[a_{21} \left(a_{11}a_{32} + a_{12}a_{31} \right) + 2a_{11}a_{22}a_{31} \right] + s_{22}a_{12} \left(a_{22} \right)^2 a_{32}$$
$$+ s_{23}a_{22}a_{23} \left(a_{13}a_{32} + a_{12}a_{33} \right)$$
$$+ s_{24}a_{22} \left[a_{22} \left(a_{13}a_{32} + a_{12}a_{33} \right) + 2a_{12}a_{23}a_{32} \right]$$
$$+ s_{25}a_{22} \left[a_{21} \left(a_{13}a_{32} + a_{12}a_{33} \right) + a_{23} \left(a_{11}a_{32} + a_{12}a_{31} \right) \right]$$
$$+ s_{26}a_{22} \left[a_{22} \left(a_{11}a_{32} + a_{12}a_{31} \right) + 2a_{12}a_{21}a_{32} \right] + s_{33}a_{13} \left(a_{23} \right)^2 a_{33}$$
$$+ s_{34}a_{23} \left[a_{23} \left(a_{12}a_{33} + a_{13}a_{32} \right) + 2a_{13}a_{22}a_{33} \right]$$
$$+ s_{35}a_{23} \left[a_{23} \left(a_{13}a_{31} + a_{11}a_{33} \right) + 2a_{13}a_{21}a_{33} \right]$$
$$+ s_{36}a_{23} \left[a_{13} \left(a_{21}a_{32} + a_{22}a_{31} \right) + a_{33} \left(a_{11}a_{22} + a_{12}a_{21} \right) \right]$$
$$+ s_{44} \left[a_{22}a_{33} \left(a_{12}a_{23} + a_{13}a_{22} \right) + a_{23}a_{32} \left(a_{12}a_{23} + a_{13}a_{22} \right) \right]$$
$$+ s_{45} \left[\left(a_{23} \right)^2 \left(a_{11}a_{32} + a_{12}a_{31} \right) + a_{23}a_{33} \left(a_{11}a_{22} + a_{12}a_{21} \right) \right.$$
$$\left. + a_{13}a_{23} \left(a_{22}a_{31} + a_{21}a_{32} \right) + 2a_{13}a_{21}a_{22}a_{33} \right]$$
$$+ s_{46} \left[\left(a_{22} \right)^2 \left(a_{11}a_{33} + a_{13}a_{31} \right) + a_{12}a_{22} \left(a_{21}a_{33} + a_{23}a_{31} \right) \right.$$
$$\left. + a_{22}a_{32} \left(a_{11}a_{23} + a_{13}a_{21} \right) + 2a_{12}a_{21}a_{23}a_{32} \right]$$
$$+ s_{55} \left[a_{11} \left(a_{23} \right)^2 a_{31} + a_{13} \left(a_{21} \right)^2 a_{33} + a_{21}a_{23} \left(a_{11}a_{33} + a_{13}a_{31} \right) \right]$$
$$+ s_{56} \left[\left(a_{21} \right)^2 \left(a_{12}a_{33} + a_{13}a_{32} \right) + a_{11}a_{21} \left(a_{22}a_{33} + a_{23}a_{32} \right) \right.$$
$$\left. + a_{21}a_{31} \left(a_{12}a_{23} + a_{13}a_{22} \right) + 2a_{11}a_{22}a_{23}a_{31} \right]$$
$$+ s_{66} \left[a_{11} \left(a_{22} \right)^2 a_{31} + a_{12} \left(a_{21} \right)^2 a_{32} + a_{21}a_{22} \left(a_{11}a_{32} + a_{12}a_{31} \right) \right], \quad (9.31)$$

$$s'_{55} = s_{11} \left(a_{11}a_{31} \right)^2 + 2s_{12}a_{11}a_{12}a_{31}a_{32} + 2s_{13}a_{11}a_{13}a_{31}a_{33}$$
$$+ 2s_{14}a_{11}a_{31} \left(a_{12}a_{33} + a_{13}a_{32} \right) + 2s_{15}a_{11}a_{31} \left(a_{11}a_{33} + a_{13}a_{31} \right)$$
$$+ 2s_{16}a_{11}a_{31} \left(a_{11}a_{32} + a_{12}a_{31} \right) + s_{22} \left(a_{12}a_{32} \right)^2 + 2s_{23}a_{12}a_{13}a_{32}a_{33}$$
$$+ 2s_{24}a_{12}a_{32} \left(a_{12}a_{33} + a_{13}a_{32} \right) + 2s_{25}a_{12}a_{32} \left(a_{11}a_{33} + a_{13}a_{31} \right)$$
$$+ 2s_{26}a_{12}a_{32} \left(a_{11}a_{32} + a_{12}a_{31} \right) + s_{33} \left(a_{13}a_{33} \right)^2 + 2s_{34}a_{13}a_{33} \left(a_{12}a_{33} + a_{13}a_{32} \right)$$

$$
\begin{aligned}
&+ 2s_{35}a_{13}a_{33}\left(a_{11}a_{33} + a_{13}a_{31}a_{33}\right) + 2s_{36}a_{13}a_{33}\left(a_{11}a_{32} + a_{12}a_{31}\right) \\
&+ s_{44}\left(a_{12}a_{33} + a_{13}a_{32}\right)^2 \\
&+ 2s_{45}\left[a_{11}a_{12}\left(a_{33}\right)^2 + \left(a_{13}\right)^2 a_{31}a_{32} + a_{13}a_{33}\left(a_{11}a_{32} + a_{12}a_{31}\right)\right] \\
&+ 2s_{46}\left[a_{11}a_{13}\left(a_{32}\right)^2 + \left(a_{12}\right)^2 a_{31}a_{33} + a_{12}a_{32}\left(a_{11}a_{33} + a_{13}a_{31}\right)\right] \\
&+ s_{55}\left(a_{11}a_{33} + a_{13}a_{31}\right)^2 \\
&+ 2s_{56}\left[\left(a_{11}\right)^2 a_{32}a_{33} + a_{12}a_{13}\left(a_{31}\right)^2 + a_{11}a_{31}\left(a_{13}a_{32} + a_{12}a_{33}\right)\right] \\
&+ s_{66}\left(a_{11}a_{32} + a_{12}a_{31}\right)^2,
\end{aligned}
\tag{9.32}
$$

$$
\begin{aligned}
s'_{56} =\ & s_{11}\left(a_{11}\right)^2 a_{21}a_{31} + s_{12}a_{11}a_{12}\left(a_{21}a_{32} + a_{22}a_{31}\right) \\
&+ s_{13}a_{11}a_{13}\left(a_{21}a_{33} + a_{23}a_{31}\right) \\
&+ s_{14}a_{11}\left[a_{12}\left(a_{21}a_{33} + a_{23}a_{31}\right) + a_{13}\left(a_{21}a_{32} + a_{22}a_{31}\right)\right] \\
&+ s_{15}a_{11}\left[a_{11}\left(a_{23}a_{31} + a_{21}a_{33}\right) + 2a_{13}a_{21}a_{31}\right] \\
&+ s_{16}a_{11}\left[a_{11}\left(a_{21}a_{32} + a_{22}a_{31}\right) + 2a_{12}a_{21}a_{31}\right] + s_{22}\left(a_{12}\right)^2 a_{22}a_{32} \\
&+ s_{23}a_{12}a_{13}\left(a_{22}a_{33} + a_{23}a_{32}\right) + s_{24}a_{12}\left[a_{12}\left(a_{22}a_{33} + a_{23}a_{32}\right) + 2a_{13}a_{22}a_{32}\right] \\
&+ s_{25}a_{12}\left[a_{11}\left(a_{23}a_{32} + a_{22}a_{33}\right) + a_{13}\left(a_{21}a_{32} + a_{22}a_{31}\right)\right] \\
&+ s_{26}a_{12}\left[a_{11}\left(a_{22}a_{32} + a_{22}a_{32}\right) + a_{12}\left(a_{21}a_{32} + a_{22}a_{31}\right)\right] + s_{33}\left(a_{13}\right)^2 a_{23}a_{33} \\
&+ s_{34}a_{13}\left[a_{12}\left(a_{23}a_{33} + a_{23}a_{33}\right) + a_{13}\left(a_{22}a_{33} + a_{23}a_{32}\right)\right] \\
&+ s_{35}a_{13}\left[a_{13}\left(a_{21}a_{33} + a_{23}a_{31}\right) + 2a_{11}a_{23}a_{33}\right] \\
&+ s_{36}a_{13}\left[a_{11}\left(a_{22}a_{33} + a_{23}a_{32}\right) + a_{12}\left(a_{21}a_{33} + a_{23}a_{31}\right)\right] \\
&+ s_{44}\left[\left(a_{12}\right)^2 a_{23}a_{33} + \left(a_{13}\right)^2 a_{22}a_{32} + a_{12}a_{13}\left(a_{22}a_{33} + a_{23}a_{32}\right)\right] \\
&+ s_{45}\left[\left(a_{13}\right)^2 \left(a_{21}a_{32} + a_{22}a_{31}\right) + a_{11}a_{13}\left(a_{22}a_{33} + a_{23}a_{32}\right)\right. \\
&\qquad\left. + a_{12}a_{13}\left(a_{21}a_{33} + a_{23}a_{31}\right) + 2a_{11}a_{12}a_{23}a_{33}\right] \\
&+ s_{46}\left[\left(a_{12}\right)^2 \left(a_{21}a_{33} + a_{23}a_{31}\right) + a_{11}a_{12}\left(a_{22}a_{33} + a_{23}a_{32}\right)\right. \\
&\qquad\left. + a_{12}a_{13}\left(a_{21}a_{32} + a_{22}a_{31}\right) + 2a_{11}a_{13}a_{22}a_{32}\right] \\
&+ s_{55}\left[\left(a_{11}\right)^2 a_{23}a_{33} + \left(a_{13}\right)^2 a_{21}a_{31} + a_{11}a_{13}\left(a_{21}a_{33} + a_{23}a_{31}\right)\right] \\
&+ s_{56}\left[\left(a_{11}\right)^2 \left(a_{22}a_{33} + a_{23}a_{32}\right) + a_{11}a_{12}\left(a_{21}a_{33} + a_{23}a_{31}\right)\right. \\
&\qquad\left. + a_{11}a_{13}\left(a_{22}a_{31} + a_{21}a_{32}\right) + 2a_{12}a_{13}a_{21}a_{31}\right] \\
&+ s_{66}\left[\left(a_{11}\right)^2 a_{22}a_{32} + \left(a_{12}\right)^2 a_{21}a_{31} + a_{11}a_{12}\left(a_{21}a_{32} + a_{22}a_{31}\right)\right],
\end{aligned}
\tag{9.33}
$$

$$
\begin{aligned}
s'_{66} =\ & s_{11}\left(a_{11}a_{21}\right)^2 + 2s_{12}a_{11}a_{12}a_{21}a_{22} + 2s_{13}a_{11}a_{13}a_{21}a_{23} \\
&+ 2s_{14}a_{11}a_{21}\left(a_{12}a_{23} + a_{13}a_{22}\right) + 2s_{15}a_{11}a_{21}\left(a_{11}a_{23} + a_{13}a_{21}\right)
\end{aligned}
$$

$$
\begin{aligned}
&+ 2s_{16}a_{11}a_{21}\left(a_{11}a_{22} + a_{12}a_{21}\right) + s_{22}\left(a_{12}a_{22}\right)^2 \\
&+ 2s_{23}a_{12}a_{13}a_{22}a_{23} + 2s_{24}a_{12}a_{22}\left(a_{12}a_{23} + a_{13}a_{22}\right) \\
&+ 2s_{25}a_{12}a_{22}\left(a_{11}a_{23} + a_{13}a_{21}\right) + 2s_{26}a_{12}a_{22}\left(a_{11}a_{22} + a_{12}a_{21}\right) \\
&+ s_{33}\left(a_{13}a_{23}\right)^2 + 2s_{34}a_{13}a_{23}\left(a_{12}a_{23} + a_{13}a_{22}\right) \\
&+ 2s_{35}a_{13}a_{23}\left(a_{11}a_{23} + a_{13}a_{21}\right) + 2s_{36}a_{13}a_{23}\left(a_{11}a_{22} + a_{12}a_{21}\right) \\
&+ s_{44}\left(a_{12}a_{23} + a_{13}a_{22}\right)^2 \\
&+ 2s_{45}\left[a_{11}a_{12}\left(a_{23}\right)^2 + \left(a_{13}\right)^2 a_{21}a_{22} + a_{13}a_{23}\left(a_{11}a_{22} + a_{12}a_{21}\right)\right] \\
&+ 2s_{46}\left[\left(a_{12}\right)^2 a_{21}a_{23} + a_{11}a_{13}\left(a_{22}\right)^2 + a_{12}a_{22}\left(a_{11}a_{23} + a_{13}a_{21}\right)\right] \\
&+ s_{55}\left(a_{11}a_{23} + a_{13}a_{21}\right)^2 \\
&+ 2s_{56}\left[\left(a_{11}\right)^2 a_{22}a_{23} + a_{12}a_{13}\left(a_{21}\right)^2 + a_{11}a_{21}\left(a_{12}a_{23} + a_{13}a_{22}\right)\right] \\
&+ s_{66}\left(a_{11}a_{22} + a_{12}a_{21}\right)^2 .
\end{aligned}
\tag{9.34}
$$

9.2 Coefficien c

The transformations concerning the subscript $q = p,e,m$ (see Eq. (9.13)) are also valid for the coefficien c_i ($i = 1, \ldots, 18$) (see Eqs. (9.35)–(9.159)), i.e. $c_i \rightarrow c_{iq}$ ($i = 1, \ldots, 18$), and accordingly $c_j \rightarrow c_{jq}$ ($j = 19, \ldots, 8143$).

The coefficien c_i ($i = 1, \ldots, 8143$) is derived as

$$
c_i = -\frac{s'_{46}\left(s'_{i2} - s'_{i3}\right) - s'_{i4}\left(s'_{26} - s'_{36}\right)}{s'_{45}\left(s'_{36} - s'_{26}\right) + s'_{46}\left(s'_{25} - s'_{35}\right)}, \quad i = 1-3,
\tag{9.35}
$$

$$
c_{3+i} = \frac{s'_{45}\left(s'_{i2} - s'_{i3}\right) - s'_{i4}\left(s'_{25} - s'_{35}\right)}{s'_{45}\left(s'_{36} - s'_{26}\right) + s'_{46}\left(s'_{25} - s'_{35}\right)}, \quad i = 1-3,
\tag{9.36}
$$

$$
c_{6+i} = s'_{1i} + c_i s'_{15} + c_{3+i} s'_{16}, \quad i = 1-3,
\tag{9.37}
$$

$$
c_{9+i} = s'_{2i} + c_i s'_{25} + c_{3+i} s'_{26}, \quad i = 1-3,
\tag{9.38}
$$

$$
c_{12+i} = s'_{5i} + c_i s'_{55} + c_{3+i} s'_{56}, \quad i = 1-3,
\tag{9.39}
$$

$$
c_{15+i} = s'_{6i} + c_i s'_{56} + c_{3+i} s'_{66}, \quad i = 1-3,
\tag{9.40}
$$

$$
\begin{aligned}
c_{18+i} &= \frac{1}{c_7 c_{12} - c_9 c_{10}} \times \\
&\left[2c_{3+i}\left(c_8 c_{12} - c_{11} c_9\right) - c_{12}\frac{\partial c_{6+i}}{\partial \varphi} + c_9\frac{\partial c_{9+i}}{\partial \varphi} + c_{15+i}\left(c_{12} - c_9\right)\right], \\
i &= 1-3,
\end{aligned}
\tag{9.41}
$$

$$
c_{21+i} = \frac{c_{3+i}\left(c_8 c_{12} - c_{11} c_9\right) + c_{15+i} c_{12}}{c_7 c_{12} - c_9 c_{10}}, \quad i = 1-3,
\tag{9.42}
$$

$$c_{24+i} = \frac{1}{c_7 c_{12} - c_9 c_{10}} \times$$
$$\left[2c_{3+i} \left(c_7 c_{11} - c_8 c_{10} \right) + c_{10} \frac{\partial c_{6+i}}{\partial \varphi} - c_7 \frac{\partial c_{9+i}}{\partial \varphi} + c_{15+i} \left(c_7 - c_{10} \right) \right],$$
$$i = 1 - 3, \tag{9.43}$$

$$c_{27+i} = \frac{c_{3+i} \left(c_7 c_{11} - c_8 c_{10} \right) - c_{15+i} c_{10}}{c_7 c_{12} - c_9 c_{10}}, \quad i = 1 - 3, \tag{9.44}$$

$$c_{30+i} = \frac{1}{c_7 c_{12} - c_9 c_{10}} \times$$
$$\left[2c_i \left(c_9 c_{11} - c_8 c_{12} \right) - c_{11} \frac{\partial c_{6+i}}{\partial \nu} + c_8 \frac{\partial c_{9+i}}{\partial \nu} + c_{12+i} \left(c_{11} - c_8 \right) \right],$$
$$i = 1 - 3, \tag{9.45}$$

$$c_{33+i} = \frac{c_i \left(c_9 c_{11} - c_8 c_{12} \right) + c_{12+i} c_{11}}{c_7 c_{12} - c_9 c_{10}}, \quad i = 1 - 3, \tag{9.46}$$

$$c_{36+i} = \frac{1}{c_7 c_{12} - c_9 c_{10}} \times$$
$$\left[2c_i \left(c_7 c_{12} - c_9 c_{10} \right) + c_{10} \frac{\partial c_{6+i}}{\partial \nu} - c_7 \frac{\partial c_{9+i}}{\partial \nu} + c_{12+i} \left(c_7 - c_{10} \right) \right],$$
$$i = 1 - 3, \tag{9.47}$$

$$c_{39+i} = \frac{c_i \left(c_7 c_{12} - c_9 c_{10} \right) - c_{12+i} c_{10}}{c_7 c_{12} - c_9 c_{10}}, \quad i = 1 - 3, \tag{9.48}$$

$$c_{42+i+3j} = c_4 c_{18+i+3j} + c_6 c_{24+i+3j} + c_1 c_{30+i+3j} + c_2 c_{36+i+3j}$$
$$- \left(2\delta_{0j} + \delta_{1j} \right) \left(c_5 c_{3+i} + c_3 c_i - \delta_{1i} \right) + \delta_{0j} \left[\frac{\partial c_{3+i}}{\partial \varphi} + \frac{\partial c_i}{\partial \nu} - \left(\delta_{2i} + \delta_{3i} \right) \right],$$
$$i = 1 - 3; \quad j = 0, 1, \tag{9.49}$$

$$c_{48+i} = -2c_{18} c_i + \frac{\partial c_{15+i}}{\partial \nu} + c_{16} c_{30+i} + c_{17} c_{36+i}$$
$$- \frac{\partial}{\partial \varphi} \left(-2c_9 c_i \frac{\partial c_{6+i}}{\partial \nu} + c_1 c_{30+i} + c_8 c_{36+i} \right)$$
$$- \sum_{j=1}^{3} \left(-2c_9 c_j + \frac{\partial c_{6+j}}{\partial \nu} + c_7 c_{30+j} + c_8 c_{36+j} \right) \left(-2c_{3+i} \delta_{2j} + c_{18+i} \delta_{1j} + c_{24+i} \delta_{3j} \right),$$
$$i = 1 - 3, \tag{9.50}$$

$$c_{51+i} = -4c_{18} c_i \frac{\partial c_{15+i}}{\partial \nu} + c_{16} \left(c_{30+i} + 2c_{33+i} \right) + c_{17} \left(c_{36+i} + 2c_{39+i} \right)$$

$$- \frac{\partial}{\partial \varphi} \left(-c_9 c_i c_7 c_{33+i} + c_8 c_{39+i} \right)$$

$$- \sum_{j=1}^{3} \left(\frac{\partial c_{6+j}}{\partial \nu} + -2c_9 c_j + c_7 c_{30+j} + c_8 c_{36+j} \right) \left(c_{21+i} \delta_{1j} - c_{3+i} \delta_{2j} + c_{27+i} \delta_{3j} \right)$$

$$- \sum_{j=1}^{3} \left[\left(c_{18+i} + c_{21+i} \right) \delta_{1j} - 3 c_{3+i} \delta_{2j} + \left(c_{24+i} + c_{27+i} \right) \delta_{3j} \right]$$

$$\times \left(-c_9 c_j + c_7 c_{33+j} + c_8 c_{39+j} \right), \quad i = 1 - 3, \tag{9.51}$$

$$c_{54+i} = -c_{18} c_i + c_{16} c_{33+i} + c_{17} c_{39+i}$$

$$- \sum_{j=1}^{3} \left(-c_9 c_j + c_7 c_{33+j} + c_8 c_{39+j} \right) \left(c_{21+i} \delta_{1j} - c_{3+i} \delta_{2j} + c_{27+i} \delta_{3j} \right),$$

$$i = 1 - 3, \tag{9.52}$$

$$c_{57+i} = -2c_{14} c_{3+i} + \frac{\partial c_{12+i}}{\partial \varphi} + c_{13} c_{18+i} + c_{15} c_{24+i}$$

$$- \frac{\partial}{\partial \nu} \left(-2c_8 c_{3+i} + \frac{\partial c_{6+i}}{\partial \varphi} + c_7 c_{18+i} + c_9 c_{24+i} \right)$$

$$- \sum_{j=1}^{3} \left(-2c_8 c_{3+j} + \frac{\partial c_{6+j}}{\partial \varphi} + c_7 c_{18+j} + c_9 c_{24+j} \right) \left(c_{30+i} \delta_{1j} + c_{36+i} \delta_{2j} - 2c_i \delta_{3j} \right),$$

$$i = 1 - 3, \tag{9.53}$$

$$c_{60+i} = -4c_{14} c_{3+i} + \frac{\partial c_{12+i}}{\partial \varphi} + c_{13} \left(c_{18+i} + 2c_{21+i} \right) + c_{15} \left(c_{24+i} + 2c_{27+i} \right)$$

$$- \frac{\partial}{\partial \nu} \left(-c_8 c_{3+i} + c_7 c_{21+i} + c_9 c_{27+i} \right)$$

$$- \sum_{j=1}^{3} \left(-2c_8 c_{3+j} + \frac{\partial c_{6+j}}{\partial \varphi} + c_7 c_{18+j} + c_9 c_{24+j} \right) \left(c_{33+i} \delta_{1j} + c_{39+i} \delta_{2j} - c_i \delta_{3j} \right)$$

$$- \sum_{j=1}^{3} \left[\left(c_{30+i} + c_{33+i} \right) \delta_{1j} + \left(c_{36+i} + c_{39+i} \right) \delta_{2j} - 3 c_i \delta_{3j} \right]$$

$$\times \left(-c_8 c_{3+j} + c_7 c_{21+j} + c_9 c_{27+j} \right), \quad i = 1 - 3, \tag{9.54}$$

$$c_{63+i} = -c_{14} c_{3+i} + c_{13} c_{21+i} + c_{15} c_{27+i}$$

$$- \sum_{j=1}^{3} \left(-c_8 c_{3+j} + c_7 c_{21+j} + c_9 c_{27+j} \right) \left(c_{33+i} \delta_{1j} + c_{39+i} \delta_{2j} - c_i \delta_{3j} \right),$$

$$i = 1 - 3, \tag{9.55}$$

$$c_{66+i} = -2c_{18}c_i + \frac{\partial c_{15+i}}{\partial \nu} + c_{16}c_{30+i} + c_{17}c_{36+i}$$
$$- \frac{\partial}{\partial \varphi}\left(-2c_{12}c_i + \frac{\partial c_{9+i}}{\partial \nu} + c_{10}c_{30+i} + c_{11}c_{36+i}\right)$$
$$- \sum_{j=1}^{3}\left(-2c_{12}c_j + \frac{\partial c_{9+j}}{\partial \nu} + +c_{10}c_{30+j} + c_{11}c_{36+j}\right)$$
$$\times \left(c_{18+i}\delta_{1j} - 2c_{3+i}\delta_{2j} + c_{24+i}\delta_{3j}\right), \quad i = 1-3, \tag{9.56}$$

$$c_{69+i} = -c_{18}c_i + c_{16}c_{33+i} + c_{17}c_{39+i} - \frac{\partial}{\partial \varphi}\left(-c_{12}c_i + c_{10}c_{33+i} + c_{11}c_{39+i}\right)$$
$$- \sum_{j=1}^{3}\left(-2c_{12}c_j + \frac{\partial c_{9+j}}{\partial \nu} + c_{10}c_{30+j} + c_{11}c_{36+j}\right)\left(c_{21+i}\delta_{1j} - c_{3+i}\delta_{2j} + c_{27+i}\delta_{3j}\right)$$
$$- \sum_{j=1}^{3}\left[\left(c_{18+i} + c_{21+i}\right)\delta_{1j} - 3c_{3+i}\delta_{2j} + \left(c_{24+i} + c_{27+i}\right)\delta_{3j}\right]$$
$$\times \left(-c_{12}c_j + c_{10}c_{33+j} + c_{11}c_{39+j}\right), \quad i = 1-3, \tag{9.57}$$

$$c_{72+i} = -\sum_{j=1}^{3}\left(-c_{12}c_j + c_{10}c_{33+j} + c_{11}c_{39+j}\right)\left(c_{21+i}\delta_{1j} - c_{3+i}\delta_{2j} + c_{27+i}\delta_{3j}\right),$$
$$i = 1-3, \tag{9.58}$$

$$c_{75+i} = -2c_{14}c_{3+i} + \frac{\partial c_{12+i}}{\partial \varphi} + c_{13}c_{18+i} + c_{15}c_{24+i}$$
$$- \frac{\partial}{\partial \nu}\left(-2c_{11}c_{3+i} + \frac{\partial c_{9+i}}{\partial \varphi} + c_{10}c_{18+i} + c_{12}c_{24+i}\right)$$
$$- \sum_{j=1}^{3}\left(-2c_{11}c_{3+j} + \frac{\partial c_{9+j}}{\partial \varphi} + c_{10}c_{18+j} + c_{12}c_{24+j}\right)$$
$$\times \left(c_{30+i}\delta_{1j} + c_{36+i}\delta_{2j} - 2c_i\delta_{3j}\right), \quad i = 1-3, \tag{9.59}$$

$$c_{78+i} = -c_{14}c_{3+i} + c_{13}c_{21+i} + c_{15}c_{27+i} - \frac{\partial}{\partial \nu}\left(-c_{11}c_{3+i} + c_{10}c_{21+i} + c_{12}c_{27+i}\right)$$
$$- \sum_{j=1}^{3}\left(-2c_{11}c_{3+j} + \frac{\partial c_{9+j}}{\partial \varphi} + c_{10}c_{18+j} + c_{12}c_{24+j}\right)\left(c_{33+i}\delta_{1j} + c_{39+i}\delta_{2j} - c_i\delta_{3j}\right)$$
$$- \sum_{j=1}^{3}\left[\left(c_{30+i} + c_{33+i}\right)\delta_{1j} + \left(c_{36+i} + c_{39+i}\right)\delta_{2j} - 3c_i\delta_{3j}\right]$$
$$\times \left(-c_{11}c_{3+j} + c_{10}c_{21+j} + c_{12}c_{27+j}\right), \quad i = 1-3, \tag{9.60}$$

$$c_{81+i} = -\sum_{j=1}^{3} \left(-c_{11}c_{3+j} + c_{10}c_{21+j} + c_{12}c_{27+j}\right)\left(c_{33+i}\delta_{1j} + c_{39+i}\delta_{2j} - c_i\delta_{3j}\right),$$
$$i = 1 - 3, \tag{9.61}$$

$$c_{85+j} = \frac{c_{45-2j}\left(c_{7+j} - c_{10+j}\right) - c_{43+j}\left(c_{9-2j} - c_{12-2j}\right)}{c_{44}\left(c_9 - c_{12}\right) - c_{45}\left(c_8 - c_{11}\right)}, \quad j = 0, 1, \tag{9.62}$$

$$c_{86+i+3j} = \frac{c_{45+i}\left(c_{9-j} - c_{12-j}\right) + c_{9+i}c_{45-j}}{c_{44}\left(c_9 - c_{12}\right) - c_{45}\left(c_8 - c_{11}\right)}, \quad i = 1 - 3; \quad j = 0, 1, \tag{9.63}$$

$$c_{93+i+2j} = -c_{62+9j}c_{88+i} + c_{69+9j}c_{91+i} + c_{65+i+9j}, \quad i, j = 0, 1, \tag{9.64}$$

$$c_{97+i+2j} = c_{61+9j} + c_{62+9j}c_{4+2i}\left(\delta_{0i} - \delta_{1i}\right) + c_{69+9j}c_{5-2i}, \quad i, j = 0, 1, \tag{9.65}$$

$$c_{101+i+2j} = \frac{c_{93+i}c_{99+j} - c_{95+i}c_{97+j}}{c_{94}c_{95} - c_{93}c_{96}}, \quad i, j = 0, 1, \tag{9.66}$$

$$c_{104+i+3j} = \frac{c_{81+i-9j}c_{93+3j} - c_{72+i+9j}c_{95-j}}{c_{94}c_{95} - c_{93}c_{96}}, \quad i = 1 - 3; \quad j = 0, 1, \tag{9.67}$$

$$c_{111+i} = c_{49+9i} + c_{50+9i}c_{85} + c_{51+9i}c_{86} - c_{47+9i}c_{102} + c_{54+9i}c_{101}, \quad i = 0, 1, \tag{9.68}$$

$$c_{112+i+3j} = \delta_{1i}\left(c_{46+9j} - c_{47+9j}c_{104} + c_{54+9j}c_{103}\right) - c_{50+9j}c_{2+i} + c_{51+9j}c_{89+i},$$
$$i = 1 - 3; \quad j = 0, 1, \tag{9.69}$$

$$c_{118+i+3j} = c_{47+9j}c_{107+i} + c_{54+9j}c_{104+i} + c_{54+i+9j}, \quad i = 1 - 3; \quad j = 0, 1, \tag{9.70}$$

$$c_{125+i+2j} = c_{101+2j}c_{115+3i} - c_{102+2j}c_{114+3i} + \delta_{0j}c_{111+i} + \delta_{1j}c_{113+3i}, \quad i, j = 0, 1, \tag{9.71}$$

$$c_{128+i+3j} = c_{104+i}c_{115+3j} + c_{107+i}c_{114+3j} + c_{118+i+3j}, \quad i = 1 - 3; \quad j = 0, 1, \tag{9.72}$$

$$c_{135+i} = c_{4+i}\left(\delta_{1i} - \delta_{0i}\right) + c_{1+6i}c_{141} + c_{2+6i}c_{142}, \quad i = 0, 1, \tag{9.73}$$

$$c_{137+i} = c_{6-3i} + c_{1+6i}c_{143} + c_{2+6i}c_{144}, \quad i = 0, 1, \tag{9.74}$$

$$c_{139+i} = c_{1+6i}c_{145} + c_{2+i}c_{146}, \quad i = 0, 1, \tag{9.75}$$

$$c_{141+i+2j} = c_{102-i+2j}\left(\delta_{1i} - \delta_{0i}\right) + c_{109-3i}c_{147+2j} - c_{110-3i}c_{148+2j}, \quad i, j = 0, 1, \tag{9.76}$$

$$c_{145+i} = c_{108-3i} + c_{109-3i}c_{151} - c_{110-3i}c_{152}, \quad i = 0, 1, \tag{9.77}$$

$$c_{147+i} = \frac{c_{126}c_{131-i} - c_{125}c_{134-i}}{c_{130}c_{134} - c_{131}c_{133}}, \quad i = 0, 1, \tag{9.78}$$

$$c_{149+i} = \frac{c_{128}c_{131-i} - c_{127}c_{134-i}}{c_{130}c_{134} - c_{131}c_{133}}, \quad i = 0, 1, \tag{9.79}$$

$$c_{151+i} = \frac{c_{132}c_{131-i} - c_{129}c_{134-i}}{c_{130}c_{134} - c_{131}c_{133}}, \quad i = 0, 1, \tag{9.80}$$

$$c_{153+i} = c_{1+3i} - c_{2+3i}c_{135} + c_{3+3i}c_{136}, \quad i = 0, 1, \tag{9.81}$$

$$c_{155+i+2j} = -c_{2+3i}c_{137+2j} + c_{3+3i}c_{138+2j}, \quad i, j = 0, 1, \tag{9.82}$$

$$c_{158+i+3j} = (2c_{9+i} - c_{6+i})\,\delta_{0j} + (c_{42+i} + c_{45+i})\,\delta_{1j}, \quad i = 1-3; \quad j = 0,1, \quad (9.83)$$

$$c_{164+i+18j} = \frac{\partial\,(c_{9+i} - c_{6+i})}{\partial\varphi}\,\delta_{0j} + \frac{\partial\,(c_{9+i} - c_{6+i})}{\partial\nu}\,\delta_{1j} - 3\,(c_{11+j} - c_{8+j})\,c_{i+3j}$$
$$+\;(c_{10} - c_7)\,c_{18+i+12j} + (c_{12-j} - c_{9-j})\,c_{24+i+12j},$$
$$i = 1-3; \quad j = 0,1, \qquad\qquad\qquad\qquad (9.84)$$

$$c_{167+i+18j} = \frac{\partial c_{9+i}}{\partial\varphi}\,\delta_{0j} + \frac{\partial c_{9+i}}{\partial\nu}\,\delta_{1j} - (5c_{11+j} - c_{8+j})\,c_{i+3j} + c_{10}c_{18+i+12j}$$
$$+\;(2c_{10} - c_7)\,c_{21+i+12j} + c_{12-j}c_{24+i+12j} + (2c_{12-j} - c_{9-j})\,c_{27+i+12j},$$
$$i = 1-3; \quad j = 0,1, \qquad\qquad\qquad\qquad (9.85)$$

$$c_{170+i+18j} = -c_{11+j}c_{i+3j} + c_{10}c_{21+i+12j} + c_{12-j}c_{27+i+12j}, \quad i = 1-3; \quad j = 0,1, \quad (9.86)$$

$$c_{173+i+18j} = \frac{\partial c_{42+i}}{\partial\varphi}\,\delta_{0j} + \frac{\partial c_{42+i}}{\partial\nu}\,\delta_{1j} - 3c_{44+j}c_{i+3j} + c_{43}c_{18+i+12j}$$
$$+c_{45-j}c_{24+i+12j}, \quad i = 1-3; \quad j = 0,1, \qquad\qquad (9.87)$$

$$c_{176+i+18j} = \frac{\partial c_{45+i}}{\partial\varphi}\,\delta_{0j} + \frac{\partial c_{45+i}}{\partial\nu}\,\delta_{1j} - (c_{44+j} + 4c_{47+j})\,c_{i+3j} + c_{46}c_{21+i+12j}$$
$$+\;(c_{43} + c_{46})\,c_{21+i+12j} + c_{48-j}c_{24+i+12j} + (c_{45-j} + c_{48-j})\,c_{27+i+12j},$$
$$i = 1-3; \quad j = 0,1, \qquad\qquad\qquad\qquad (9.88)$$

$$c_{179+i+18j} = -c_{47+j}c_{i+3j} + c_{46}c_{21+i+12j} + c_{48-j}c_{27+i+12j}, \quad i = 1-3; \quad j = 0,1, \quad (9.89)$$

$$c_{200+i+3j} = c_{159+3j}\delta_{2i} + c_{10+36j}\delta_{3i} + c_{160+3j}c_{139+2i} + c_{161+3j}c_{140+2i}$$
$$+\;c_{11+36j}c_{145+2i} - c_{12+36j}c_{146+2i}, \quad i = 1-3; \quad j = 0,1, \qquad (9.90)$$

$$c_{206+i+3j} = c_{165+9j}\delta_{1i} + c_{168+9j}\delta_{2i} + c_{171+9j}\delta_{3i} - c_{166+9j}c_{133+2i} + c_{167+9j}c_{134+2i}$$
$$+\;c_{169+9j}c_{139+2i} + c_{170+9j}c_{140+2i} + c_{172+9j}c_{145+2i} - c_{173+9j}c_{146+2i},$$
$$i = 1-3; \quad j = 0-3, \qquad\qquad\qquad\qquad (9.91)$$

$$c_{218+i+9j} = \sum_{k=0}^{1} c_{158+i+3j}\delta_{jk} + \sum_{k=2}^{5} \left[c_{164+i+9(j-2)} + c_{167+i+9(j-2)} \right] \delta_{jk}$$
$$+\sum_{k=6}^{9} \left[c_{48+i+9(j-6)} + c_{51+i+9(j-6)} \right] \delta_{jk}, \quad i = 1-3; \quad j = 0-9, \qquad (9.92)$$

$$c_{221+i+9j} = \sum_{k=0}^{1} \left[c_{158+i+3j} + 2c_{9+i+36j} \right] \delta_{jk}$$

$$+ \sum_{k=2}^{5} \left[c_{167+i+9(j-2)} + 2c_{170+i+9(j-2)} \right] \delta_{jk}$$

$$+ \sum_{k=6}^{9} \left[c_{51+i+9(j-6)} + 2c_{54+i+9(j-6)} \right] \delta_{jk}, \quad i = 1 - 3; \quad j = 0 - 9, \quad (9.93)$$

$$c_{224+i+9j} = \sum_{k=0}^{1} c_{9+i+36j} \delta_{jk} + \sum_{k=2}^{5} c_{170+i+9(j-2)} \delta_{jk} + \sum_{k=6}^{9} c_{54+i+9(j-6)} \delta_{jk},$$
$$i = 1 - 3; \quad j = 0 - 9, \quad (9.94)$$

$$c_{308+i+9(j+2k)} = \frac{\partial c_{158+i+3j}}{\partial \varphi} \delta_{0k} + \frac{\partial c_{158+i+3j}}{\partial \nu} \delta_{1k} - 4c_{160+3j+k} c_{i+3k}$$
$$+ c_{159+3j} \left(c_{18+i+12k} + c_{21+i+12k} \right) + c_{161+3j-k} \left(c_{24+i+12k} + c_{27+i+12k} \right),$$
$$i = 1 - 3; \quad j, k = 0, 1, \quad (9.95)$$

$$c_{311+i+9(j+2k)} = \frac{\partial c_{9+i+36j}}{\partial \varphi} \delta_{0k} + \frac{\partial c_{9+i+36j}}{\partial \nu} \delta_{1k} - \left(c_{160+3j+k} + 5c_{11+36j+k} \right) c_{i+3k}$$
$$+ c_{10+36j} c_{18+i+12k} + \left(c_{159+3j} + 2c_{10+36j} \right) c_{21+i+12k} + c_{12+36j-k} c_{24+i+12k}$$
$$+ \left(c_{161+3j-k} + 2c_{12+36j-k} \right) c_{27+i+12k}, \quad i = 1 - 3; \quad j, k = 0, 1, \quad (9.96)$$

$$c_{314+i+9(j+2k)} = -c_{11+36j+k} c_{i+3k} + c_{10+36j} c_{21+i+12k} + c_{12+36j-k} c_{27+i+12k},$$
$$i = 1 - 3; \quad j, k = 0, 1, \quad (9.97)$$

$$c_{344+i+12(j+4k)} = \frac{\partial c_{164+i+9j}}{\partial \varphi} \delta_{0k} + \frac{\partial c_{164+i+9j}}{\partial \nu} \delta_{1k} - 3c_{166+9j+k} c_{i+3k}$$
$$+ c_{165+9j} c_{18+i+12k} + c_{167+9j-k} c_{24+i+12k},$$
$$i = 1 - 3; \quad j = 0 - 3; \quad k = 0, 1, \quad (9.98)$$

$$c_{347+i+12(j+4k)} = \frac{\partial c_{167+i+9j}}{\partial \varphi} \delta_{0k} + \frac{\partial c_{167+i+9j}}{\partial \nu} \delta_{1k}$$
$$- \left(c_{166+9j+k} + 4c_{169+9j+k} \right) c_{i+3k} + c_{168+9j} c_{18+i+12k}$$
$$+ \left(c_{165+9j} + c_{186+9j} \right) c_{21+i+12k} + c_{170+9j-k} c_{24+i+12k}$$
$$+ \left(c_{167+9j-k} + c_{170+9j-k} \right) c_{27+i+12k},$$
$$i = 1 - 3; \quad j = 0 - 3; \quad k = 0, 1, \quad (9.99)$$

$$c_{350+i+12(j+4k)} = \frac{\partial c_{170+i+9j}}{\partial \varphi} \delta_{0k} + \frac{\partial c_{170+i+9j}}{\partial \nu} \delta_{1k}$$
$$- \left(c_{169+9j+k} + 5c_{172+9j+k}\right) c_{i+3k} + c_{171+9j} c_{18+i+12k}$$
$$+ \left(c_{168+9j} + 2c_{171+9j}\right) c_{21+i+12k} + c_{173+9j-k} c_{24+i+12k}$$
$$+ \left(c_{170+9j-k} + 2c_{173+9j-k}\right) c_{27+i+12k},$$
$$i = 1 - 3; \quad j = 0 - 3; \quad k = 0, 1, \tag{9.100}$$

$$c_{353+i+12(j+4k)} = -c_{172+9j+k} c_{i+3k} + c_{171+9j} c_{21+i+12k} + c_{173+9j-k} c_{27+i+12k},$$
$$i = 1 - 3; \quad j = 0 - 3; \quad k = 0, 1, \tag{9.101}$$

$$c_{440+i+12(j+4k)} = \frac{\partial c_{48+i+9j}}{\partial \varphi} \delta_{0k} + \frac{\partial c_{48+i+9j}}{\partial \nu} \delta_{1k} - 3c_{50+9j+k} c_{i+3k}$$
$$+ c_{49+9j} c_{18+i+12k} + c_{51+9j-k} c_{24+i+12k},$$
$$i = 1 - 3; \quad j = 0 - 3; \quad k = 0, 1, \tag{9.102}$$

$$c_{443+i+12(j+4k)} = \frac{\partial c_{51+i+9j}}{\partial \varphi} \delta_{0k} + \frac{\partial c_{51+i+9j}}{\partial \nu} \delta_{1k} - \left(c_{50+9j+k} + 4c_{53+9j+k}\right) c_{i+3k}$$
$$+ c_{52+9j} c_{18+i+12k} + \left(c_{49+9j} + c_{52+9j}\right) c_{21+i+12k} + c_{54+9j-k} c_{24+i+12k}$$
$$+ \left(c_{51+9j-k} + c_{54+9j-k}\right) c_{27+i+12k}, \quad i = 1 - 3; \quad j = 0 - 3; \quad k = 0, 1, \tag{9.103}$$

$$c_{446+i+12(j+4k)} = \frac{\partial c_{54+i+9j}}{\partial \varphi} \delta_{0k} + \frac{\partial c_{54+i+9j}}{\partial \nu} \delta_{1k} - \left(c_{53+9j+k} + 5c_{56+9j+k}\right) c_{i+3k}$$
$$+ c_{55+9j} c_{18+i+12k} + \left(c_{52+9j} + 2c_{55+9j}\right) c_{21+i+12k} + c_{57+9j-k} c_{24+i+12k}$$
$$+ \left(c_{54+9j-k} + 2c_{57+9j-k}\right) c_{27+i+12k}, \quad i = 1 - 3; \quad j = 0 - 3; \quad k = 0, 1, \tag{9.104}$$

$$c_{449+i+12(j+4k)} = -c_{56+9j+k} c_{i+3k} + c_{55+9j} c_{21+i+12k} + c_{57+9j-k} c_{27+i+12k},$$
$$i = 1 - 3; \quad j = 0 - 3; \quad k = 0, 1, \tag{9.105}$$

$$c_{537+4i} = c_{220+9i} c_{141} + c_{221+9i} c_{142} + \left(c_{223+9i} - 2c_{226+9i}\right) c_{147}$$
$$- \left(c_{224+9i} - 2c_{227+9i}\right) c_{148}, \quad i = 0 - 13, \tag{9.106}$$

$$c_{538+4i} = c_{219+9i} + c_{220+9i} c_{143} + c_{221+9i} c_{144} + c_{226+9i} c_{147} - c_{227+9i} c_{148}$$
$$+ \left(c_{223+9i} - c_{226+9i}\right) c_{149} - \left(c_{224+9i} - c_{227+9i}\right) c_{150}, \quad i = 0 - 13, \tag{9.107}$$

$$c_{539+4i} = c_{222+9i} + c_{220+9i} c_{145} + c_{221+9i} c_{146} + c_{226+9i} c_{149} - c_{227+9i} c_{150}$$
$$+ c_{223+9i} c_{151} - c_{224+9i} c_{152}, \quad i = 0 - 13, \tag{9.108}$$

$$c_{540+4i} = c_{225+9i} + c_{226+9i}c_{151} - c_{227+9i}c_{152}, \quad i = 0 - 13, \tag{9.109}$$

$$c_{593+4i} = c_{345+12i} - c_{346+12i}c_{135} + c_{347+12i}c_{136} + c_{349+12i}c_{141} + c_{350+12i}c_{142}$$
$$+ \left(c_{352+12i} - 2c_{355+12i}\right) c_{147} - \left(c_{353+12i} - 2c_{356+12i}\right) c_{148}, \quad i = 0 - 15, \tag{9.110}$$

$$c_{594+4i} = c_{348+12i} - c_{346+12i}c_{137} + c_{347+12i}c_{138} + c_{349+12i}c_{143} + c_{350+12i}c_{144}$$
$$+ c_{355+12i}c_{147} - c_{356+12i}c_{148} + \left(c_{352+12i} - c_{355+12i}\right) c_{149}$$
$$- \left(c_{353+12i} - c_{356+12i}\right) c_{150}, \quad i = 0 - 15, \tag{9.111}$$

$$c_{595+4i} = c_{351+12i} - c_{346+12i}c_{139} + c_{347+12i}c_{140} + c_{349+12i}c_{145} + c_{350+12i}c_{146}$$
$$+ c_{355+12i}c_{149} - c_{356+12i}c_{150} + c_{352+12i}c_{151} - c_{353+12i}c_{152},$$
$$i = 0 - 15, \tag{9.112}$$

$$c_{596+4i} = c_{354+12i} + c_{355+12i}c_{151} - c_{356+12i}c_{152}, \quad i = 0 - 15, \tag{9.113}$$

$$c_{656+i+12j} = \sum_{k=0}^{13} c_{218+i+9j}\delta_{jk} + \sum_{k=14}^{29} \left[c_{344+i+12(j-14)} + c_{347+i+12(j-14)}\right] \delta_{jk},$$
$$i = 1 - 3; \quad j = 0 - 29, \tag{9.114}$$

$$c_{659+i+12j} = \sum_{k=0}^{13} \left(c_{218+i+9j} + 2c_{221+i+9j}\right) \delta_{jk}$$
$$+ \sum_{k=14}^{29} \left[c_{347+i+12(j-14)} + 2c_{350+i+12(j-14)}\right] \delta_{jk},$$
$$i = 1 - 3; \quad j = 0 - 29, \tag{9.115}$$

$$c_{662+i+12j} = \sum_{k=0}^{13} \left(c_{221+i+9j} + 3c_{224+i+9j}\right) \delta_{jk}$$
$$+ \sum_{k=14}^{29} \left[c_{350+i+12(j-14)} + 3c_{353+i+12(j-14)}\right] \delta_{jk},$$
$$i = 1 - 3; \quad j = 0 - 29, \tag{9.116}$$

$$c_{664+i+12j} = \sum_{k=0}^{13} c_{224+i+9j}\delta_{jk} + \sum_{k=14}^{29} c_{353+i+12(j-14)}\delta_{jk},$$
$$i = 1 - 3; \quad j = 0 - 29, \tag{9.117}$$

$$c_{998+i+12(j+14k)} = \frac{\partial c_{218+i+9j}}{\partial \varphi} \delta_{0k} + \frac{\partial c_{218+i+9j}}{\partial \nu} \delta_{1k} - 4c_{220+9j+k} c_{i+3k}$$
$$+ c_{219+9j} \left(c_{18+i+12k} + c_{21+i+12k} \right) + c_{221+9j-k} \left(c_{24+i+12k} + c_{27+i+12k} \right),$$
$$i = 1 - 3; \quad j = 0 - 13; \quad k = 0, 1, \tag{9.118}$$

$$c_{1001+i+12(j+14k)} = \frac{\partial c_{221+i+9j}}{\partial \varphi} \delta_{0k} + \frac{\partial c_{221+i+9j}}{\partial \nu} \delta_{1k}$$
$$- \left(c_{220+9j+k} + 5c_{223+9j+k} \right) c_{i+3k} + c_{222+9j} c_{18+i+12k}$$
$$+ \left(c_{219+9j} + 2c_{222+9j} \right) c_{21+i+12k} + c_{224+9j-k} c_{24+i+12k}$$
$$+ \left(c_{221+9j-k} + 2c_{224+9j-k} \right) c_{27+i+12k},$$
$$i = 1 - 3; \quad j = 0 - 13; \quad k = 0, 1, \tag{9.119}$$

$$c_{1004+i+12(j+14k)} = \frac{\partial c_{224+i+9j}}{\partial \varphi} \delta_{0k} + \frac{\partial c_{224+i+9j}}{\partial \nu} \delta_{1k}$$
$$- \left(c_{223+9j+k} + 6c_{226+9j+k} \right) c_{i+3k} + c_{225+9j} c_{18+i+12k}$$
$$+ \left(c_{222+9j} + 3c_{225+9j} \right) c_{21+i+12k} + c_{227+9j-k} c_{24+i+12k}$$
$$+ \left(c_{224+9j-k} + 3c_{227+9j-k} \right) c_{27+i+12k},$$
$$i = 1 - 3; \quad j = 0 - 13; \quad k = 0, 1, \tag{9.120}$$

$$c_{1007+i+12(j+14k)} = -c_{226+9j+k} c_{i+3k} + c_{225+9j} c_{21+i+12k} + c_{227+9j-k} c_{27+i+12k},$$
$$i = 1 - 3; \quad j = 0 - 13; \quad k = 0, 1, \tag{9.121}$$

$$c_{1352+i+15(j+16k)} = \frac{\partial c_{344+i+12j}}{\partial \varphi} \delta_{0k} + \frac{\partial c_{344+i+12j}}{\partial \nu} \delta_{1k} - 3c_{346+12j+k} c_{i+3k}$$
$$+ c_{345+12j} c_{18+i+12k} + c_{347+12j-k} c_{24+i+12k},$$
$$i = 1 - 3; \quad j = 0 - 15; \quad k = 0, 1, \tag{9.122}$$

$$c_{1355+i+15(j+16k)} = \frac{\partial c_{347+i+12j}}{\partial \varphi} \delta_{0k} + \frac{\partial c_{347+i+12j}}{\partial \nu} \delta_{1k}$$
$$- \left(c_{346+12j+k} + 4c_{349+12j+k} \right) c_{i+3k} + c_{348+12j} c_{18+i+12k}$$
$$+ \left(c_{345+12j} + c_{348+12j} \right) c_{21+i+12k} + c_{350+12j-k} c_{24+i+12k}$$
$$+ \left(c_{347+12j-k} + c_{350+12j-k} \right) c_{27+i+12k},$$
$$i = 1 - 3; \quad j = 0 - 15; \quad k = 0, 1, \tag{9.123}$$

$$c_{1358+i+15(j+16k)} = \frac{\partial c_{350+i+12j}}{\partial \varphi} \delta_{0k} + \frac{\partial c_{350+i+12j}}{\partial \nu} \delta_{1k}$$
$$- \left(c_{349+12j+k} + 5c_{352+12j+k} \right) c_{i+3k} + c_{351+12j} c_{18+i+12k}$$
$$+ \left(c_{348+12j} + 2c_{351+12j} \right) c_{21+i+12k} + c_{353+12j-k} c_{24+i+12k}$$
$$+ \left(c_{350+12j-k} + 2c_{353+12j-k} \right) c_{27+i+12k},$$
$$i = 1 - 3; \quad j = 0 - 15; \quad k = 0, 1, \tag{9.124}$$

$$c_{1361+i+15(j+16k)} = \frac{\partial c_{353+i+12j}}{\partial \varphi} \delta_{0k} + \frac{\partial c_{353+i+12j}}{\partial \nu} \delta_{1k}$$
$$- \left(c_{352+12j+k} + 6c_{355+12j+k} \right) c_{i+3k} + c_{354+12j} c_{18+i+12k}$$
$$+ \left(c_{351+12j} + 3c_{354+12j} \right) c_{21+i+12k} + c_{356+12j-k} c_{24+i+12k}$$
$$+ \left(c_{353+12j-k} + 3c_{356+12j-k} \right) c_{27+i+12k},$$
$$i = 1 - 3; \quad j = 0 - 15; \quad k = 0, 1, \tag{9.125}$$

$$c_{1364+i+15(j+16k)} = -c_{355+12j+k} c_{i+3k} + c_{354+12j} c_{21+i+12k} + c_{356+12j-k} c_{27+i+12k},$$
$$i = 1 - 3; \quad j = 0 - 15; \quad k = 0, 1, \tag{9.126}$$

$$c_{1833+5i} = c_{658+12i} c_{141} + c_{659+12i} c_{142} + \left[c_{661+12i} - 2 \left(c_{664+12i} - 3c_{667+12i} \right) \right] c_{147}$$
$$- \left[c_{662+12i} - 2 \left(c_{664+12i} - 3c_{668+12i} \right) \right] c_{148}, \quad i = 0 - 57, \tag{9.127}$$

$$c_{1834+5i} = c_{657+12i} + c_{658+12i} c_{143} + c_{659+12i} c_{144} + \left(c_{664+12i} - 4c_{667+12i} \right) c_{147}$$
$$- \left(c_{664+12i} - 4c_{668+12i} \right) c_{148} + \left(c_{661+12i} - c_{664+12i} + 2c_{667+12i} \right) c_{149}$$
$$- \left(c_{662+12i} - c_{664+12i} + 2c_{668+12i} \right) c_{150}, \quad i = 0 - 57, \tag{9.128}$$

$$c_{1835+5i} = c_{660+12i} + c_{658+12i} c_{145} + c_{659+12i} c_{146} + c_{667+12i} c_{147} - c_{668+12i} c_{148}$$
$$+ \left(c_{664+12i} - 2c_{667+12i} \right) c_{149} - \left(c_{664+12i} - 2c_{668+12i} \right) c_{150} + c_{661+12i} c_{151}$$
$$- c_{662+12i} c_{152}, \quad i = 0 - 57, \tag{9.129}$$

$$c_{1836+5i} = c_{663+12i} + c_{667+12i} c_{149} - c_{668+12i} c_{150} + c_{664+12i} c_{151} - c_{665+12i} c_{152},$$
$$i = 0 - 57, \tag{9.130}$$

$$c_{1837+5i} = c_{666+12i} + c_{667+12i} c_{151} - c_{668+12i} c_{152}, \quad i = 0 - 57, \tag{9.131}$$

$$c_{2123+5i} = c_{1353+15i} - c_{1354+15i} c_{135} + c_{1355+15i} c_{136} + c_{1357+15i} c_{141} + c_{1358+15i} c_{142}$$
$$+ \left[c_{1360+15i} - 2 \left(c_{1363+15i} - 3c_{1366+15i} \right) \right] c_{147}$$
$$- \left[c_{1361+15i} - 2 \left(c_{1364+15i} - 3c_{1367+15i} \right) \right] c_{148}, \quad i = 0 - 31, \tag{9.132}$$

$$c_{2124+5i} = c_{1356+15i} - c_{1354+15i} c_{137} + c_{1355+15i} c_{138} + c_{1357+15i} c_{143}$$
$$+ c_{1358+15i} c_{144} + \left(c_{1363+15i} - 4c_{1366+15i} \right) c_{147} - \left(c_{1364+15i} - 4c_{1367+15i} \right) c_{148}$$
$$+ \left(c_{1360+15i} - c_{1363+15i} + 2c_{1366+15i} \right) c_{149}$$
$$- \left(c_{1361+15i} - c_{1364+15i} + 2c_{1367+15i} \right) c_{150}, \quad i = 0 - 31, \tag{9.133}$$

$$c_{2125+5i} = c_{1359+15i} - c_{1354+15i}c_{139} + c_{1355+15i}c_{140} + c_{1357+15i}c_{145}$$
$$+ c_{1358+15i}c_{146} + c_{1366+15i}c_{147} - c_{1367+15i}c_{148} + \left(c_{1363+15i} - 2c_{1366+15i}\right)c_{149}$$
$$- \left(c_{1364+15i} - 2c_{1367+15i}\right)c_{150} + c_{1360+15i}c_{151} - c_{1361+15i}c_{152},$$
$$i = 0 - 31, \tag{9.134}$$

$$c_{2126+5i} = c_{1362+15i} + c_{1366+15i}c_{149} - c_{1367+15i}c_{150} + c_{1363+15i}c_{151}$$
$$- c_{1364+15i}c_{152}, \quad i = 0 - 31, \tag{9.135}$$

$$c_{2127+5i} = c_{1365+15i} + c_{1366+15i}c_{151} - c_{1367+15i}c_{152}, \quad i = 0 - 31, \tag{9.136}$$

$$c_{2282+i+15j} = \sum_{k=0}^{57} c_{656+i+12j}\delta_{jk} + \sum_{k=58}^{89} \left(c_{1352+i+15j} + c_{1355+i+15j}\right)\delta_{jk}$$
$$+ \sum_{k=90}^{147} \left[\frac{\partial c_{656+i+12j}}{\partial \varphi} + c_{657+12j}\left(c_{18+i} + c_{21+i}\right)\right.$$
$$\left. - 4c_{658+12j}c_{3+i} + c_{659+12j}\left(c_{24+i} + c_{27+i}\right)\right]\delta_{jk}$$
$$+ \sum_{k=148}^{205} \left[\frac{\partial c_{656+i+12j}}{\partial \nu} + c_{657+12j}\left(c_{30+i} + c_{33+i}\right)\right.$$
$$\left. + c_{658+12j}\left(c_{36+i} + c_{39+i}\right) - 4c_{659+12j}c_i\right]\delta_{jk},$$
$$i = 1 - 3; \quad j = 0 - 205, \tag{9.137}$$

$$c_{2285+i+15j} = \sum_{k=0}^{57} \left(c_{656+i+12j} + 2c_{659+i+12j}\right)\delta_{jk}$$
$$+ \sum_{k=58}^{89} \left(c_{1355+i+15j} + 2c_{1358+i+15j}\right)\delta_{jk}$$
$$+ \sum_{k=90}^{147} \left[\frac{\partial c_{659+i+12j}}{\partial \varphi} + c_{657+12j}c_{21+i} - c_{658+12j}c_{3+i}\right.$$
$$+ c_{659+12j}c_{27+i} + c_{660+12j}\left(c_{18+i} + 2c_{21+i}\right)$$
$$\left. - 5c_{661+12j}c_{3+i} + c_{662+12j}\left(c_{24+i} + 2c_{27+i}\right)\right]\delta_{jk}$$
$$+ \sum_{k=148}^{205} \left[\frac{\partial c_{659+i+12j}}{\partial \nu} + c_{657+12j}c_{33+i} + c_{658+12j}c_{39+i}\right.$$
$$\left. - c_{659+12j}c_i + c_{660+12j}\left(c_{30+i} + 2c_{33+i}\right)\right.$$

$$+ c_{661+12j} \left(c_{36+i} + 2c_{39+i} \right) - 5c_{662+12j}c_i \Big] \delta_{jk},$$
$$i = 1 - 3; \quad j = 0 - 205, \tag{9.138}$$

$$c_{2288+i+15j} = \sum_{k=0}^{57} \left(c_{659+i+12j} + 3c_{662+i+12j} \right) \delta_{jk}$$

$$+ \sum_{k=58}^{89} \left(c_{1358+i+15j} + 3c_{1361+i+15j} \right) \delta_{jk}$$

$$+ \sum_{k=90}^{147} \left[\frac{\partial c_{662+i+12j}}{\partial \varphi} + c_{660+12j}c_{21+i} - c_{661+12j}c_{3+i} \right.$$
$$+ c_{662+12j}c_{27+i} + c_{663+12j} \left(c_{18+i} + 3c_{21+i} \right)$$
$$\left. - 6c_{664+12j}c_{3+i} + c_{665+12j} \left(c_{24+i} + 3c_{27+i} \right) \right] \delta_{jk}$$

$$+ \sum_{k=148}^{205} \left[\frac{\partial c_{662+i+12j}}{\partial \nu} + c_{660+12j}c_{33+i} + c_{661+12j}c_{39+i} \right.$$
$$- c_{662+12j}c_i + c_{663+12j} \left(c_{30+i} + 3c_{33+i} \right)$$
$$\left. + c_{664+12j} \left(c_{36+i} + 3c_{39+i} \right) - 6c_{665+12j}c_i \right] \delta_{jk},$$
$$i = 1 - 3; \quad j = 0 - 205, \tag{9.139}$$

$$c_{2291+i+15j} = \sum_{k=0}^{57} \left(c_{662+i+12j} + 4c_{665+i+12j} \right) \delta_{jk}$$

$$+ \sum_{k=58}^{89} \left(c_{1361+i+15j} + 4c_{1364+i+15j} \right) \delta_{jk}$$

$$+ \sum_{k=90}^{147} \left[\frac{\partial c_{665+i+12j}}{\partial \varphi} + c_{663+12j}c_{21+i} - c_{664+12j}c_{3+i} \right.$$
$$+ c_{665+12j}c_{27+i} + c_{666+12j} \left(c_{18+i} + 4c_{21+i} \right)$$
$$\left. - 7c_{667+12j}c_{3+i} + c_{668+12j} \left(c_{24+i} + 4c_{27+i} \right) \right] \delta_{jk}$$

$$+ \sum_{k=148}^{205} \left[\frac{\partial c_{665+i+12j}}{\partial \nu} + c_{663+12j}c_{33+i} + c_{664+12j}c_{39+i} \right.$$
$$- c_{665+12j}c_i + c_{666+12j} \left(c_{30+i} + 4c_{33+i} \right)$$
$$\left. + c_{667+12j} \left(c_{36+i} + 4c_{39+i} \right) - 7c_{668+12j}c_i \right] \delta_{jk},$$
$$i = 1 - 3; \quad j = 0 - 205, \tag{9.140}$$

$$c_{2294+i+15j} = \sum_{k=0}^{57} c_{665+i+12j}\delta_{jk} + \sum_{k=58}^{89} c_{1364+i+15j}\delta_{jk}$$

$$+ \sum_{k=90}^{147} \left(c_{666+12j}c_{21+i} - c_{667+12j}c_{3+i} + c_{668+12j}c_{27+i} \right) \delta_{jk}$$

$$+ \sum_{k=148}^{205} \left(c_{666+12j}c_{33+i} + c_{667+12j}c_{39+i} - c_{668+12j}c_i \right) \delta_{jk},$$

$$i = 1 - 3; \quad j = 0 - 205, \tag{9.141}$$

$$c_{5372+i+18j} =$$

$$\sum_{k=0}^{31} \left(\frac{\partial c_{1352+i+15j}}{\partial \varphi} + c_{1353+15j}c_{18+i} - 3c_{1354+15j}c_{3+i} + c_{1355+15j}c_{24+i} \right) \delta_{jk}$$

$$+ \sum_{k=32}^{63} \left(\frac{\partial c_{1352+i+15j}}{\partial \nu} + c_{1353+15j}c_{30+i} + c_{1354+15j}c_{36+i} - 3c_{1355+15j}c_i \right) \delta_{jk},$$

$$i = 1 - 3; \quad j = 0 - 63, \tag{9.142}$$

$$c_{5375+i+18j} =$$

$$\sum_{k=0}^{31} \left[\frac{\partial c_{1355+i+15j}}{\partial \varphi} + c_{1353+15j}c_{21+i} - c_{1354+15j}c_{3+i} \right.$$

$$+ c_{1355+15j}c_{27+i} + c_{1356+15j} \left(c_{18+i} + c_{21+i} \right)$$

$$\left. - 4c_{1357+15j}c_{3+i} + c_{1358+15j} \left(c_{24+i} + c_{27+i} \right) \right] \delta_{jk}$$

$$+ \sum_{k=32}^{63} \left[\frac{\partial c_{1355+i+15j}}{\partial \nu} + c_{1353+15j}c_{33+i} + c_{1354+15j}c_{39+i} \right.$$

$$- c_{1355+15j}c_i + c_{1356+15j} \left(c_{30+i} + c_{33+i} \right)$$

$$\left. + c_{1357+15j} \left(c_{36+i} + c_{39+i} \right) - 4c_{1358+15j}c_i \right] \delta_{jk},$$

$$i = 1 - 3; \quad j = 0 - 63, \tag{9.143}$$

$$c_{5378+i+18j} =$$

$$\sum_{k=0}^{31} \left[\frac{\partial c_{1358+i+15j}}{\partial \varphi} + c_{1356+15j}c_{21+i} - c_{1357+15j}c_{3+i} \right.$$

$$+ c_{1358+15j}c_{27+i} + c_{1359+15j} \left(c_{18+i} + 2c_{21+i} \right)$$

$$\left. - 5c_{1360+15j}c_{3+i} + c_{1361+15j} \left(c_{24+i} + 2c_{27+i} \right) \right] \delta_{jk}$$

$$+ \sum_{k=32}^{63} \left[\frac{\partial c_{1358+i+15j}}{\partial \nu} + c_{1356+15j}c_{33+i} + c_{1357+15j}c_{39+i} \right.$$
$$- c_{1358+15j}c_i + c_{1359+15j}\left(c_{30+i} + 2c_{33+i}\right)$$
$$\left. + c_{1360+15j}\left(c_{36+i} + 2c_{39+i}\right) - 5c_{1361+15j}c_i \right] \delta_{jk},$$
$$i = 1 - 3; \quad j = 0 - 63, \tag{9.144}$$

$$c_{5381+i+18j} =$$
$$\sum_{k=0}^{31} \left[\frac{\partial c_{1361+i+15j}}{\partial \varphi} + c_{1359+15j}c_{21+i} - c_{1360+15j}c_{3+i} \right.$$
$$+ c_{1361+15j}c_{27+i} + c_{1362+15j}\left(c_{18+i} + 3c_{21+i}\right)$$
$$\left. - 6c_{1363+15j}c_{3+i} + c_{1364+15j}\left(c_{24+i} + 3c_{27+i}\right) \right] \delta_{jk}$$
$$+ \sum_{k=32}^{63} \left[\frac{\partial c_{1361+i+15j}}{\partial \nu} + c_{1359+15j}c_{33+i} + c_{1360+15j}c_{39+i} \right.$$
$$- c_{1361+15j}c_i + c_{1362+15j}\left(c_{30+i} + 3c_{33+i}\right)$$
$$\left. + c_{1363+15j}\left(c_{36+i} + 3c_{39+i}\right) - 6c_{1364+15j}c_i \right] \delta_{jk},$$
$$i = 1 - 3; \quad j = 0 - 63, \tag{9.145}$$

$$c_{5384+i+18j} =$$
$$\sum_{k=0}^{31} \left[\frac{\partial c_{1364+i+15j}}{\partial \varphi} + c_{1362+15j}c_{21+i} - c_{1363+15j}c_{3+i} \right.$$
$$+ c_{1364+15j}c_{27+i} + c_{1365+15j}\left(c_{18+i} + 4c_{21+i}\right)$$
$$\left. - 7c_{1366+15j}c_{3+i} + c_{1367+15j}\left(c_{24+i} + 4c_{27+i}\right) \right] \delta_{jk}$$
$$+ \sum_{k=32}^{63} \left[\frac{\partial c_{1364+i+15j}}{\partial \nu} + c_{1362+15j}c_{33+i} + c_{1363+15j}c_{39+i} \right.$$
$$- c_{1364+15j}c_i + c_{1365+15j}\left(c_{30+i} + 4c_{33+i}\right)$$
$$\left. + c_{1366+15j}\left(c_{36+i} + 4c_{39+i}\right) - 7c_{1367+15j}c_i \right] \delta_{jk},$$
$$i = 1 - 3; \quad j = 0 - 63, \tag{9.146}$$

$$c_{5387+i+18j} = \sum_{k=0}^{31} \left(c_{1365+15j}c_{21+i} - c_{1366+15j}c_{3+i} + c_{1367+15j}c_{27+i}\right) \delta_{jk}$$

$$+ \sum_{k=32}^{63} \left(c_{1365+15j} c_{33+i} + c_{1366+15j} c_{39+i} - c_{1367+15j} c_i \right) \delta_{jk},$$
$$i = 1 - 3; \quad j = 0 - 63, \tag{9.147}$$

$$c_{6525+6i} = \sum_{j=0}^{1} c_{2284+15i+j} c_{141+j}$$
$$+ \sum_{j=0}^{1} c_{147+j} \left\{ c_{2287+15i+j} - 2 \left[c_{2290+15i+j} - 3 \left(c_{2293+15i+j} - 4 c_{2296+15i+j} \right) \right] \right\}$$
$$\times \left(\delta_{0j} - \delta_{1j} \right), \quad i = 0 - 205, \tag{9.148}$$

$$c_{6526+6i} = c_{2283+15i} + \sum_{j=0}^{1} c_{2284+15i+j} c_{143+j}$$
$$+ \sum_{j=0}^{1} \left\{ 2 \left[c_{2293+15i+j} \left(c_{149+j} - 2 c_{147+j} \right) + 3 c_{2296+15i+j} \left(3 c_{147+j} - c_{149+j} \right) \right] \right.$$
$$\left. + c_{2287+15i+j} c_{149+j} + c_{2290+15i+j} \left(c_{147+j} - c_{149+j} \right) \right\} \left(\delta_{0j} - \delta_{1j} \right),$$
$$i = 0 - 205, \tag{9.149}$$

$$c_{6527+6i} = c_{2286+15i} + \sum_{j=0}^{1} c_{2284+15i+j} c_{145+j}$$
$$+ \sum_{j=0}^{1} \left[c_{2287+15i+j} c_{151+j} + c_{2290+15i+j} c_{149+j} + c_{2293+15i+j} \left(c_{147+j} - 2 c_{149+j} \right) \right.$$
$$\left. - 6 c_{2296+15i+j} \left(c_{147+j} - c_{149+j} \right) \right] \left(\delta_{0j} - \delta_{1j} \right), \quad i = 0 - 205, \tag{9.150}$$

$$c_{6528+6i} = c_{2289+15i}$$
$$+ \sum_{j=0}^{1} \left[c_{2290+15i+j} c_{151+j} + c_{2293+15i+j} c_{149+j} \right.$$
$$\left. + c_{2296+15i+j} \left(c_{147+j} - 3 c_{149+j} \right) \right] \left(\delta_{0j} - \delta_{1j} \right), \quad i = 0 - 205, \tag{9.151}$$

$$c_{6529+6i} = c_{2292+15i} + \sum_{j=0}^{1} \left(\delta_{0j} - \delta_{1j} \right) \left(c_{2293+15i+j} c_{151+j} + c_{2296+15i+j} c_{149+j} \right),$$
$$i = 0 - 205, \tag{9.152}$$

$$c_{6530+6i} = c_{2295+15i} + \sum_{j=0}^{1} \left(\delta_{0j} - \delta_{1j}\right) c_{2296+15i+j} c_{151+j}, \quad i = 0 - 205, \qquad (9.153)$$

$$c_{7761+6i} = c_{5373+18i} + \sum_{j=0}^{1} c_{5377+18i+j} c_{141+j}$$

$$+ \sum_{j=0}^{1} \left(\left\{c_{5380+18i+j} - 2\left[c_{5383+18i+j} - 3\left(c_{5386+18i+j} - 4c_{5389+18i+j}\right)\right]\right\} c_{147+j}\right.$$
$$\left. - c_{5374+18i+j} c_{135+j}\right) \left(\delta_{0j} - \delta_{1j}\right), \quad i = 0 - 63, \qquad (9.154)$$

$$c_{7762+6i} = c_{5376+18i} + \sum_{j=0}^{1} c_{5377+18i+j} c_{143+j}$$

$$+ \sum_{j=0}^{1} \left[c_{5380+18i+j} c_{149+j} + c_{5383+18i+j} \left(c_{147+j} - c_{149+j}\right)\right.$$
$$+ 2c_{5386+18i+j} \left(c_{149+j} - 2c_{147+j}\right) + 6c_{5389+18i+j} \left(3c_{147+j} - c_{149+j}\right)$$
$$\left. - c_{5374+18i+j} c_{137+j}\right] \left(\delta_{0j} - \delta_{1j}\right), \quad i = 0 - 63, \qquad (9.155)$$

$$c_{7763+6i} = c_{5379+18i} + \sum_{j=0}^{1} c_{5377+18i+j} c_{145+j}$$

$$+ \sum_{j=0}^{1} \left[c_{5380+18i+j} c_{151+j} + c_{5383+18i+j} c_{149+j} + c_{5386+18i+j} \left(c_{147+j} - 2c_{149+j}\right)\right.$$
$$\left. - 6c_{5389+18i+j} \left(c_{147+j} - c_{149+j}\right) - c_{5374+18i+j} c_{139+j}\right] \left(\delta_{0j} - \delta_{1j}\right),$$
$$i = 0 - 63, \qquad (9.156)$$

$$c_{7764+6i} = c_{5382+18i}$$
$$+ \sum_{j=0}^{1} \left[c_{5383+18i+j} c_{151+j} + c_{5386+18i+j} c_{149+j} + c_{5389+18i+j} \left(c_{147+j} - 3c_{149+j}\right)\right]$$
$$\times \left(\delta_{0j} - \delta_{1j}\right), \quad i = 0 - 63, \qquad (9.157)$$

$$c_{7765+6i} = c_{5385+18i} + \sum_{j=0}^{1} \left(\delta_{0j} - \delta_{1j}\right) \left(c_{5386+18i+j} c_{151+j} + c_{5389+18i+j} c_{149+j}\right),$$
$$i = 0 - 63, \qquad (9.158)$$

$$c_{7766+6i} = c_{5388+18i} + \sum_{j=0}^{1} \left(\delta_{0j} - \delta_{1j}\right) c_{5389+18i+j} c_{151+j}, \quad i = 0 - 63. \qquad (9.159)$$

Bibliography

[1] Eshelby, J.D. *Proc. Royal Soc. London A* 1957, *241*, 376–396.

[2] Li, S.; Sauer, R.A.; Wang, G. *J. Appl. Mech.* 2007, *74*, 770–783.

[3] Li, S.; Sauer, R.A.; Wang, G. *J. Appl. Mech.* 2007, *74*, 784–797.

[4] Skočovský, P.; Bokůvka, O.; Palček, P. *Materials Science*; EDIS: Žilina, SK, 1996 (in Slovak).

[5] Kuba, F. *Theory of Elasticity and Selected Applications*; SNTL/Alfa: Prague, CZ, 1982 (in Czech).

[6] Hearmon, R.F.S. *Introduction to Applied Anisotropic Elasticity*; SNTL: Prague, CZ, 1965 (in Czech); translation from English: The Clarendon Press: Oxford, UK, 1961.

[7] Brdička, M.; Samek, L.; Sopko, B. *Mechanics of Continuum*; Academia: Prague, CZ, 2000 (in Czech).

[8] Rektorys, K. *Review of Applied Mathematics*; SNTL: Prague, CZ, 1973 (in Czech).

[9] Ceniga, L. *Analytical Models of Thermal Stresses in Composite Materials I*; Nova Science Publishers, New York, US, 2008.

[10] Ceniga, L. *Analytical Models of Thermal Stresses in Composite Materials II*; Nova Science Publishers, New York, US, 2007.

[11] Hajko, V.; Potocký, L.; Zentko, A. *Magnetization Processes*; Alfa: Bratislava, SK, 1982.

[12] Diko, P. *Supercond. Sci. Technol.* 1998, *11*, 68–72.

[13] Bhadeshia, H.K.D.H.; Strang, A.; Gooch, D.J. *Int. Mater. Rev.* 1998, *43*, 45-53.

[14] Dimmler, G.; Weinert, P.; Kozeschnik, E.; Cerjak, H. *Mater. Character.* 2003, *51*, 341-352.

[15] Výrostková, A.; Kroupa, A.; Janovec, J.; Svoboda, M. *Acta Mater.* 1998, *46*, 31-38.

[16] Výrostková, A.; Homolová, V.; Pecha, J.; Svoboda, M. *Mater. Sci. Eng. A* 2008, *480*, 289–298.

[17] Janovec, J.; Svoboda, M.; Kroupa, A.; Výrostková, A. *J. Mater. Sci.* 2006, *41*, 3425-3433.

[18] Novikova, S.I. *Measurem. Techniq.*, 1984, *27*, 933–938.

Index

A

anisotropy, x, 29, 42, 170
applied mathematics, 6
atoms, 43, 44, 163

B

base, 43

C

carbides, 163
Cartesian system, 1, 2, 10, 12, 16, 21, 22, 23, 24, 29, 44, 52, 87, 88, 91, 96, 146, 164, 169, 171
cell surface, 11, 12, 16, 23, 24, 51, 52, 53
central spherical particle, 1, 10, 95
ceramic, 86, 109
compatibility, 26, 30, 31, 33
composite materials, 1, 4, 5, 9, 12
compression, 47, 48
computation, 102
computer, 6, 7
cooling, 3, 11, 12, 41, 43, 44, 48
cooling process, 3, 11, 12, 41, 44, 48
crack formation, x, 5, 85, 87, 90, 91, 92, 93, 94, 95, 96, 98, 101, 110, 111, 112, 113, 114, 115, 116, 119, 122, 123, 124, 125, 127, 131, 132, 133, 135, 136, 138, 139
cracks, 87, 100, 101, 102, 103, 104
critical value, 89, 161, 162
cubic cells, 14, 16, 85, 88, 159

D

deformation, 2, 11, 13, 21, 26, 38, 51, 85, 86, 87, 88, 94, 108, 147, 152
derivatives, 6
differential equations, 33, 34, 35
discontinuity, 14
displacement, vi, 2, 4, 11, 12, 21, 22, 23, 24, 25, 49, 51, 54, 55
distribution, 1, 9, 11, 12, 20, 95, 149, 153, 160

E

energy, 2, 5, 6, 11, 12, 13, 14, 20, 21, 36, 37, 38, 39, 40, 46, 82, 85, 86, 87, 88, 91, 92, 93, 94, 96, 98, 108, 109, 116, 117, 119, 123, 127, 128, 130, 131, 132, 138, 146, 147, 148, 149, 150, 151, 152, 154, 155, 156, 157, 158
energy density, 2, 6, 37, 39, 40, 85, 86, 87, 88, 92, 94, 96, 108, 109, 116, 117, 127, 128, 130, 146, 147, 148, 150, 151, 152, 154, 155, 156, 157, 158
engineering, 4, 5, 6
equilibrium, 2, 27, 31, 33, 38, 93
exploitation, 90, 163, 164, 165

F

force, 96, 97, 98, 116, 117, 128
formation, 5, 47, 85, 86, 87, 90, 91, 92, 93, 94, 95, 96, 98, 101, 109, 110, 111, 112, 113, 114, 115, 116, 119, 121, 122, 123, 124, 125, 126, 127, 131, 132, 133, 134, 135, 136, 138, 139
formula, 6, 7, 13, 14, 85, 86
fracture toughness, 86, 93

Index

G

geometry, 21

H

height, 44, 91, 92
hexagonal-based prismatic cells, 10, 11, 12, 13
high-speed crack propagation, 109
hysteresis loop, 149

I

individual differences, 43
industry, 5
initial state, 163
initiation, 86, 98, 103, 104, 106, 108
integration, 5, 6, 7, 33, 36, 37, 39, 49, 50, 52, 53, 54, 55, 57, 58, 64, 65, 66, 69, 71, 72, 74, 75, 77, 79, 88, 99, 101, 103, 146, 147, 160
inter-particle distances, 1

L

lattices, 3, 22, 29, 41, 42, 43, 44, 170
lifetime, 5, 88, 89, 90, 159, 160, 163, 164, 165, 166, 167

M

magnetic field, 148
magnetic materials, 149
magnetic moment, 148, 149
magnetization, 149
materials, 1, 4, 5, 9, 12, 86, 87, 88, 89, 149, 160, 161, 163
mechanical stress, 5
melting, 42, 163
melting temperature, 42
microscopy, 12
microstructure, 5, 89, 163, 164
minimal thermal-stress induced elastic energy, 46
modelling, 1, 29, 42, 85, 89, 92, 159
models, 5, 6, 9, 12, 20
modulus, 29, 42, 47, 93, 146, 152, 154, 156, 169, 170, 171

O

operations, 33, 34, 35

P

parallel, 11, 147, 150, 151, 152
phase transformation, 3, 22, 41, 44, 55, 56, 82, 83
physics, 4
plastic deformation, 164
principles, 1, 3, 5, 7
programming, 6
propagation, 86, 98, 106, 108, 109

R

radius, 1, 6, 7, 10, 11, 12, 21, 22, 23, 45, 86, 87, 89, 90, 94, 95, 98, 99, 102, 105, 124, 125, 126, 127, 139, 141, 143, 144, 159, 164, 165, 166
real numbers, 6
relaxation, 3, 11, 13, 22, 41, 45, 163, 164
researchers, 4
resistance, 87, 90, 98, 123, 125, 127, 138, 141, 143, 144, 153, 161, 163, 166, 167
response, 89
root, 33, 34, 35, 36, 98, 101, 102
roots, 36, 167

S

shape, 1, 2, 4, 9, 11, 12, 21, 27, 44, 85, 86, 87, 90, 91, 92, 93, 98, 100, 101, 103, 104, 109, 147, 152, 163
shear, 2, 23, 25, 27, 28, 29, 96, 160
signs, 55, 88, 153, 155, 157, 158, 161
simulation, 89, 90, 163, 164, 165, 166, 167
solid matrix, 42
solution, 14, 54, 55, 56, 82, 83
spherical envelope, 1, 2, 4, 5, 10, 11, 14, 24, 25, 29, 36, 37, 38, 39, 41, 42, 43, 46, 53, 54, 55, 65, 69, 74, 83, 86, 89, 94, 95, 96, 97, 106, 109, 112, 113, 114, 115, 127, 129, 134, 138, 142, 146, 149, 151, 153, 156, 157, 159, 160, 161, 171
state, 2, 5, 6, 11, 13, 21, 38, 51, 86, 93, 103, 104, 106, 160, 163
steel, 163, 164, 165
stress, 2, 3, 4, 5, 11, 12, 13, 14, 21, 23, 24, 28, 31, 33, 34, 35, 38, 40, 41, 45, 46, 47, 49, 51, 52, 57, 58, 64, 65, 66, 69, 71, 72, 74, 75, 79, 81, 82, 85,

86, 87, 88, 89, 94, 96, 97, 98, 108, 116, 117, 118, 128, 147, 152, 160, 161, 162, 163, 165, 171
surface area, 15, 88, 91, 92, 93, 152, 155
surface energy, 93
symmetry, 11, 93, 95, 149, 153, 160

T

techniques, 1, 2, 6, 12, 21, 26, 29, 38, 82, 163
temperature, 3, 11, 12, 13, 22, 41, 42, 43, 44, 45, 46, 47, 48, 89, 90, 163, 164, 165, 166
tension, 48
tetragonal crystal lattice (TCL), 43
thermal expansion, 3, 9, 22, 41, 42, 43, 47, 56, 81, 82, 84, 87, 89, 118, 129, 145, 146, 159, 163, 169, 170

thermal stresses, 1, 3, 5, 6, 9, 10, 11, 12, 13, 20, 22, 36, 38, 41, 42, 43, 45, 46, 47, 48, 49, 51, 57, 58, 66, 71, 81, 89, 96, 97, 109, 110, 111, 112, 114, 115, 116, 118, 128, 129, 160, 161, 162, 163, 164, 166, 171
time periods, 90, 165
transformation, 3, 29, 39, 41, 43, 44, 82, 90, 95, 106, 118, 129, 144, 146, 150, 151,163, 166, 167, 170, 171, 180
translation, 195
variables, 56, 83, 87, 89, 90, 116, 117, 127, 130, 131, 145, 147, 162, 164, 165

Y

yield, 3, 47, 48, 164